About Professional Baking

THE ESSENTIALS

Join us on the web at

culinary.delmar.com

About Professional Baking

THE ESSENTIALS

Gail Sokol

THOMSON

DELMAR LEARNING

Australia Canada Mexico Singapore Spain United Kingdom United States

About Professional Baking, The Essentials
Gail Sokol

Vice President, Career Education Strategic Business Unit:
Dawn Gerrain

Director of Learning Solutions:
Sherry Dickinson

Managing Editor:
Robert L. Serenka, Jr.

Acquisitions Editor:
Matthew Hart

Product Manager:
Patricia M. Osborn

Editorial Assistant:
Patrick B. Horn

Director of Production:
Wendy A. Troeger

Senior Production Editor:
Kathryn B. Kucharek

Project Editor:
Maureen M.E. Grealish

Technology Project Manager:
Sandy Charette

Director of Marketing:
Wendy Mapstone

Channel Manager:
Kristin McNary

Images:
Paul Castle, Castle Photography, Inc.
except for the following:

Chapter 9.
© James Scherzi Photography, Inc. 2005.

Chef Hat courtesy of Getty Inc.

Cover Image:
© James Scherzi Photography, Inc. 2005.

Cover and Text Design:
Jeffrey Potter

For permission to use material from this text or product, submit a request online at http://www.thomsonrights.com
Any additional questions about permissions can be submitted by email to thomsonrights@thomson.com

Library of Congress Cataloging-in-Publication Data
Sokol, Gail.
About professional baking, The essentials / Gail Sokol. -- Retail ed.
    p. cm.
Includes bibliographical references and index.
ISBN-13: 978-1-4180-5143-3 (hardcover)
ISBN-10: 1-4180-5143-8 (hardcover)
1. Baking. I. Title.
TX763.S58 2006b
641.8'15--dc22

2006018613

NOTICE TO THE READER

Publisher does not warrant or guarantee any of the products described herein or perform any independent analysis in connection with any of the product information contained herein. Publisher does not assume, and expressly disclaims, any obligation to obtain and include information other than that provided to it by the manufacturer.

The reader is expressly warned to consider and adopt all safety precautions that might be indicated by the activities herein and to avoid all potential hazards. By following the instructions contained herein, the reader willingly assumes all risks in connection with such instructions.

The Publisher makes no representation or warranties of any kind, including but not limited to, the warranties of fitness for particular purpose or merchantability, nor are any such representations implied with respect to the material set forth herein, and the publisher takes no responsibility with respect to such material. The Publisher shall not be liable for any special, consequential, or exemplary damages resulting, in whole or part, from the readers' use of, or reliance upon, this material.

This book is dedicated to my mother, Carole Koblantz Deitcher, who gave me free rein in the kitchen as a young child with one stipulation: that I clean up the mess that I had made!

-AND-

To my husband, Dr. Harold Sokol, my computer guru and best friend who gave his time and energies to painstakingly help me put this textbook into the computer. One of the finest, most caring human beings on this planet, he has always been very supportive of any task I have undertaken, even when this textbook was just a pipe dream. He always encouraged me to reach for the stars even though they were a million miles away. Words are not enough to express how I feel.

Thomson Delmar Learning is excited to announce the *About Series,* the first installment in a robust line of culinary arts books from a leader in educational publishing. You'll soon discover why it's all *About Baking, Garde Manger, and Wine*! These essential culinary arts books present the tools and techniques necessary to ensure success as a culinary professional in a highly visual, accessible, and motivating format. It is truly the first culinary arts series written for today's culinary arts professionals and enthusiasts.

About Professional Baking by Gail Sokol. Over 700 full-color photographs demonstrating best practices and key techniques. Features include profiles of professional bakers; 125 fully kitchen-tested recipes written in an easy-to-comprehend format; clearly stated objectives; and hundreds of detailed step-by-step procedural photographs.

About Wine by Patrick Henderson and Dellie Rex. This introductory wine book presents practical and detailed information about wine. The five distinct sections of the book cover the basics of wine, the wine regions of the world, types of wine, and the business of wine. Special features of *About Wine* include detailed color diagrams, maps, and photographs. Useful appendices designed for use as a quick reference or as a basis for more research are also included, making this book a valuable resource.

Modern Garde Manger by Robert Garlough and Angus Campbell. This innovative and comprehensive book is designed to meet the needs of culinary arts students, enthusiasts, and experienced culinary professionals. Carefully researched content and fully tested recipes span the broad international spectrum of the modern garde manger station. Seventeen chapters are divided between five areas of instruction, each focusing on a different aspect of the garde manger chef's required knowledge and responsibilities. With over 500 color photographs, 250 recipes, and 75 beautifully illustrated graphs and charts, *Modern Garde Manger* is the most comprehensive book of its kind available for today's culinary arts student and the professional chef.

We look forward to providing you with the highest quality educational products available. Please contact us at (800) 477-3692 to preorder your desk copies.

All the Best!

Matthew Hart
Culinary Arts Acquisitions Editor
matthew.hart@thomson.com

CONTENTS

This book teaches the basic principles of baking with the most fundamental components of recipes and formulas—the ingredients. It is organized predominantly by ingredients because each ingredient plays a major role in how baked goods develop and take shape. The focus is to learn the role of ingredients in baking while realizing the science behind their interactions and how chemical reactions among ingredients create baked goods with specific characteristics. The act of baking makes much more sense when the science of why ingredients react in specific ways is incorporated into the learning process.

Throughout each chapter the role of specific ingredients in different baked goods is discussed, as are the scientific principles behind those roles. Each chapter culminates in the real world application of recipes that directly reflect the principles being taught. Most of the chapters end with recipes and because some of the methods and principles taught in a chapter will apply also to recipes taught in other chapters, certain recipes are listed in more than one chapter, with page number reference.

Lessons that demonstrate specific recipe principles taught in the chapter precede each recipe. You can then easily make a direct correlation between what has just been learned in the corresponding chapter and why the specific ingredients and methods chosen for that recipe are used. Emphasis is placed not on the quantity of recipes but rather on how each recipe provided is applicable to the knowledge of baking principles. You are then free to explore other recipes that apply the same foundations and principles.

Each recipe is written with explanations of how the ingredients should look at each stage of preparation; each recipe is accompanied by step-by-step photographs for a visual presentation to learn from. The recipes provide small yields focusing on principles and techniques while handling a smaller, more manageable quantity of dough or batter.

A chapter on healthy baking substitutes healthier ingredients for less healthy ones without detracting from the flavor or texture of the original baked good. It is important to provide healthier options for a population with a variety of health concerns. How to substitute healthier fats, adding whole-grain options to improve the nutritional profile of baked goods, and sugar substitutes are discussed in detail. What ingredients are most easily substituted for others and why is discussed. Nutritional information is given at the end of each recipe.

The final chapter is a troubleshooting chapter. The troubleshooting chapter lists individual chapters in sequential order and general problems are identified and addressed, followed by reasons for the problems and, if applicable, solutions to those problems. Finally, Appendix A provides essential information on Baking Tools and Equipment, Appendix B provides directions on how to prepare baking pans, Appendix C discusses weights and measurements and Appendix D provides information on canned goods.

An added feature is the professional profile included in every chapter. Each professional profile discusses one person who has graduated from a culinary school and who has become successful in the baking profession. Along with a personal photograph, the chef profiled answers questions about his or her life and career so that students can see the career options open to them.

A detailed glossary at the end of the book provides easy access to definitions of terms, methods, and principles.

This book is meant to lay the foundation of baking knowledge and should merely be the beginning of a lifetime of knowledge that is sought by the passionate baker.

Notes on Specific Ingredients and Techniques Used within the Text

The following is meant to clarify the text by providing specific notes on ingredients and techniques used:

- Dry ingredients are whisked together to thoroughly combine them.

- Powdered dried egg whites are substituted for fresh eggs in specific recipes that will receive no further cooking.

- A pasteurized whole egg substitute is used in the recipes in Chapter 13, Healthy Baking, or is given as an option.

- Only large fresh shell eggs are used in the recipes within this text except where mentioned.

- Only unsalted butter is used in this text unless otherwise noted in specific recipes. The taste of butter is preferred over vegetable shortening, which can leave an unpleasant film on the tongue.

- Instant active dry yeast is the yeast of choice in this text, but other types of yeast can be used in the correct amounts. Correct conversions are given in Chapter 1.

- A suggested water temperature is given in each recipe using yeast, but this temperature should be used only as a guide because conditions vary from kitchen to kitchen. Water temperature can be calculated using the method in Chapter 1.

- Many recipes for yeast breads in this text begin by mixing the dough with the paddle attachment of the mixer and then switching to the dough hook, because the ingredients combine more thoroughly if the paddle is used first. Ingredients, especially in smaller quantities, may not combine well with a dough hook during the beginning stages of mixing.

- Chapter 1, Working with Yeast in Straight Doughs may be used as a foundation of knowledge for the following two chapters: Chapter 2, Starters and Preferments, and Chapter 3, Laminated Doughs.

- Many bakers do not agree on whether or not to scald milk in the preparation of yeast doughs. Many bakers do not scald milk to destroy the enzyme protease because they believe it to be destroyed during pasteurization. However, protease may become reactivated during storage. Milk is scalded in this text to be sure the enzyme has been deactivated.

To Toby Strianese, CCE, author, and Director of the Hotel, Culinary Arts, and Tourism
Department at Schenectady County Community. A special thank you for supporting me, and
encouraging me in every step of the process. Your door has always been open to me as a
teacher, mentor, and friend, and I will never forget all that you have done for me.

To my father, Dr. Kenneth Deitcher, who taught me through his work as a pediatrician and
highly acclaimed photographer how extremely rewarding life can be when you pursue your
passion.

To my daughter, Rebekah, who always was willing to give me an editorial opinion when I
needed it the most and a palate willing to be experimented on.

To my daughter, Drue, who freely gave herself as a guinea pig and sounding board, sampling
many versions of recipes while always giving me her most honest opinion, even when it hurt.

To my maternal grandmother, Betty Koblantz, a fabulous baker herself, who encouraged my interest in baking at an early age.

To my paternal grandmother, Etta Deitcher, a wonderful baker who was the first who taught me
how to properly fold an egg white foam into a sponge cake batter, thereby sparking my passion for baking when I was just a little girl.

To Certified Master Chef Dale Miller, who taught me so much and who was always willing to try
my creations. He encouraged me to enter my very first pastry competition, which I won.

To Paul Krebs, CCE, chef and professor at Schenectady County Community College, my teacher,
colleague, and friend who reviewed this text with the eyes of an eagle to make sure I got it
right.

To Matthew Hart, Acquisitions Editor, whose vision for this book matched my own.

To Gerald O'Malley, Developmental Editor, who was always available when I needed him and
was instrumental in my achieving my goal of completing this book.

Thank you to Kathy Kucharek, Senior Production Editor, who diligently organized the photo
shoot and assisted in the final editing of this text so everything would go smoothly.

To Lisa Flatley, Editorial Assistant, for whose boundless energy, enthusiasm, and willingness to
help I will be forever grateful.

To Maureen Grealish, Project Editor, who has the patience of a saint and whose attention to detail helped to make this book more understandable.

To Kristin McNary, Channel Manager, for her efforts in helping to promote this text.

To Paul Castle, Photographer, and his assistant, Christopher Morris, who painstakingly strived
to capture my vision of each baked good.

To David Lenweaver, food stylist and recipe tester extraordinaire who, with his attention to detail and his artistic sense, helped to bring my vision for presentation to a whole new level.

To the students of Schenectady County Community College: Rebekah Milewski and Josh Gabri
who tirelessly assisted during the photo shoot and did whatever was asked of them.

REVIEWERS

I would very much like to thank the following reviewers whose reviews I read diligently and who gave me great insights into their needs as culinary educators:

DEBRA SOCHA
Milwaukee Area Technical College
Gillet, WI

JEFFERY BRICKER
Ivy Tech State College of Central Indiana
Indianapolis, IN

MATHIEU COCHENNEC
City College of San Francisco
San Francisco, CA

VINCENT DONATELLI
Asheville-Buncombe Technical
    Community College
Asheville, NC

WAYNE SMITH
Mesa State College
Grand Junction, CO

JEFFREY J. CLICK
Henry Ford Community College
Dearborn, MI

MICHELE McGRAW
Cuyahoga Community College
Cleveland, OH

MARY K. COWELL
Consultant
Belle Vernon, PA

CHAPTER 1

Working with Yeast in Straight Doughs

After reading this chapter, you should be able to:

- Describe the role carbon dioxide plays in the leavening of yeast breads.

- Explain what yeast is.

- List the different types of yeast.

- Define gluten.

- Define gluten's role in baking.

- List ways to control gluten.

- Demonstrate the 12 steps of yeast dough production.

- Show how to best handle a yeast dough when shaping.

- Demonstrate how to work with yeast in straight doughs by preparing the recipes in this chapter.

In order to prepare many breads and pastries,

it is most important for a professional baker to know how to work with yeast. Yeast is a leavening agent that helps doughs and batters rise through the incorporation of gases. These gases expand in the oven and push the batters and doughs upward and outward, increasing volume and forming texture. Yeast is the foundation for breads ranging from French bread and pizza to sweet cinnamon rolls and Danish pastry.

Yeast breads take more time to prepare than quick breads made with chemical leaveners (e.g., baking powder and baking soda). However, because of their versatility, the investment in their preparation is well worth it.

This chapter and the following two chapters deal with different types of baked goods that are leavened with yeast. This chapter is to be used as a foundation of knowledge for the two chapters that follow.

This chapter gives a basic understanding of yeast, what it is, the different types, and how it is used in straight doughs that are both savory and sweet. Straight doughs are yeast doughs in which most of the ingredients are mixed together in one bowl at the beginning of preparation.

Other types of yeast breads using sponges or preferments are discussed in Chapter 2.

The last chapter on yeast breads, Chapter 3, discusses laminated or rolled-in doughs in which the fat is incorporated into the dough through a series of folds and turns. Laminated doughs are used to prepare rich sweet doughs like Danish pastry and croissants.

Defining Yeast

Yeast is a one-celled living organism called a *fungus* (Figure 1–1). Yeast is everywhere. It is on fresh fruits and vegetables and even in the air. The more a baker bakes with yeast, the more airborne wild yeasts permeate the kitchen. Many bakers attribute this presence of yeast in the surrounding air as contributing better flavor to breads baked in those kitchens.

Yeast requires specific conditions to grow. These conditions include the presence of moisture, warmth, and sugar. Yeast is extremely sensitive to temperature. Its activity slows down or speeds up depending on the surrounding temperature. Table 1–1 lists yeast activity levels at various temperature ranges.

Bakers who prepare straight yeast doughs are interested in reliable yeast, known as *baker's yeast,* which is grown commercially. Specific strains of baker's yeast are grown in the laboratory and then sold commercially in different forms to professional bakers. Determining which type of yeast to use depends on the baker's preference and how quickly the yeast bread needs to be completed. In general, any type of commercial yeast may be used, but because all yeast is not equal, correctly converting from one type to another is essential.

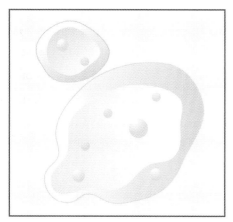

FIGURE 1–1

FIGURE 1–2

Table 1–1 Yeast Activity Levels at Various Temperature Ranges

34° to 40°F (1° to 4°C)	Yeast is dormant (refrigeration temperature).
50°F (10°C)	Yeast slowly wakes up.
60° to 70°F (16° to 21°C)	Yeast becomes more active.
70° to 80°F (21° to 27°C)	Optimum fermentation temperature.
120°F (50°C)	Fermentation slows significantly.
140°F (60°C)	Yeast is killed.

There are many different types of commercial yeast. They include fresh or compressed yeast, active dry yeast, instant active dry yeast, and osmotolerant instant active dry yeast (Figure 1–2).

Fresh or Compressed Yeast

Fresh or compressed yeast usually comes in 1-pound (450-g) cakes shaped like a brick. It is also available in smaller quantities as small cubes. Fresh yeast is also available crumbled in bags.

Fresh yeast is light tan in color and crumbles easily. It does not require the higher water temperatures to dissolve it like active dry or instant active dry yeast. It cannot be killed with cold water. Fresh yeast dissolves easily in water or other liquid ingredients. Most bakers dissolve it first in approximately two times its weight in warm water (100°F; 38°C). Although fresh yeast does not require mixing it with water before adding it to a dough, it is wise to do so. Mixing it in the dry state risks uneven distribution within the dough.

Fresh or compressed yeast lasts for approximately 2 weeks when wrapped airtight and refrigerated. It can be frozen for 3 to 4 months.

Active Dry Yeast

Active dry yeast comes in the form of tiny pellets or granules. It has been dried or dehydrated to increase its shelf life and is sold in vacuum-packed jars or packets. The container should have a date of expiration on it. If unopened, active dry yeast has a shelf life of up to 1 year. Once opened, it is best stored in the refrigerator where it will last for at least 2 months.

Once dried, the yeast becomes inactive and is said to be dormant (in a sleeping state) until it is mixed with warm water and rehydrated.

Because of the severity of the drying process, a small percentage of the yeast may die. Although the dead yeast can provide some flavor to the finished bread, these dead yeast cells produce a substance (glutathione) that relaxes and weakens gluten. This weakening of gluten may be a plus in specific types of yeast dough such as pizza, because weakened gluten allows the dough to be stretched and shaped more easily—but generally it has a negative effect on yeast breads. (See Ingredients That Negatively Affect Yeast Breads.)

Active dry yeast must be combined with approximately four times its weight in warm water that is approximately 110°F (43°C) to become activated. Cold water can damage or kill active dry yeast.

Many recipes call for proofing active dry yeast, which is a quick test to make sure the yeast is indeed alive. Although not necessary, proofing ensures that other ingredients will not be wasted by adding them to yeast that is dead. To proof the yeast, the yeast granules are placed in warm water and mixed together to rehydrate. The mixture is allowed to sit at room temperature for 5 to 10 minutes. Some recipes may call for a pinch of sugar to be added as an encouragement for the yeast to begin the fermentation process. The mixture should produce carbon dioxide gas and become foamy, proving that the yeast is alive and feeding on the sugar. If the mixture does not foam, the yeast should be discarded.

Instant Active Dry Yeast

Instant active dry yeast is also referred to as *instant yeast* or *fast-rising yeast.* Like active dry yeast, it, too, goes through a drying process. Because this process is more gentle, it produces fewer dead yeast cells.

Because of its porous nature, instant active dry yeast absorbs water instantly without having to sit for several minutes. Because it also produces more carbon dioxide gas per yeast cell, a smaller amount can be used to achieve the same results.

If a baker uses the same amount of instant active dry yeast as active dry yeast, then the yeast doughs made with instant active dry yeast will ferment more quickly. This is why this type of yeast is sometimes known as fast-rising yeast.

Because of its vigorous nature, instant active dry yeast is best used with yeast doughs that require short fermentation times. However, instant active dry yeast can be used successfully for yeast doughs with longer fermentation times if a lesser amount is used.

Instant active dry yeast does not require proofing and can be sprinkled into dry ingredients directly or dissolved into a portion of the liquid ingredients, which should be at approximately 110°F (43°C) or warmer. It should not be added directly into cold water or into a cold dough that is below 70°F (21°C) because this temperature can damage or slow down the yeast's activity. The ideal dough temperature for a dough using instant active dry yeast should be approximately 70° to 90°F (21° to 32°C).

Instant active dry yeast has a longer shelf life than active dry yeast. It can last at least 1 year if left unopened at room temperature. Once opened, it can be refrigerated for several months but keeps best wrapped airtight and frozen.

Osmotolerant Instant Active Dry Yeast

Osmotolerant instant active dry yeast is a special type of yeast used in rich, sweet doughs that contain greater amounts of sugar, fats, and eggs. Because the ingredients contain little water, active dry and instant active dry yeast cells have trouble hydrating themselves and must fight to maintain any moisture within them. By using osmotolerant instant active dry yeast, the yeast cells are able to withstand the lack of moisture without being damaged; they are specifically

Equivalent Amounts of Various Yeasts

- To convert from fresh compressed yeast to active dry yeast, use one half of the amount of fresh or compressed yeast called for in the recipe.

- To convert from active dry yeast to fresh or compressed yeast, use double the amount of active dry yeast called for in the recipe.

- To convert from active dry yeast to instant active dry yeast, use three fourths of the amount of active dry yeast called for in the recipe.

- To convert from fresh compressed yeast to instant active dry yeast, use one third of the amount of fresh or compressed yeast called for in the recipe.

selected when they are grown to withstand conditions with little moisture. Osmotolerant yeast is able to tolerate the osmotic changes within the dough.

Water tends to go from a higher concentration of itself to a lower concentration of itself across a semipermeable membrane in order to create an equilibrium. This is known as osmosis. The semipermeable membrane in this case is the cell membrane of the yeast. With little water in a rich dough, any available water in the yeast cell will move through the cell membrane and be attracted to the sugar in the dough because of the sugar's hygroscopic nature (attracting moisture). This movement of water is made in an attempt to equalize the concentration of water. For this reason, non-osmotolerant yeast cells (e.g., active dry yeast and instant active dry yeast) would dry out and die.

Yeast and Fermentation

Yeast live on simple sugars. As long as the correct living conditions are combined with an adequate supply of oxygen, the yeast will divide and multiply. Flour contains starches that are comprised of long chains of sugars. When chemically broken down, the starches become simple sugars. Enzymes that break down the starches into simple sugars are present naturally within the flour or may be added during the milling process. Enzymes within the yeast also break down the starches in the flour to simple sugars. The yeast then use these simple sugars as food.

The process whereby yeast eat sugar and give off waste products that include carbon dioxide gas and alcohol is referred to as fermentation. Fermentation begins as soon as the yeast and other ingredients are mixed together and does not end until the yeast die in the oven during baking (Figure 1–3).

Of the many waste products that are produced during fermentation, bakers are very interested in the carbon dioxide gas bubbles given off by the yeast. It is important for these gas bubbles to become trapped within the yeast dough to ensure that the product will rise (see How Gluten Traps Carbon Dioxide Gas). Once in the oven, these gases expand, forcing the dough to rise. The alcohol evaporates during and immediately after baking.

The Correct Temperature to Ferment Yeast Doughs

It is important for yeast doughs to begin the fermentation process at the correct temperature. Professional bakers use heated cabinets called *proof boxes* that are temperature and humidity controlled to ferment their doughs when the temperature of the kitchen is not ideal for fermentation to occur. However, yeast doughs can ferment at room temperature. The best temperature to ferment a yeast dough is the same temperature at which it was mixed. The ideal temperature range for fermentation is approximately 70° to 80°F (21° to 27°C).

Success with yeast dough depends on the temperature of three factors: the flour, the surrounding air, and the water. The temperature of the dough determines the rate of

FIGURE 1–3

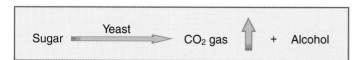

fermentation. Because the temperature of the flour and air in the kitchen cannot be as easily controlled, the water temperature is the easiest way to manipulate the temperature of the other two factors. These manipulations help achieve the desired final dough temperature. For example, if the kitchen is too cold, the water temperature can be raised to compensate, or if the kitchen is too warm, the water temperature can be lowered accordingly. Old World European bakers can sense the temperature just by touch, but less experienced bakers rely on formulas and thermometers to calculate the correct temperature of the water. Some bakers even make the assumption that the flour and the kitchen are the same temperature.

Friction

The sheer act of mixing produces energy, which is transferred to the dough as heat and increases its temperature. Bakers call this *friction*. Friction needs to be taken into consideration when calculating the desired water temperature. Friction can be measured exactly by taking the temperature of the dough before mixing and then after mixing. The difference between the two numbers is the temperature increase due to friction. This is known as the friction factor and can be calculated using the formula that follows. Friction factors vary from kitchen to kitchen, depending on how aggressively the baker mixes the dough or what type of mixer is used. The friction factor is the highest when doughs are mixed in large commercial machines. (*Note:* The friction factor needs to be calculated only once for any particular mixer.)

The friction factor (FF) can be calculated for a particular mixer with the following calculation that takes into account the temperature of the room (RT), the temperature of the flour (FT), and the temperature of the water used (WT). A measurement of the dough temperature after it has been mixed is also needed (ADT; actual dough temperature).

The following formula can be used to determine the friction factor (FF) for a particular mixer:

$$FF = (ADT \times 3) - (RT + FT + WT)$$

For example, if RT = 75°F (24°C), FT = 75°F (24°C), WT = 64°F (18°C), and measured ADT was 78°F (26°C):

Using the Fahrenheit system:

Step 1	RT	75°F
	FT	75°F
	WT	64°F
		214°F
Step 2	ADT × 3	78 × 3 = 234°F
Step 3	Subtract	234 − 214 = 20°F, the FF.

Using the Celsius system:

Step 1	RT	24°C
	FT	24°C
	WT	18°C
		66°C
Step 2	ADT × 3	26 × 3 = 78°C
Step 3	Subtract	78 − 66 = 12°C, the friction factor.

Temperature

Every bread baker has his or her ideal desired dough temperature (DDT) to ferment yeast doughs. If the desired dough temperature is known along with the room temperature (RT), flour temperature (FT), and friction factor (FF), we can now find the calculated water temperature (CWT) to be used.

The following formula can be used to determine the calculated water temperature (CWT) to be used to prepare the dough:

$$CWT = (DDT \times 3) - (RT + FT + FF)$$

For example, if RT = 75°F (24°C), FT = 75°F (24°C), FF = 20°F (11°C), and DDT = 78°F (26°C).

Using the Fahrenheit system:

Step 1	RT	75°F
	FT	75°F
	FF	20°F
		170°F
Step 2	DDT × 3	78 × 3 = 234°F.
Step 3	Subtract	234 − 170 = 64°F, the CWT.

Using the Celsius system:

Step 1	RT	24°C
	FT	24°C
	FF	11°C
		59°C
Step 2	ADT × 3	26 × 3 = 78°C
Step 3	Subtract	78 − 59 = 19°C, the CWT.

How Gluten Traps Carbon Dioxide Gas

To understand how yeast breads rise, you must understand the role of gluten. Gluten is produced from the proteins in wheat flour and gives strength and structure to baked goods.

Although many different types of flour are used in baking, bakers are most interested in flour that comes from the wheat plant because wheat flours contain two proteins called *glutenin* and *gliadin.* Different wheat flours contain different amounts of glutenin and gliadin.

Gluten develops when water is mixed into the flour. The more gluten-producing proteins there are in a wheat flour, the more water it can absorb. As gluten develops within a yeast dough, a web-like structure forms. This structure has nooks and crannies where carbon dioxide gas bubbles become entrapped (Figure 1–4). As the yeast continue to feed on the sugars within the broken down starches in the flour, fermentation continues. As fermentation continues, even more gases are produced, causing the gluten fibers to stretch out and hold even more gases.

Eventually, the dough is put in the oven to bake and these gases plus the alcohol vapors expand in the dough while any moisture within the dough turns to steam. This combination of gases and steam expanding inside the dough causes the network of gluten to stretch and push against the dough, forcing it to rise and increase in volume.

FIGURE 1–4

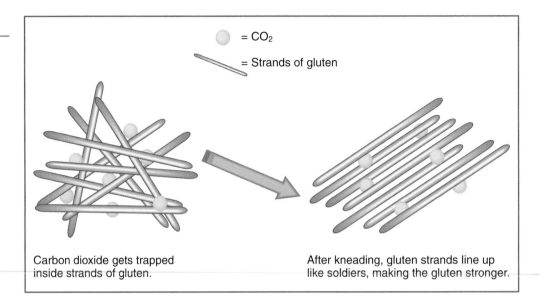

= CO$_2$

= Strands of gluten

Carbon dioxide gets trapped inside strands of gluten.

After kneading, gluten strands line up like soldiers, making the gluten stronger.

Developing Flavor in Yeast Breads

There are many factors that contribute to the flavor of yeast breads. The fermentation process itself provides a great deal of flavor to yeast breads. One of the reasons for this is the work of enzymes (in the flour and yeast) that help release simple sugars from within the starch molecules in flour.

Proteins called *enzymes* can speed up chemical reactions by breaking down starch molecules in the flour into simple sugars that yeast can eat. Without these important enzymes, the fermentation process would never begin, and flavorful by-products would not be produced. Thus, these enzymes also act as flavor regulators.

How quickly these flavorful by-products of fermentation are produced is affected by other factors, including the temperature and amount of time that fermentation is allowed to occur. Ingredients also play a role in providing flavor for yeast breads.

Temperature

The temperature at which the dough is fermented affects the flavor. Fermentation speeds up at higher temperatures of 90°F (32°C) and above. These higher temperatures cause off-flavors to develop while different acids are produced at the increased rate of fermentation and the yeast multiply more quickly. This eventually causes the yeast to run out of food because they have consumed most of the sugar within the flour.

Fermentation at too low a temperature also has an effect on flavor. Doughs fermented at too low a temperature, below 70°F (21°C), have little flavor because the yeast is not active enough to produce enough flavorful by-products. The best temperature at which to ferment a yeast dough is the temperature at which it was mixed.

Time

The amount of time allowed for fermentation plays a part in flavor. The longer the fermentation time, the more flavor the bread has because enzymes have more time to break down starches into sugar feeding the fermentation process. Longer fermentation times also allow for more flavorful organic compounds (by-products of fermentation) to be produced.

However, if fermentation continues for too long, the yeast run out of food and begin to die off. This produces yeast breads with little flavor and poor volume. Because there is no residual or leftover sugar, the crust color will be too light. Any residual sugar within the dough provides flavor and enhances crust color and browning. Browning of the crust provides great flavor caused by the Maillard reaction, a chemical reaction between the proteins and sugars within the dough. For more information on browning and the Maillard reaction, see step 10, Baking, in The 12 Steps of Yeast Dough Production.

Ingredients

Even the ingredients in a yeast dough provide flavor. The flour itself provides a flavor component, especially for lean doughs in which there are no rich ingredients. The flavor in rich straight doughs used to make such baked goods as cinnamon rolls or coffee cakes comes from rich ingredients. These ingredients may include sugar, eggs, fat, and milk.

How the Amount of Water in a Yeast Dough Can Affect Hole Structure

It is true that gluten cannot develop unless water is mixed with the proteins in the wheat flour. However, too much water (over 70 percent using baker's percentages compared to 50 to 60 percent for other lean doughs) can overpower the proteins and actually weaken gluten. This is used to the advantage of bread bakers who bake rustic breads that traditionally are prized for their large, irregular hole structure. The dough is prepared with a large quantity of water so that it is a cross between a batter and a dough. As the bread bakes in the oven, the excess water within the dough produces large quantities of steam which force the dough to expand. Because the gluten is weakened, large holes develop within the structure of the bread. Some bakers prepare these rustic breads using flours with a lower protein content in addition to large quantities of water.

Ingredients That Negatively Affect Yeast Breads

There are a few ingredients that a baker should be aware of that can negatively affect yeast breads. They include expired or dead yeast, water, salt, and protease.

Expired or Dead Yeast

Using dead yeast greatly affects yeast bread production. Do not use yeast after the expiration date has passed. Dead yeast produces a substance called glutathione that can weaken gluten in yeast doughs. (See Water in this section.) Bread baked with dead yeast is heavy and flat with little flavor. When in doubt of the freshness of the yeast, proof it in a bowl of warm water with a pinch of granulated sugar. If after 5 to 10 minutes it has not become foamy, discard it and start over with a fresh package.

Water

Water is one of the necessary ingredients in yeast bread production. However, water that is either too cold or too hot can kill yeast. Fresh compressed yeast and natural wild yeast starters are not harmed when mixed with cold water. However, active dry yeast and instant active dry yeast are damaged when exposed to water temperatures below 100°F (38°C). Both active dry yeast and instant active dry yeast are typically rehydrated in warm water, although instant active dry yeast may be added directly to the dry ingredients.

Cool water temperatures can damage or kill yeast cells by weakening the cell's membrane, which allows the cell's inner contents to leak out. As this occurs, glutathione is released from the yeast's damaged cell wall. Glutathione, an amino acid also found in milk that has not been pasteurized, can weaken gluten in yeast doughs, causing the dough to become soft and sticky. This is one reason why some recipes for yeast breads call for the milk to be scalded. For another reason to scald milk, see protease. (*Note:* See Defining Yeast in this chapter for the specific water temperature intervals needed for rehydrating specific types of yeast and Table 1-1, Yeast Activity Levels at Various Temperature Ranges, for the temperature at which yeast is killed.)

The amount of minerals that are in the water (how hard the water is) also affect yeast breads by slowing down fermentation and creating a tough dough. Soft water, with few minerals, can make yeast doughs sticky and too soft. If water is too hard or too soft, use bottled water.

Salt

Salt is a very important ingredient in yeast breads. Salt not only adds flavor but also strengthens gluten, making it "stretchy" and elastic. Care must be taken as to when the salt is added.

Salt can also decrease fermentation and, because of osmosis, has the potential to kill the yeast. Osmosis is the tendency of water to go from a higher concentration of itself to a lower concentration of itself in order to find an equilibrium or balance. If yeast is surrounded with salt, the salt draws any water away from the yeast by sucking the water through the cell membrane of the yeast, which prevents the yeast from rehydrating. This causes the yeast to dry out and die. Some bakers add the salt directly in with the other dry ingredients like flour without any problems. The flour and other dry ingredients protect the yeast by buffering it against the negative effects of the salt. The salt can also be dissolved in the liquid ingredients first.

Some bakers prepare the dough (with the yeast) and then, after a short rest period but before kneading, the salt is added. Because salt tightens gluten, it is easier to mix the ingredients of the dough before adding it. If salt is added in the beginning with the other ingredients it can prevent the flour from absorbing as much water as it needs. If the salt is added after the rest period, however, be sure it is mixed in thoroughly.

Protease

There is a controversy over a substance known as protease when preparing rich yeast doughs. Protease, found in milk and other dairy products, is an enzyme that breaks down proteins. Because the gluten network within yeast doughs contains proteins, some bakers believe that protease can negatively affect gluten formation and yeast doughs that contain it produce breads with less volume and a coarser texture. Protease can also slow down yeast growth. That is why, in some recipes for yeast doughs, the milk is scalded to destroy protease before the milk is added to the other ingredients.

Scalding significantly reduces the protease level. To completely inactivate protease, it would be necessary to heat milk to an extremely high temperature. For bread making purposes, scalding the milk to 180°F (82°C) is sufficient. This is also true for dry milk. Not all dry milk solids have been exposed to a high enough heat to destroy the protease; however, dry milk solids that are labeled as *high-heat treated*, have been.

Some bakers believe protease is destroyed when fresh and dry milk are pasteurized, so it is not necessary to scald the milk. Still others believe that storing pasteurized milk leads to the reactivation of the protease over time and, in order to completely inactivate it, scalding the milk is still necessary.

The 12 Steps of Yeast Dough Production

There is an order to making a yeast dough with several steps to follow. Each step builds on the one before it. Each step also accomplishes specific goals that culminate in a superior finished yeast bread. There are 12 steps of yeast dough production.

1. Scaling Ingredients

The exact measurement of ingredients is crucial. Most ingredients in a professional kitchen are weighed with the exception of some thin liquids such as water, eggs, and milk, which can be weighed or measured by volume. The act of weighing out ingredients is referred to as scaling. Scaling ingredients is done on a baker's scale or a digital scale.

2. Mixing

Once all the ingredients are scaled and brought to the work area, the mixing of ingredients begins. Mixing brings the ingredients together into a uniform dough. The act of mixing distributes ingredients throughout the dough so they are uniformly dispersed. Mixing also develops gluten. Once the initial mixing of ingredients is completed, kneading begins. Kneading is the act of pushing and folding the dough against a work surface to develop a network of gluten. Kneading can also be accomplished in an electric mixer using a dough hook. As kneading continues, the dough transforms from a shaggy mass to a smooth, elastic dough.

Some recipes call for a short rest period just after the flour and water have been mixed together but before kneading is begun. This short rest period gives the flour time to absorb all the liquid ingredients and form gluten. The more gluten that forms, the more areas in which carbon dioxide bubbles can become trapped.

This rest is called the autolyse. As the flour takes time to thoroughly absorb the water that has been added, the gluten that is beginning to form is disorganized (the network of fibers are randomly arranged). After the autolyse and once kneading has begun, the gluten sheets are able to line up and face the same direction. This helps form stronger sheets of gluten fibers that are better aligned to give strength and structure to the bread. The autolyse also reduces the time needed to knead the dough, bringing it to a soft, elastic mass more quickly; and the less kneading required, the better the color and flavor of the finished bread. Not every recipe uses the autolyse.

In some recipes, the salt is added with the other ingredients before the autolyse. In other recipes, the salt is added after the autolyse. Whichever method is followed, it is extremely important not to have the salt come into direct contact with the yeast. See Ingredients That Negatively Affect Yeast Breads.

There are two mixing methods for yeast doughs. They include the straight dough method and the sponge method.

STRAIGHT DOUGH METHOD

The straight dough method is the simplest. All the ingredients are added to one bowl and mixed. Because the yeast may not get evenly distributed using this method, many bakers mix the yeast with some of the water called for in the recipe before blending it with other ingredients. Instant active dry yeast can be mixed in with the dry ingredients without first mixing it with water.

There are variations to the straight dough method. Some recipes call for the fat and sugar to be blended together first before adding other ingredients to ensure that they are evenly distributed throughout the dough. This is true for richer, sweet yeast doughs. Variations to the straight dough method are referred to as the modified straight dough method.

The 12 Steps of Yeast Dough Production

1. Scaling ingredients
2. Mixing
3. Fermentation
4. Punching
5. Scaling to size
6. Rounding
7. Resting or benching
8. Makeup and panning
9. Proofing
10. Baking
11. Cooling
12. Storing

Note: Although it is traditional to mix just the flour and water before the rest period, other ingredients may be added.

SPONGE METHOD

The sponge method refers to yeast doughs that are prepared in two stages. A portion of the flour, water, and yeast are mixed together before the actual dough is prepared. It is then allowed to ferment for hours or overnight before the actual dough is mixed. This sponge or "pre-dough" is referred to as a *preferment* and imparts great flavor to breads. For detailed information on preferments, see Chapter 2.

3. Fermentation

Fermentation is the process during which yeast eat the sugar from the broken down starches within the flour and then release carbon dioxide gas and alcohol. This is the first rising that the dough, as a whole, undergoes before being divided and shaped. Once fermentation begins, it continues until the time the dough goes into the oven and the temperature reaches 140°F (60°C), the temperature at which yeast is killed.

Once the mixing and kneading is complete, the dough is placed into a greased bowl or other container large enough to handle the dough, even after it has expanded. It is flipped over so that the greased side is facing up, preventing it from drying out. It is then covered and allowed to ferment. Once fermentation is complete, most doughs should be doubled in volume. This takes approximately 1 to 2 hours.

Commercial bakeries use *proof boxes* for fermentation. Proof boxes are cabinets that are temperature and humidity controlled where both fermentation and proofing can take place. For those establishments that do not have a proof box, a warm area of the kitchen will also work. The covered bowl of dough can be left to rise in a warm area, or many bakers make their own makeshift proof box by using a clean, unused garbage bag. The bag should be shaken out to open it up fully. A bowl of dough can then be slid into the bag. The bag is twisted shut repeatedly, which inflates the interior with air. The bag is then secured shut with a tie or is knotted. The bowl of dough can be left in the bag at room temperature to ferment.

Rich sweet doughs tend to be fermented for less time and they may not double in volume. This is due to the addition of rich ingredients like fat, eggs, and sugar, which tend to slow fermentation. Generally, the fermentation process is complete when the impression left from a finger gently pushed into the dough remains and springs back slowly. For more on fermentation, see Yeast and Fermentation and Developing Flavor in Yeast Breads.

4. Punching

Punching is not as it sounds. The edges of the dough are pulled over into the center, which deflates the dough and expels much of the built up gases. Punching is also referred to as degassing. Deflating the dough in this way helps redistribute the yeast, preparing the dough for its final rise before it is placed in the oven. Punching helps even out the temperature of the dough, because the warmest place after fermentation is in the middle. Punching also helps redistribute the yeast and help it find fresh food, because it probably ingested all the sugars in the flour that was surrounding it during the fermentation period. This increases the yeast activity, ultimately yielding a better risen yeast bread.

Some yeast breads are not fully deflated and are treated more gently than other doughs, because in certain yeast breads, it is desirable for the dough to maintain as much gas as possible to ensure that the existing holes within the dough will enlarge even more during the proofing stage. Many French breads use this gentle punching to create large, irregular holes within the final bread.

5. Scaling to Size

The dough is now ready to be divided into pieces. Scaling should be done quickly because the process of fermentation is continuing. The pieces are weighed on a scale and their size is determined by the product being made.

There are machines that are manufactured to cut large quantities of dough at once for commercial bakeries. One such device is called a *dough divider.* The dough is pushed into a pan and a press cuts into the dough to create uniform pieces of dough that can be used to form three dozen rolls at once.

6. Rounding

The pieces of dough are now rounded into smooth balls, which will make the final shaping of the dough easier. Rounding forms a smooth, elastic skin of gluten around the outside of the dough. This elastic skin keeps needed carbon dioxide gas within the dough.

7. Resting or Benching

The rounded pieces of dough are lightly covered and allowed to rest for a short period of time, usually 10 to 20 minutes. Resting or benching relaxes gluten just enough so the pieces of dough can be easily shaped into what they will ultimately become: rolls, loaves, braids, or rounds. Fermentation is still continuing.

8. Makeup and Panning

The pieces of dough are now ready to be formed or molded into the final shape they will have going into the oven. This step is known as makeup and panning. The shaped pieces are placed on sheet pans or in loaf pans. Some pieces are placed in flour bannetons or baskets so the dough will have attractive markings when unmolded.

9. Proofing

The final step before baking is referred to as proofing, and the shaped dough is allowed to continue the fermentation process one final time before it will be baked. This final fermentation causes the shaped dough to further increase in volume.

In general, proofing temperatures are higher than fermentation temperatures. For lean doughs, with no fat or sugar, the temperature range is approximately 80 to 85°F (27 to 30°C). The dough should double in volume. For rich, sweet doughs, the temperature range is approximately 75° to 77°F (24° to 25°C). Rich, sweet doughs such as laminated doughs for Danish pastry or croissants use a lower proofing temperature to prevent the fat within the dough from melting and oozing out from between the layers of dough.

Lean doughs that tend to use fewer ingredients, like French bread, tend to rely on long proofing times to develop flavor within the bread. Longer proofing times help the gluten strands stretch and hold more gases, which help create an open hole structure within the bread.

Rich, sweet doughs tend to be given shorter proofing times. The fat within the dough coats gluten strands, which weakens them so the dough does not benefit from a longer proofing time. Flavor for rich, sweet doughs generally comes from the rich ingredients within them and not as much from the fermentation and proofing stages.

The proofing process is complete when an indentation made by a finger gently pressed into the dough springs back slowly.

Tips to Follow When Shaping Yeast Doughs

1. Handle the dough as little as possible during the shaping step. Overhandling develops too much gluten, making the dough harder to shape.

2. Use as little flour as possible when shaping to prevent the dough from becoming too dry.

3. Spritz the work surface lightly with water or a drop of oil to allow the dough to stick slightly to the work surface for better shaping.

10. Baking

The dough is now ready for baking.

Breads that were proofed in baskets are unmolded by flipping them out onto a peel that has been dusted with flour or cornmeal and placed onto the floor of the oven to bake.

SCORING OR SLASHING

Some breads, especially lean doughs such as French bread, are cut with a razor just before being placed into the oven. This is referred to as scoring or slashing. The cuts are shallow and prevent the bread from splitting in the oven as it continues to rise, before the crust has formed. Shapes can be cut into the dough which can form into attractive patterns after the bread is baked.

WASHES

Just before the dough is placed into the oven, some bread doughs are brushed with a liquid known as a wash. Much like glazing a clay pot before it goes into a kiln, a wash can give breads a rich brown color, provide a shiny exterior, or prevent the dough from drying out and forming a crust too quickly. Washes can also help a topping of seeds or nuts to adhere to the surface of the dough. Liquids that are used as washes include whole eggs, egg yolks, egg whites, milk, cream, butter, water, or a paste made from starch and water.

STEAM

Steam is used frequently within the first 5 to 10 minutes of baking for lean doughs with hard crusts like French bread. Steam can be brought into the oven by spritzing the inside of the oven and the dough with a water bottle or by placing a pan of ice cubes on the bottom of the oven. Many commercial bread ovens have the capability to inject steam directly into the oven. The moisture produced from the steam keeps the dough soft enough so that it has time to expand and rise quickly and evenly without forming a crust too soon.

THE BAKING PROCESS

The shaped dough is placed in the oven, allowing the yeast its final rise before dying at 140°F (60°C).

There is much activity during the baking process. For instance, the carbon dioxide gas bubbles trapped throughout the strands of gluten in the dough begin to expand as the temperature rises, pushing the dough upward. The moisture within the dough forms steam that also expands, aiding in this process of leavening the dough. This rapid rising of the dough is known as oven spring.

Once the dough has risen, the gluten within the dough sets and becomes firm, giving strength and structure to the bread. During the baking process, the starches within the flour in and on the surface of the dough absorb moisture. The starches then swell and become firm. The firming of the starches helps maintain the structure of the bread so it does not collapse after baking. A thin, crisp crust also forms. This process is referred to as the gelatinization of starches.

MAILLARD REACTION

As the temperature in the oven climbs to between 300° and 500°F (149° and 260°C), the moisture on the surface of the dough combines with the sugars (broken down starches) and proteins within the flour to form a crisp, flavorful brown crust with complex caramel-like flavors. This chemical reaction is known as the Maillard reaction. It is named after the scientist who first observed it.

COMPLETION OF BAKING

As yeast breads finish their time baking in the oven, flavorful aromas permeate the kitchen. These aromas are due to the evaporation of alcohols and other volatile (easily evaporated) organic compounds from the freshly baked bread.

Breads are done when they sound hollow when thumped with a fist or an instant read thermometer inserted into the bread reads between 200° and 212°F (94° and 100°C).

11. Cooling

The freshly baked bread is removed from the oven and allowed to cool on cooling racks. Breads baked in pans are removed from them and placed on the racks. The cooling racks allow maximum air circulation, which aids in the speedy evaporation of any excess moisture.

12. Storing

Storing yeast breads properly is most important in maintaining their quality. Hard-crusted breads that will be eaten within an 8- to 10-hour period may be left uncovered at room temperature. To store breads for longer periods of time, it is important that they be cooled completely before being wrapped for storage.

For longer storage, a yeast bread should be wrapped airtight in plastic wrap and then in aluminum foil. It should then be placed in a plastic bag, which is again sealed airtight with a tie, or knotted. Before sealing, remove as much air as possible from the bag. Then store the bread in the freezer.

Wrapping and freezing hard-crusted breads will soften the crusts. After thawing, they can be placed in a hot oven to regain crispness. Yeast breads with a softer crust like rich coffee cake or sweet rolls can also be placed in the oven to regain a "just baked" texture. It is not recommended to refrigerate yeast breads because staling is encouraged at those temperatures.

Two Basic Types of Yeast Doughs

There are two basic types of yeast doughs: lean doughs and rich doughs.

Lean Doughs

A lean dough is a yeast dough that uses little or no fat or sugar. They include breads that are prepared using few ingredients and tend to have hard crusts. Lean dough breads include French, Italian, and pizza.

Recipes that fall in between a lean and a rich bread are sometimes called *enriched*. For example, white bread or whole wheat bread recipes may contain milk, fat, sugar, and eggs, but the percentages of these ingredients are small enough to still label them as lean doughs. Challah bread falls into this category.

RECIPES

BBQ Chicken Pizza (This chapter, page 20)
Double Braided Challah (This chapter, page 30)
French Bread (This chapter, page 17)
Reuben Braid (This chapter, page 23)
Walnut Lemon Buttermilk Bread (This chapter, page 27)

Rich Doughs

A rich dough, in general, contains greater amounts of fat and sugar. Rich doughs may also include eggs. The crust of breads made with rich doughs tends to be softer. Rich yeast breads include brioche, coffee cakes, and cinnamon rolls.

LAMINATED DOUGHS

Another type of specialty rich yeast dough is called a *laminated dough*, also referred to as a rolled-in dough. Laminated doughs are doughs that do not have the fat incorporated with the other ingredients of the dough; instead, the fat is rolled into the dough using a series of folds and turns, which produces hundreds of thin layers of dough and fat. When baked, these rich doughs increase in volume to make very light, flaky pastries. Laminated doughs produce such pastries as croissants and Danish pastry. Puff pastry does not contain yeast. They are discussed in greater depth in Chapter 3.

RECIPES

Almond Danish Braid (Chapter 3, page 94)
Croissants (Chapter 3, page 97)
Pain au Chocolat Croissant Coffee Cake (Chapter 3, page 102)
Raspberry Danish Spirals (Chapter 3, page 88)
Rich Cinnamon Rolls (This chapter, page 34)

Determining Which Flour Is Best

Because many recipes combine different types of flour to obtain specific textures in baked goods, it is important to use the specific flour stated in the recipe. Bakers can remake the same recipe using different combinations of flours to experiment with altering the texture of the finished product. Figure 1–5 shows different types of wheat and their uses.

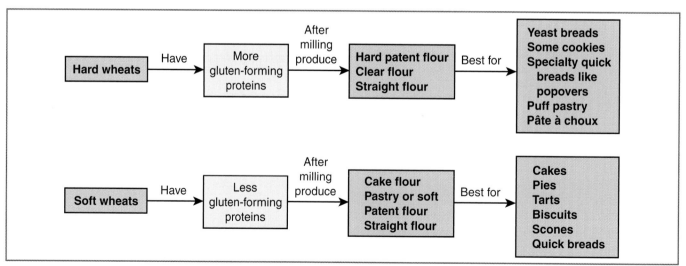

FIGURE 1–5

FRENCH BREAD

Makes three 12¾-ounce
(360 g), 14-inch (35-cm)
baguettes

Lean Dough Recipes

Lessons demonstrated in this recipe:

- How to prepare a lean yeast dough using the straight dough method.
- The addition of wheat germ adds texture and flavor to an otherwise plain bread.
- An additional fermentation period helps flavors develop even further.
- Spritzing the baguettes (long thin loaves) before baking keeps the dough moist, allowing it to attain plenty of oven spring. The moisture also helps the starches on the surface to firm up and create a crisp, outer crust.

MEASUREMENTS				INGREDIENTS
U.S.		**METRIC**	**BAKER'S %**	
1 pound + 4 ounces	4 cups	570 g	92%	bread flour
2 ounces	½ cup	50 g	8%	raw wheat germ
⅛ ounce	1¾ teaspoons	5 g	0.8%	instant active dry yeast
14 fluid ounces	1¾ cups	420 mL	68%	warm water (110° to 115°F; 43° to 46°C)
½ ounce	2 teaspoons	12 g	1.9%	kosher salt
				cornmeal for dusting
			170.7%	Total French Bread percentage

1. In the bowl of an electric mixer place the flour, wheat germ, and yeast. Using a spoon or hands, combine the flour mixture until well mixed.

2. Using the dough hook attachment on medium speed, add the water to the flour and yeast mixture and mix until a dough forms. Turn the machine off and cover the mixer with a kitchen towel and allow the dough to rest for 10 minutes.

3. On low speed, add the salt and mix until the salt is incorporated.

4. On a work surface, knead the dough until smooth and elastic, adding as little dusting flour as possible to the work surface.

5. Spray a large mixing bowl with nonstick cooking spray and place the dough in it, turning it over so the greased side is up. Cover the bowl with plastic wrap and let the dough rise for 1 to 2 hours or until it has doubled in volume.

6. Punch the dough down by pulling the edges up and down into the center and then turn it over. Place the dough back into the bowl, covering it, and allow it to rise for approximately 1 hour or until it has doubled again. This extra fermentation helps develop flavor. If there is no time, skip to step 7 but proof the loaves in step 8 for 1 to 1½ hours, not 30 minutes.

7. Preheat an oven lined with baking tiles or with a baking stone on the bottom to 500°F (260°C). Divide the dough into thirds. Shape each piece into a rough ball and allow them to rest, covered, for 10 minutes.

8. Shape each ball into a baguette (long cylinder with tapered ends) about 14 inches (35 cm) long (Figure 1–6). Place the baguettes on a wooden board dusted with cornmeal, leaving a space between each baguette. The shaped dough can, alternatively, be placed into a greased baguette pan. Cover with a clean kitchen towel or greased plastic wrap and allow to proof at room temperature for about 30 minutes. The baguettes will look puffy but will not have doubled. (*Note:* Sometimes bakers use a piece of heavy canvas or linen couche. During proofing, baguettes and other yeast breads can be nestled in a couche, which is used to hold their shape during the proofing process.)

9. Make 3 to 5 angled cuts down the length of each baguette using a razor (Figure 1–7). Spritz each baguette with a water bottle and slide it into the oven on a peel or bake in baguette pans. Immediately turn the temperature down to 450°F (230°C). Bake for 5 minutes, quickly spritzing the inside of the oven three more times, waiting 1 to 1½ minutes between each spraying.

10. Bake for another 15 to 20 minutes or until the loaves are golden brown and feel firm to the touch. They should sound hollow when tapped with a finger. Place the baguettes on a rack to cool.

FIGURE 1–6

FIGURE 1–7

French Bread

BBQ CHICKEN PIZZAS

Makes six 8- to 9-inch (20- to 23-cm) individual pizzas

Lessons demonstrated in this recipe:

- How to prepare a lean yeast bread using the straight dough method.
- A short rest or autolyse allows the gluten fibers to line up and develop properly, producing better gluten.
- The dough can be baked on a grill or in an oven.

STEP A: PREPARE THE DOUGH

MEASUREMENTS				INGREDIENTS FOR TOPPING
U.S.		**METRIC**	**BAKER'S %**	
1 pound + ¼ ounce	3¼ cups	460 g	90%	bread flour, plus extra for dusting the work surface
1¾ ounces	¼ cup	50 g	10%	yellow cornmeal
	½ teaspoon	1 g	0.2%	garlic powder
⅛ ounce	1¾ teaspoons	5 g	1%	instant active dry yeast
11 fluid ounces	1⅓ cup	330 mL	65%	lukewarm water (110°F; 43°C)
2 fluid ounces	¼ cup	60 mL	12%	olive oil
	1½ teaspoons	10 g	2%	kosher salt
				nonstick cooking spray
				extra olive oil (or cooking spray) for brushing during grilling
			180.2%	Total Pizza Dough (Step A) percentage

1. In a large mixing bowl, whisk together the flour, cornmeal, garlic powder, and yeast. Make a well in the center of the mixture and add the water and olive oil (Figure 1–8). Mix well with your hands or a spoon until the dough forms a ball (Figure 1–9). Cover the ball with a clean kitchen towel and allow it to rest for 10 minutes. Uncover the dough. Sprinkle the salt over the dough and knead it in.

FIGURE 1–8

FIGURE 1–9

2. Place the dough on a flat work surface and knead it for approximately 1 to 2 minutes or until it becomes elastic.

3. Place the dough in a large bowl sprayed with nonstick cooking spray. Flip the dough over so the greased side is up. Cover the bowl with plastic wrap. Let the dough rise for about 1 hour or until it has doubled in volume.

4. Place the dough onto a lightly floured work surface. Cut the dough into 6 equal pieces using a dough cutter or a knife. Knead each piece briefly to deflate it and cover with a damp cloth or kitchen towel. Allow the dough to rest for about 10 minutes. Resting allows the gluten to relax a bit so you can better shape the pizzas.

STEP B: ASSEMBLY

| MEASUREMENTS | | | | INGREDIENTS |
U.S.		METRIC	BAKER'S %	
7 ounces	⅔ cup	200 g	39%	barbecue sauce
9 ounces	2 cups	255 g	50%	roasted, skinned, boned chicken, cut into small dice
4¼ ounces	1 cup	120 g	23%	mozzarella cheese, shredded
5 ounces	1½ cups	140 g	27%	smoked Gouda cheese, shredded
1¾ ounces	¾ cup	50 g	10%	sliced scallions
½ ounce	½ cup	15 g	2.7%	finely chopped cilantro
			151.7%	total topping (Step B) percentage
			331.9%	Total BBQ Chicken Pizza percentage

1. Start the grill on medium-high heat. Roll out each piece of dough into a thin rough circle approximately 8 to 9 inches (20 to 22.5 cm) in diameter (Figure 1–10). Brush each circle with a little olive oil and grill on both sides until brown and puffed, about 4 minutes per side (Figure 1–11). Use tongs to turn the pizzas. (The grilling can be done several hours ahead or a few days before. After grilling, allow the crust to cool to room temperature and cover each grilled pizza with plastic wrap and refrigerate until needed.) The dough can also be grilled ahead of time, cooled, wrapped in plastic wrap, and placed in a freezer bag to be frozen for up to 3 to 4 months.

FIGURE 1–10

FIGURE 1-11

TIP If a grill is not available, the dough rounds can be baked in a preheated 450°F (230°C) oven on a parchment-lined sheet pan or directly on the bottom of an oven lined with baking tiles. Bake for 10 to 15 minutes or until puffy and lightly browned.

(*Note:* Instead of making individual pizzas, one large pizza can be made.)

2. In a bowl, lightly mix one fourth of the barbecue sauce with the chicken. Transfer the pizza crusts to a parchment or foil-lined baking sheet and spread each crust with some of the remaining barbecue sauce, the cheeses, and the chicken. Sprinkle some scallions and cilantro over each one (Figure 1–12). Place the pizzas in a preheated 450°F (230°C) oven for about 10 to 15 minutes or until the cheeses have melted and are bubbly. Serve at once.

FIGURE 1–12

BBQ Chicken Pizza

REUBEN BRAID

Makes one approximately 17-inch (43-cm) braid serving 6 people

This recipe is based on a sandwich known as a Reuben—a sandwich made of corned beef, Swiss cheese, and sauerkraut on rye bread that is traditionally fried in butter.

Lessons demonstrated in this recipe:

- How to prepare a lean yeast bread using the straight dough method.
- Waiting to add the salt after the dough has been mixed allows better distribution of ingredients before the gluten tightens up the dough.
- The finished braid is not allowed to proof to prevent the finished bread from getting too big and overpowering the filling.

STEP A: PREPARE THE DOUGH

MEASUREMENTS				INGREDIENTS FOR DOUGH
U.S.		**METRIC**	**BAKER'S %**	
18 fluid ounces	2¼ cups	530 mL	69%	warm water (110°F; 43°C)
1¼ fluid ounces	2 tablespoons + ¾ teaspoon	35 mL	4.6%	canola or olive oil
1 pound + 6½ ounces	4½ cups	640 g	83%	bread flour plus extra for dusting the work surface
4½ ounces	1 cup	130 g	17%	rye flour
⅛ ounce	1¾ teaspoons	5 g	0.7%	instant active dry yeast
	2¼ teaspoons	15 g	2%	kosher salt
				nonstick cooking spray
			176.3%	Total Dough (Step A) percentage

1. In the bowl of an electric mixer, add the warm water and the oil.

2. In another bowl, whisk together 18 ounces (4 cups; 570 g) of the bread flour, all of the rye flour, and the yeast. With the paddle attachment on low speed, slowly add the flour mixture to the liquids in the bowl and blend well. Clean off the paddle and change to the dough hook. Add up to 3 ounces (⅔ cup; 85 g) of the remaining bread flour until a soft dough forms. The entire amount may not be needed. Stop the machine and cover the bowl with a clean kitchen towel. Let the dough rest for 10 minutes. Uncover the bowl and on medium speed add the salt and mix until well combined. This will take approximately 1 minute.

3. Place the dough onto a floured work surface and knead it to develop the gluten for about 3 to 5 minutes until the dough feels smooth and elastic. The dough should not feel too sticky.

4. Spray a large bowl with nonstick cooking spray and place the dough in the bowl. Flip the dough over so the greased side is up. Cover the bowl tightly with plastic wrap. Let the dough rise for 1 hour.

Note: The dough is rolled out directly on an inverted sheet pan so it does not have to be moved after filling and shaping.

5. Preheat the oven to 375°F (190°C). Punch the dough down in the bowl by folding the outer edges into the middle to collapse the dough. Let the dough rest (covered with a clean kitchen towel or plastic wrap) in the bowl for 10 minutes.

FIGURE 1–13A

FIGURE 1–13B

6. Turn a half sheet pan upside down and spray the bottom with nonstick cooking spray. Place the dough on the baking sheet and roll it into a rectangle 15 inches by 10 inches (37.5 cm by 25 cm) long (Figure 1–13A and B). Alternatively, the dough can be rolled out onto a floured work surface and, after it is shaped, it can then be transferred to a parchment lined sheet pan.

STEP B: PREPARE THE FILLING AND ASSEMBLE THE BRAID

MEASUREMENTS				INGREDIENTS FOR FILLING
U.S.		METRIC	BAKER'S %	
2½ ounces	⅓ cup	70 g	9%	mayonnaise
2½ ounces	¼ cup	70 g	9%	chili sauce
8 ounces		225 g	29%	thinly sliced corned beef or smoked turkey
4 ounces		115 g	15%	sliced Swiss cheese
1 8-ounce		225 g	29%	can sauerkraut, well-drained (squeeze out excess liquid)
1 each		28 g	3.7%	large egg white, beaten
				caraway seeds, optional
			94.7%	Total Filling (Step B) percentage
			271%	Total Reuben Braid percentage

FIGURE 1–14

FIGURE 1–15

1. Mix the mayonnaise and the chili sauce together in a small bowl. Spread the mixture down the center of the dough, lengthwise, leaving 2.5 inches (6.5 cm) of dough uncovered on either side (Figure 1–14). Top with layers of corned beef or smoked turkey, cheese, and sauerkraut (Figure 1–15).

2. Cut the uncovered dough along the sides of the filling into 2-inch (5-cm) long strips, using a small knife or scissors (Figure 1–16). This will resemble fringe. Starting at one end and alternating sides, fold the strips in at an angle across the filling to look like they are being braided (Figure 1–17). The braided dough will not be proofed. Brush the finished braid with the beaten egg white. Sprinkle with caraway seeds (Figure 1–18).

3. Bake the bread at 375°F (190°C) for 35 to 40 minutes, or until nicely browned on the top and the sides. Cool for 5 minutes, cut into slices, and serve warm or at room temperature.

FIGURE 1–16

FIGURE 1–17

FIGURE 1–18

Reuben Braid

WALNUT LEMON BUTTERMILK BREAD

Makes one 9-inch by 5-inch by 3-inch (22.5-cm by 12.5-cm by 7.5-cm) loaf

Lessons demonstrated in this recipe:

- How to prepare a lean yeast dough using the straight dough method.
- Giving the dough a short rest or autolyse before adding the salt allows for better gluten development.
- The addition of buttermilk gives the bread a tangy flavor. As an acid, the buttermilk reduces gluten formation, producing a more tender crumb.
- A small amount of walnut oil intensifies the walnut flavor of the bread.
- Wheat bran and wheat germ add flavor and nutritional value and help produce a crunchy texture that complements the walnuts.
- The lemon zest adds a pleasant citrus aroma throughout the bread.

MEASUREMENTS				INGREDIENTS
U.S.		METRIC	BAKER'S %	
15 ounces	3 cups	425 g	85%	bread flour plus extra for dusting the work surface
1 ounce	½ cup	25 g	6%	wheat bran
1¾ ounces	½ cup	50 g	10%	raw wheat germ
5 ounces	1¼ cups	145 g	29%	finely chopped walnuts, toasted
½ ounce	1½ tablespoons	15 g	2.8%	lemon zest
⅜ ounce	3½ teaspoons	10½ g	2.1%	instant active dry yeast
4 fluid ounces	½ cup	120 mL	24%	warm water at 110° to 115°F (43° to 46°C)
1¼ fluid ounces	3 tablespoons	45 mL	9%	walnut oil
½ ounce	1 tablespoon	20 g	4%	honey
12 fluid ounces	1½ cups	340 mL	68%	buttermilk
½ ounce	2 teaspoons	12 g	2.4%	kosher salt
			242.3%	Total Walnut Lemon Buttermilk Bread percentage

1. In the bowl of an electric mixer using the paddle attachment on low speed, blend the bread flour, wheat bran, wheat germ, walnuts, lemon zest, and yeast until well combined.

2. In a mixing bowl, whisk together the water, the walnut oil, honey, and buttermilk. Blend well. On low speed add the water and buttermilk mixture to the dry ingredients. Remove the paddle and change to the dough hook. Mix on medium speed for 1 minute. Cover the mixer with a kitchen towel and allow the dough to rest for 10 minutes.

3. On low speed, add the salt and mix for 1 minute until thoroughly incorporated.

4. Remove the dough from the bowl and knead it on a lightly floured surface until smooth, satiny, and still slightly sticky.

5. Place the dough into a large bowl sprayed with nonstick cooking spray. Flip it over so the greased side is facing up. Cover the bowl tightly with plastic wrap and allow it to rise for about 1 to 1½ hours or until it has doubled in volume.

6. Punch the dough down and shape it into a rough ball. Cover and allow the dough to rest for 10 minutes.

7. Spray a 9-inch by 5-inch by 3-inch (22.5-cm by 12.5-cm by 7.5-cm) loaf pan with nonstick cooking spray. Shape the dough into a rectangle the same length as the loaf pan (Figure 1–19). Roll the dough lengthwise into a tight spiral (Figure 1–20). Gather it under each end and gently pick it up and place it into the greased pan, making sure that the seam side is down.

8. Cover the filled loaf pan lightly with greased plastic wrap or a clean kitchen towel and allow it to proof until the dough doubles in volume, about 1 hour (Figure 1–21A and B).

9. Preheat the oven to 375°F (190°C). Bake the loaf for about 40 minutes or until golden brown. Remove the bread from the pan while it is still hot and place it, on its side back in the oven on a stone, or baking tiles, if possible, for another 6 to 8 minutes to create a crisp crust. Cool on a rack.

FIGURE 1–19

FIGURE 1–20

FIGURE 1–21A

FIGURE 1–21B

Walnut Lemon Buttermilk Bread

DOUBLE BRAIDED CHALLAH

Makes one large double braid, approximately 3 pounds (1⅓ kg)

Challah is a yeast bread made yellow in color by the addition of eggs and egg yolks to the dough. Challah bread is a leaner version of brioche, a tender French yeast bread made richer by the addition of butter. Challahs are often shaped into braided loaves, which is traditional in the Jewish religion. Challah bread is great for sandwiches and makes delicious French toast!

Lessons demonstrated in this recipe:

- How to prepare a lean yeast dough using the straight dough method.
- The browning of the dough is intensified by the yolks in the egg wash.

MEASUREMENTS				INGREDIENTS
U.S.		METRIC	BAKER'S %	
1 pound + 9 ounces	5 cups	710 g	100%	bread flour
⅜ ounce	3½ teaspoons	10½ g	1.5%	instant active dry yeast
3½ ounces	¼ cup + 2 tablespoons	105 g	15%	granulated sugar
2 fluid ounces	¼ cup	60 mL	8%	canola oil
12 fluid ounces	1½ cups	360 mL	51%	warm water 110°F (43°C)
1 pinch		1 pinch	0%	saffron, optional
2 each		94 g	13%	large eggs
1 each		19 g	2.7%	large egg yolk
	2 teaspoons	12 g	1.7%	kosher salt
				extra flour for dusting
				nonstick cooking spray
				1 large egg, beaten in a small bowl to be used as an egg wash
			192.9%	Total Double Braided Challah percentage

1. In a medium bowl, whisk together the flour, the yeast, and the sugar and set aside. In the bowl of an electric mixer using the paddle attachment, blend the canola oil, the water, the saffron (optional), the eggs, and the egg yolk, on low speed.

2. Add one half of the flour mixture and blend well. Then stop the machine and change the paddle to the dough hook attachment. While mixing on medium speed, add the remaining flour mixture. A ball of dough will start to form. Stop the machine, pull the dough off the hook to turn it over and put it back into the bowl. The dough should be somewhat sticky. Run the mixer for 1 to 2 minutes on medium speed until well mixed. Turn the mixer off and cover the bowl with a kitchen towel. Allow the dough to rest for 10 minutes.

3. Remove the kitchen towel and with the mixer running on medium speed, add the salt and continue mixing for another minute to thoroughly incorporate the salt.

4. Take the ball of dough out of the bowl and place it on a lightly floured surface to knead it for a few minutes and develop the gluten. Do not use too much flour or the challah will be dry. The dough should feel nice and soft, yet smooth and elastic.

Saffron is an expensive spice that is actually a part of the crocus's purple flower. Saffron comes from the stigma, or inner part, that looks like small threads but are actually sacs that contain pollen. It is used as a flavoring in the Middle East and to give a yellowish-orange color to foods.

5. Spray a large bowl with nonstick cooking spray and place the ball of dough in it. Flip the dough over so that the greased side is facing up. Cover the bowl tightly with plastic wrap. Allow the dough to rise for about 1½ hours, or until it has doubled in volume.

6. Uncover the dough and punch down by pulling the outer edges of the dough into the middle to gently deflate it. Spray one full sheet pan with nonstick cooking spray or cover it with parchment paper and set nearby. Have the beaten egg wash and a pastry brush ready. Shape the dough into a rough rectangle and place it on a large, smooth, clean work surface. Divide the dough into two pieces, one twice as large as the other (Figure 1–22).

7. Shape both pieces into flattened rectangles (Figure 1–23). Using a dough cutter, cut each rectangle lengthwise into three ropes (Figure 1–24).

FIGURE 1–22

FIGURE 1–23

FIGURE 1–24

8. Roll each of the three larger ropes, separately, into 15- to 16-inch (38.5- to 41-cm) lengths. Set them aside. Roll each of the three smaller ropes into 15- to 16-inch (38.5- to 41-cm) lengths. Braid the three larger ropes together, starting in the middle and working out to the ends. This will help to keep the braid uniform in size. Braid the three smaller ropes together in the same manner (Figure 1–25).

9. Place the larger braid onto the prepared full sheet pan.

10. Using a pastry brush, paint the larger braid with egg wash and place the smaller braid on top of it (Figure 1–26). Push down slightly to make sure the two braids stick together.

11. Cover the challah loosely with a clean kitchen towel or a piece of lightly greased plastic wrap. Allow the challah to rise, until doubled in size (about 1 hour).

12. Preheat the oven to 375°F (190°C). Brush the entire challah with egg wash and bake for 15 minutes. Remove the challah from the oven and brush with egg wash again. Continue baking for another 10 to 15 minutes or until golden brown. Cool on a rack.

Variation:

Instead of one large challah, approximately two dozen knotted challah rolls can be made with this recipe. Follow the above recipe until step 6. Instead of dividing the dough into pieces, scale out 24 2-ounce (55-g) pieces of dough. Roll out each piece of dough into a 9- to 10-inch (22.5- to 25-cm) rope. Shape each rope into a knot and place on a full sheet pan that has been greased or covered with parchment paper. Allow some space between each knot. Allow the knots to rise for about 1 hour or until they are doubled in size. Brush each knot with egg wash and bake at 375°F (190°C) for 10 minutes. Remove the knots from the oven and brush each knot with egg wash again. Return the rolls to the oven and continue baking for another 10 to 15 minutes, or until golden brown. Cool on a rack.

FIGURE 1–25

FIGURE 1–26

Double Braided Challah

RICH CINNAMON ROLLS

Makes 32 rolls

Rich Dough Recipe

Lessons demonstrated in this recipe:

- How to prepare a rich, sweet yeast dough using the modified straight dough method.
- The milk is scalded to destroy protease, an enzyme that can decrease volume in yeast breads.
- The flavor of the cinnamon rolls comes directly from cinnamon and the rich ingredients, such as butter, eggs, and sugar.
- Because instant yeast is more lively than either compressed or active dry yeast, it is best used for rich, sweet yeast doughs in which fermentation times are short, such as rich cinnamon rolls.
- Because flavor is derived from the rich ingredients, the dough is shaped without any bulk fermentation, going directly to shaping and proofing, which saves time.

STEP A: MAKING THE DOUGH

MEASUREMENTS				INGREDIENTS
U.S.		METRIC	BAKER'S %	
8 fluid ounces	1 cup	240 mL	19%	milk
6 fluid ounces	¾ cup	180 mL	15%	room temperature water
6 ounces	¾ cup	170 g	14%	unsalted butter, softened
7½ ounces	2 cups	215 g	17%	cake flour
2 pounds + 4¼ ounces	8¼ cups	1030 g	83%	bread flour and more, if needed
½ ounce	5 teaspoons	15 g	1.2%	instant active dry yeast
11 ounces	1½ cups	310 g	25%	granulated sugar
5 each		235 g	19%	large eggs, beaten and warmed to room temperature in a warm water bath
¼ ounce	1 teaspoon	6 g	0.5 %	salt
			193.7%	Total Dough (Step A) percentage

1. In a medium saucepan scald the milk and remove it from the heat. Add the water and the butter. Stir until the butter is melted. Allow the mixture to cool to 115° to 120°F (46° to 49°C).

2. In the bowl of an electric mixer using the paddle attachment, blend the cake flour, approximately one half of the bread flour, the yeast, and the sugar on low speed to mix well. Reserve the remaining bread flour to be added later.

3. Using the paddle attachment, add the warm milk mixture on low speed, followed by the eggs. Stop the machine and switch to the dough hook.

4. On low speed, continue mixing and add as much of the reserved bread flour as necessary to make a soft dough. The dough should be shiny and a bit sticky. If it is too sticky, add a small amount of bread flour. Turn the mixer off, cover the bowl with the kitchen towel, and allow the dough to rest for 10 to 15 minutes.

FIGURE 1–27

FIGURE 1–28

5. On medium speed, add the salt and mix well. Remove the dough from the bowl and knead it on a work surface until smooth and elastic. Do not dust the work surface with too much flour or the dough will become dry.

6. Shape the dough into a rough rectangle and using a dough cutter, divide the dough into two equal parts (Figure 1–27). Roll out each half to an 11- by 20-inch (28- by 51-cm) rectangle (Figure 1–28).

Makes 1 pound, 8¼ ounces, 690 gm; 2¾ cups

STEP B: CINNAMON PASTE FILLING

MEASUREMENTS				INGREDIENTS
U.S.		METRIC	BAKER'S %	
15 ounces	2 cups	425 g	34%	light brown sugar (packed, only if measured by volume)
7¼ ounces	1 cup	205 g	17%	granulated sugar
½ ounce	2 tablespoons	15 g	1.2%	ground cinnamon
1 each		28 g	2.3%	large egg white
1 fluid ounce	2 tablespoons	30 mL	2.4%	water
1 each		28 g	2.3%	large egg white, to be used as glue
			59.2%	Total Filling (Step B) percentage
			252.9%	Total Cinnamon Rolls percentage

1. Mix all ingredients except the last egg white (which will be used as glue later on) until blended and a spreadable paste forms.

FIGURE 1–29

FIGURE 1–30

STEP C: ASSEMBLY

1. Using a pastry brush, paint each rectangle with the egg white that was reserved to be used as glue. Spread each rectangle with one half of the cinnamon paste filling using a palette knife or offset spatula. Starting at the long side, roll each rectangle into a tight cylinder (Figure 1–29).

2. Using a dough cutter or a sharp knife, cut each cylinder into 1-inch (2.5-cm) wide rolls (Figure 1–30). There should be 16 rolls per cylinder, 32 total.

3. Place the rolls cut side down on a parchment-lined full sheet pan leaving a space between each one. Use a second sheet pan, if necessary.

4. Cover the rolls lightly with greased plastic wrap or a clean kitchen towel and allow them to proof until they have doubled in size, approximately 1 to $1\frac{1}{2}$ hours.

5. Preheat oven to 350°F (175°C) and bake for 25 to 30 minutes or until golden brown.

6. Drizzle or spread with simple icing (Chapter 8) while still warm.

Variation:

Combine $10\frac{1}{4}$ ounces (2 cups; 290 g) raisins and $8\frac{1}{2}$ ounces (2 cups; 240 g) chopped pecans in a bowl. Scatter one half of the mixture over the cinnamon paste filling on each rectangle of dough before rolling into a cylinder (see Step C: Assembly Step 1).

Rich Cinnamon Rolls

Professional Profile

BIOGRAPHICAL INFORMATION

Beverly A. Bates, ACF
Pastry Chef Instructor and Coordinator of Culinary Arts
Baltimore International College
Baltimore, MD

1. Question: *When did you realize that you wanted to pursue a career in baking and pastry?*

Answer: *I realized I wanted to pursue a career in baking and pastry when I was a lowly cook in Watertown, NY, and the pastry team quit. I was the only prep person left who wanted to do it. It was love from the very beginning.*

2. Question: *Was there a person or event that influenced you to go into this line of work?*

Answer: *As a child I loved looking through the cookbooks at the color pictures of food as a lot of poor kids do, and I always was drawn to the beautiful wedding cakes, thinking "How do they do that?" Because we had eight children in my family, my mother was always baking and cooking, so she always had plenty of cookbooks around.*

3. Question: *What did you find most challenging when you first began working in baking and pastry?*

Answer: *The biggest challenge in the field when I first started was to get a handle on the large number of basic skills a pastry chef needs to know each very different from the other. Pastry cream is nothing like banana bread or a baguette or an ice cream or a chocolate mousse, yet you must speak the language of each fluently.*

4. Question: *Where and when was your first practical experience in a professional baking setting?*

Answer: *My first professional experience in baking was about 18 years ago at the Black River Valley Club in Watertown, NY, working for Vince Rose. That is where and when the pastry team walked out.*

5. Question: *How did this experience affect your later professional development?*

Answer: *Being able to rise to the challenge and perform as well as I did gave me the confidence and cockiness I needed to compete to work in a highly competitive industry filled with very talented people.*

6. Question: *Who were your mentors when you were starting out?*

Answer: *My first mentor was a culinary instructor named Floyd Misek and later a pastry chef at The Culinary Institute of America named Frank Vollkommer.*

7. Question: *What would you list as the greatest rewards in your professional life?*

Answer: *Among the greatest rewards in my professional life is the ability to travel a little while doing what I love to do.*

8. Question: *What traits do you consider essential for anyone entering the field?*

Answer: *An essential trait for anyone entering the field is to have high personal standards. Don't worry about what everyone else is doing, just do what you know is your best, and never look back!*

9. Question: *If there was one message you would impart to all students in this field what would that be?*

Answer: *I would prepare and caution students to expect to work harder than they think they will need to. It is a competitive field and it is full of very talented people. So, keep your eyes open and never stop learning.*

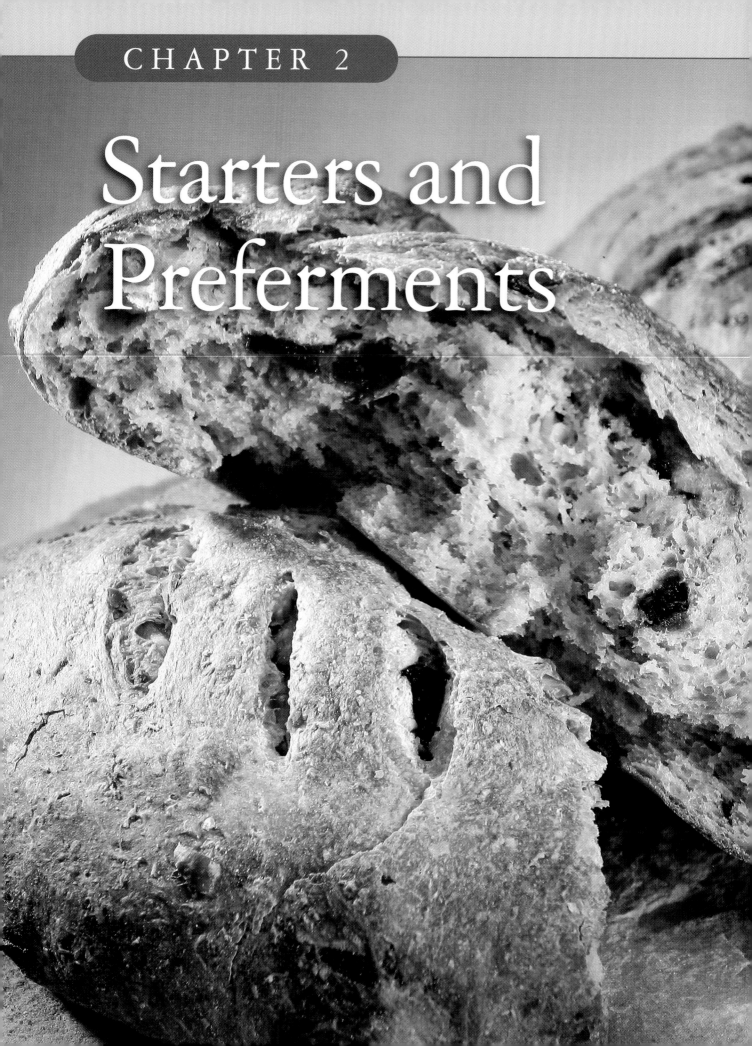

CHAPTER 2

Starters and Preferments

After reading this chapter, you should be able to:

- Define a preferment.

- List the benefits of using a preferment.

- Understand the difference between a straight dough and a dough that uses a sponge or starter.

- Recall the differences between the various types of sponges and sourdoughs.

- Work with preferments and prepare the recipes at the end of this chapter.

In the previous chapter, you learned how yeast is used to create breads from straight yeast doughs. This chapter deals with very flavorful yeast breads that are made with preferments. The word preferment means "to ferment before." Doughs made with preferments are prepared in stages. The yeast and some of the flour and water are mixed together first and allowed to ferment anywhere from 30 minutes to several days at cool room temperature or in the refrigerator. This mixture is then used as a foundation with which to build a dough. Using a preferment gives the fermentation process a head start, contributing great flavor and/or leavening to the finished bread.

There are several different types of preferments. Some are thick and stiff like a dough, whereas others are thin like a batter. Some are prepared using commercial yeast, whereas others are prepared using wild yeast.

All yeast doughs that use some form of a preferment add a great deal of flavor to a bread. Just like a good wine gets better over time, so does a bread with a long, slow fermentation. As you learned in the previous chapter, there are straight yeast doughs and doughs made using preferments. Chapter 1 deals exclusively with straight doughs and this chapter deals exclusively with preferments. At times, references will be made to straight doughs for comparison.

This chapter explores the different types of preferments and their applications in yeast doughs.

Straight Doughs

To review, straight doughs are very simple yeast doughs in which most of the ingredients are mixed together at the beginning of bread making. Straight doughs are used to make breads such as white, whole wheat, and challah, among others. They are given a relatively quick fermentation and proofing and then are off to the oven. This quick process does not add a great deal of flavor to the finished bread.

Preferments

Preferments, on the other hand, are yeast mixtures that are fermented before the actual dough is made. The yeast and some of the flour and water are mixed together first and allowed to ferment anywhere from 30 minutes to several days. Preferments are used for two reasons: (1) to increase flavor in breads and (2) to provide leavening.

ARTISAN BREADS

Preferments are associated with artisan breads, which are prepared by bakers who manipulate the dough with their hands with great care and skill using traditional methods. Traditional methods can be traced back to the Old World European style of bread baking wherein breads were baked by hand in wood-fired ovens using little machinery such as mixers and proof boxes. Although modern commercial bakeries do use mixers and some other equipment to prepare enough bread to meet demand, for the most part, commercial operations that prepare artisan breads tend to remain small-scale so the baker can feel and craft each loaf.

Artisan breads tend to share similar characteristics:

■ They tend to include only natural ingredients with few preservatives.

■ They tend to be prepared using a preferment that contributes great flavor.

■ They tend to be manipulated and crafted by hand at some stage during their preparation.

Artisan bakers choose to follow the Old World bakers of years ago. Proof boxes are not used; instead, the dough is fermented at room temperature for a longer period of time than that of a straight dough. These longer fermentation periods enhance the flavor of the finished bread considerably. Because some commercial yeast may not survive a long, cool fermentation period, some artisan bakers prefer to use natural or wild yeasts instead to prepare preferments such as sourdough starters.

Two Categories of Preferments

Preferments can be broken down into two categories: sponges and sourdoughs. (See Table 2–1, Various Types of Preferments.) Before discussing the various types of preferments, it should be noted that there are times when the terminology used by some chefs to refer to the different types of preferments is not used consistently and may be confusing. For example, some bakers use the terms *sponges* and *sourdoughs* interchangeably. Technically both sponges and sourdoughs are preferments. However, there are differences between them. One difference is that sponges tend to be made from commercial yeast while sourdoughs are made with wild yeast.

> Preferments are used for two reasons:
>
> **1.** To increase flavor in breads
>
> **2.** As a leavening agent

Sponges

Sponges generally are a mixture of flour, water, and yeast that is mixed together and allowed to ferment anywhere from 30 minutes to several hours. The sponge is then mixed with other ingredients to form a dough and allowed to ferment one more time. (*Note:* The sponge is completely used up in the recipe.) The dough is then formed and proofed before being baked. Sponges tend to be a quicker way to add flavor and a mild sour taste to a bread. Sponges are also used when a rich, sweet dough containing a large quantity of sugar is made. High quantities of sugar and fat can slow down fermentation, so forming a sponge before the sugar and fat are added gives the yeast a head start. Sponges can also decrease overall fermentation time.

Although there are many variations of sponges, there are two major types: those that are thin and batter-like and those that are thick and dough-like. They vary mainly in the amount of water that is added. Thinner sponges tend to ferment more quickly, whereas thicker sponges ferment more slowly, giving the baker more time before it needs to be used. Examples of thinner

Table 2–1 **Various Types of Preferments**

SPONGES		SOURDOUGH STARTERS
Use commercial yeast		Use wild or natural yeast
Completely used up in a recipe		Can be maintained for long periods of time
Poolish	Thin	Chef
Levain-levure	Thin or thick	Levain
Biga	Thick	Mother
Pâte fermentée	Thick	Barm
Altus brat	Thick	Desem

sponges include poolish and levain-levure. Examples of thicker types of sponges include biga, pâte fermentée, and altus brat.

POOLISH

Poolish, the French word for "Polish," was so named because Polish bakers are thought to have originated this type of sponge technique. It is often simply referred to as a sponge. A poolish is typically prepared with equal parts of flour and water by weight.

The amount of commercial yeast used can vary. The longer the poolish is expected to ferment, the less yeast should be used. When a poolish is fermented fully, it should have risen and then fallen back onto itself so that it has a wrinkled appearance on the top. A poolish is fermented at room temperature from 3 to 15 hours. It can also be refrigerated overnight and brought to room temperature before using it.

RECIPES

Buttermilk, Dried Plum, and Hazelnut Bread (This chapter, page 53)
Overnight Poolish Bread (This chapter, page 56)

LEVAIN-LEVURE

A levain-levure is a French term for a preferment using commercial yeast that means "leaven of the yeast." It is typically a stiff sponge, but can be thinner like a poolish. A levain-levure should not be confused with a levain, a sourdough starter using wild yeast.

BIGA

A biga is an Italian preferment so thick that it resembles a bread dough. A typical biga contains approximately 50 to 60 percent water as compared to the weight of the flour and up to one half of a percent of instant yeast. Because of its stiff consistency, it can be prepared several hours in advance and left at cool room temperature for up to 3 days. A stiff sponge like a biga is the preferment of choice when preparing yeast breads with a high water content, because the tight gluten network that forms within the biga provides great structure to the dough. A biga tends to contain more yeast to make up for its thick texture, which can slow down fermentation.

PÂTE FERMENTÉE

Pâte fermentée is a French term that simply means "old dough." It is simply a piece of dough that has been saved from a previous batch of bread dough. This piece of dough is added into a new batch of dough toward the end of mixing. The old dough is a quick way to add the flavor of a preferment to a new batch of bread because it has already gone through the fermentation process.

Some bakers prepare a separate dough just to be used as a pâte fermentée so the fermentation time can be monitored carefully. Because a pâte fermentée contains all the ingredients of a bread dough, including salt, which can inhibit fermentation, additional yeast may be added to make up for this.

Care must be taken not to overferment a pâte fermentée because the extra yeast can produce an overly acidic dough as a result of excess by-products of fermentation. Therefore, these doughs tend to be fermented at cool room temperature (approximately 65°F; 19°C).

ALTUS BRAT

Altus brat is a German expression meaning "old bread." It is similar to a pâte fermentée except, instead of a piece of old dough being added to a batch of new dough, old bread is soaked in water, squeezed dry, and allowed to ferment. This fermented old bread is then added to a new batch of dough.

Sourdoughs

Sourdoughs or sourdough starters are similar to sponges but tend to need more of a time commitment, requiring several days before they are ready to use. Besides the yeast sponge, a sourdough also contains bacteria. Two species of bacteria that exist in a sourdough are Lactobacillus and Acetobacillus. As sugar within the flour undergoes fermentation by yeast, the bacteria in the starter also undergo a slightly different kind of fermentation and give off lactic and acetic acids. It is these acids that give sourdough breads a pleasant sour flavor. Sourdoughs, like sponges, can use commercially available yeast or they can take advantage of wild yeast that exist naturally in the air, or on the skins of fruits or vegetables without using any commercial yeast at all. For example, grapes with wild yeast on their skins can be added to a flour and water mixture to begin a sourdough starter.

Sometimes fruits or vegetables are used merely as an attractant on which airborne wild yeast can feed. For example, a vegetable, such as potato, can be boiled in water. The starches and natural sugars are released from the potato into the water. This potato water is then used to prepare the sourdough starter. These starters are left at cool room temperature to attract wild yeast that can survive at the same cool temperatures.

A sourdough starter can be created in many ways. Typically, a sourdough starter starts with equal parts of flour and water by weight. The amount of water may vary depending upon the thickness desired. The type of flour used can also vary. Some bakers prefer to start with an organic whole grain rye or whole wheat flour that contains a great deal of natural yeast and bacteria already in it. As the starter becomes strong and active over time, white flour can be added gradually while decreasing the amount of rye or whole wheat flour. Eventually, this would turn the culture into what is referred to as an *all white starter*. However, an all white starter can be made successfully without starting with amount of rye or whole wheat flours using just unbleached white bread flour. An example of this is seen in the potato starter recipe in this chapter.

Sourdoughs have been kept for centuries and were used exclusively before commercial yeast became available. Sourdoughs that use natural or wild yeast are called natural starters or sourdough cultures (sourdough starters). There are as many flavors of sourdough breads as there are species of yeast and bacteria. A sourdough bread made in San Francisco will taste different than a sourdough bread made in New York. This is because there are so many different varieties of yeast. Certain species of yeast may thrive in certain locations of the world but may die in others. Each area has its own wild yeast native to that area. Sourdoughs are like bird feeders. Give them the right food, and they will come. Strains of wild yeast eat sugars in the flour. A starter attracts natural yeast from a particular area. Those yeast tend to survive because they are already used to the environment. A kitchen where there is a great deal of yeast bread baking will already be plentiful with natural yeast endemic to the area.

A sourdough starter in its beginning stages is referred to as a chef or seed culture. After the chef is fed over a period of time, it becomes strong enough to bake bread. It is then referred to as a *sourdough culture* or a *sourdough starter*. Some other names for sourdough starters include *barm, desem,* and *mother.* The sourdough breads from France are known as *pain au levain.* They are prepared using a sourdough starter known as levain. Bakers in France develop their own special formula to make their pain au levain unique in flavor.

Sourdough starters can be thin like a batter or thick like a dough. Thinner, more fluid starters tend to impart more of a sour taste to a bread than a thicker one does. Thinner starters also ferment at a faster rate.

A healthy sourdough starter can be used successfully as the sole leavening agent for bread. However, sometimes a small amount of commercial yeast is added to the starter or the bread dough to ensure a well-risen loaf. This is known as *spiking*.

RECIPES
Basil and Rosemary Pesto Sourdough Bread (This chapter, page 64)
Garlic-Infused Olive, Onion, and Sweet Pepper Focaccia (This chapter, page 60)

Yeast and Bacteria Living Together

During the beginning stages of developing a natural starter, an interesting phenomenon begins to occur. As a natural starter grows, yeast and bacteria begin to live harmoniously in a symbiotic relationship within the starter. Natural or wild yeast that are grown in sourdough starters thrive in an acidic environment unlike commercially grown yeast, which may not survive.

In sponges made using commercially grown yeast, sugars broken down within the flour are completely consumed by the yeast, leaving no available food for any bacteria to grow. So breads made with these sponges will not be very sour.

In starters made using natural or wild yeast, not all the available sugars are eaten by the yeast. Because the wild yeast are not able to digest certain types of sugar that only commercial yeast can, the natural bacteria present feed eagerly on these leftover sugars. The result is a starter in which bacteria begin to grow and live happily side-by-side with the yeast.

Through the process of fermentation, the yeast give off alcohol and carbon dioxide. The bacteria give off lactic and acetic acids. Other organic compounds are also produced. All of these organic compounds add to the flavor of the finished bread.

Sourdough starters need to be fed regularly to maintain the crucial balance between yeast and bacteria. The feedings usually consist of flour and water.

(Note: Bread made with natural starters only get better over time. Breads made with a new starter, as good as it may be, will never taste as wonderful as bread made with an older, more mature one.)

As previously mentioned, one difference between sponges and starters is that sponges are completely used up in a recipe, whereas only a portion of the starter is actually used at any one time. This leaves plenty of starter that can be fed and maintained almost indefinitely. After all the trouble a baker goes to, to grow and nurture a starter, especially one with wild yeast, discarding it is not an option.

Stirring a healthy starter sounds like bubble paper being popped. It is full of life and, upon close inspection, gas bubbles can be seen forming and popping constantly. It can be very rewarding to create a sourdough starter and then bake incredible bread from it. All this effort yields a great sense of accomplishment for the baker and satisfaction for the customer.

Reliability and Hardiness of Starters

Starters using commercial yeast tend to be much more reliable at leavening yeast breads than natural sourdough starters. If natural or wild yeast are not plentiful when made, the starter becomes overrun with bacteria and may never become healthy and strong. A natural balance must exist between yeast and bacteria that is necessary to maintain the starter's health.

This is the main reason so many bakers add at least a pinch of commercial yeast to their natural starters as a "jump starter." As the wild yeast take hold and multiply, eventually they overrun the commercial yeast and become healthy and strong. Otherwise, unless wild yeast exist in the kitchen, success of a natural sourdough cannot be guaranteed.

Well-maintained sourdough starters are very hardy. A healthy starter can last for weeks and even months in the refrigerator without being fed. However, this is not recommended because the balance of yeast and bacteria may be altered.

Yeast lie dormant between 34° and 40°F (1° and 4°C) without being harmed under adverse conditions like a lack of food or water or too cool a temperature. In this cooler temperature range, fermentation still occurs but at a much slower rate.

The starter can be successfully "reawakened" or reactivated by allowing it to warm up to room temperature and then beginning feedings to get the yeast back to a healthy, active state. A healthy starter that has been properly maintained can be kept for many years.

Preferments, especially sourdough starters, tend to fend off bad bacterial growth through the natural acids they produce. This acidity inhibits the development of molds and staling, giving breads made with these starters a longer shelf life and an antibacterial quality.

Healthy sourdough starters are so hardy that if an otherwise healthy older starter became contaminated in some way, it is likely that it could be brought back to good health through proper feedings with no deleterious effects.

Developing Flavor in Sourdough Breads

Chapter 1 describes how enzymes and fermentation help develop flavor in yeast breads.

Developing flavor in sourdough breads made using wild yeast starters introduces another organism into the mix—bacteria. Bacteria, like yeast, also eat sugars that are released through enzymatic activity. Bacteria also give off their own by-products, including lactic and acetic acids. The temperature at which fermentation occurs determines how much of each acid will form. The particular combination of acidic by-products that forms in a starter imparts a pleasant sour taste to the sourdough bread baked with them.

Acetic acid, which is the more sour of the two acids, forms at lower temperatures of between 40° and 55°F (4° and 13°C). The milder lactic acid is formed at warmer temperatures of between 55° and 90°F (13° and 32°C).

The strain of wild yeast and the specific types of bacteria present will also determine how sour the starter and ultimately the bread baked with it will be.

Sourdough starters using wild instead of commercial yeast can be fermented for longer periods of time at cooler temperatures without being overproofed because yeast activity is slowed by the acidic environment. These long fermentation times at cooler temperatures produce a greater depth of flavor because the by-products of fermentation have more time to be produced.

Retarding the Dough

Many preferments can be fermented in the refrigerator overnight for a longer, cooler fermentation to bring out more flavors. This is known as retarding. Yeast activity slows down at refrigerated temperatures, but continues nonetheless.

Another reason bakers retard the dough is for purposes of scheduling. Because preparing breads using preferments can be so time-consuming, the long fermentation time slows down bread baking. The unattended time necessary for a preferment to ferment breaks up the work involved and what might have been too laborious in 1 day becomes more manageable over a period of a few days.

Retarding yeast doughs can be successfully accomplished with some straight doughs as well as those using preferments.

Differences between Sponges and Sourdoughs

Sponges—Tend to be short-lived. They ferment anywhere from 30 minutes to several hours and are then completely used up in a recipe. Sponges can leaven while imparting complex flavors to breads.

Sourdoughs—The fermentation process for a sourdough tends to take more time to mature, at least 24 hours and up to several days. Sourdough starters can be maintained for years. They need to be fed regularly. Only a portion is used in a recipe, while the rest is saved. It can be refrigerated or frozen. Sourdough starters can leaven bread and give them a complex flavor with a pleasant sour taste.

OVERNIGHT POOLISH SPONGE

Lessons demonstrated in this recipe:

- How to prepare a thin sponge using commercial yeast.
- Less yeast is used to compensate for a long fermentation time.
- The poolish is retarded overnight in the refrigerator to slow down fermentation and increase flavor.
- Allowing the poolish to retard in the refrigerator overnight also helps scheduling become more manageable over 2 days.
- A variety of flours are provided as food for the yeast to encourage fermentation.
- Semolina, the flour of choice for pastas, comes from durum wheat. It is so hard that, during the milling process, as semolina is cracked open, the starches in it are damaged. These starches break down easily and become excellent food for the yeast.
- The sponge is used up completely in the recipe.

MEASUREMENTS				INGREDIENTS
U.S.		**METRIC**	**BAKER'S %**	
4 fluid ounces	½ cup	120 mL	109%	water at 78°F (26°C)
	½ teaspoon	2½ mL	2.3%	honey
	¼ teaspoon	1¼ g	1.1%	instant active dry yeast
¾ ounce	⅛ cup	20 g	18%	semolina flour
½ ounce	⅛ cup	15 g	14%	whole wheat flour
2½ ounces	½ cup	75 g	68%	bread flour
			212.4%	Total Overnight Poolish Sponge percentage

1. In the bowl of an electric mixer using the paddle attachment, blend the water, honey, yeast, and flours on low speed (Figure 2–1).

2. Stop the machine, scrape down the bowl using a rubber spatula, and mix on low speed until thoroughly combined.

3. Turn the machine on medium speed and mix for 1 minute to develop the gluten.

4. Cover the mixing bowl with plastic wrap and place it in the refrigerator overnight for 12 to 15 hours (Figure 2–2). The next morning, remove the poolish from the refrigerator and allow it to warm up to room temperature for 1 hour before using it in a recipe (Figure 2–3). (See the Overnight Poolish Bread recipe that uses this sponge.)

TIP If the bread has to be made within 1 day, the poolish can be made and covered with plastic wrap and allowed to sit at room temperature for 2 hours before using it in a recipe. The poolish should look puffy and be used just as it is beginning to fall. It will look dimpled and slightly pulled inward.

FIGURE 2–1

FIGURE 2–2

FIGURE 2–3

POTATO SOURDOUGH STARTER

Lessons demonstrated in this recipe:

- How to prepare a natural sourdough starter.
- The potato water supplies the wild yeast with more sugars and starches with which to feed on.
- A natural starter takes several days to mature and a portion of it can be saved and maintained for future baking.

MEASUREMENTS				INGREDIENTS
U.S.		METRIC	BAKER'S %	
16 fluid ounces	2 cups	475 mL	191%	potato water
8¾ ounces	1¾ cups	250 g	100%	unbleached white bread flour
	1 teaspoon	6 g	2.4%	granulated sugar
			293.4%	Total Potato Sourdough Starter percentage

Potato Water

MEASUREMENTS				INGREDIENTS
U.S.		METRIC		
1 each		1 each		large potato, peeled and cut into 1-inch (2½-cm) chunks
28 fluid ounces	3½ cups	830 mL		bottled water

1. Place the potato chunks and water into a medium saucepan over medium-high heat. Cook for approximately 12 to 15 minutes or until the potato chunks can be easily pierced with a fork. Remove from the heat.

2. Allow the potato and water mixture to cool until almost room temperature, about 78°F (25°C). Measure out 2 cups (500 mL) of the potato water, discarding the rest, and reserving the potato for another use.

The Starter

3. In a very clean 2-gallon (7½-liter) plastic container, mix the potato water, flour, and sugar, using a wooden spoon (Figure 2–4).

4. Cover the container with plastic wrap secured with a rubber band (Figure 2–5). Let it sit at cool room temperature (between 70° and 75°F; 21° and 24°C) for about 24 hours.

FIGURE 2–4

FIGURE 2–5

FIGURE 2–6

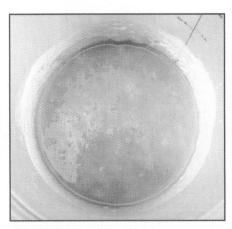

FIGURE 2–7

FIGURE 2–8

5. Stir the starter once every day, allowing it to sit for 3 days while making observations about how it looks and smells (Figure 2–6).

6. On the fourth day, the starter will start to separate and smell quite sour (Figure 2–7). The sour smell is from the bacteria beginning to outnumber the yeast. Feed the starter 9 ounces (1¾ cups; 255 g) unbleached white bread flour and 16 fluid ounces (2 cups; 480 mL) water at 78°F (25°C) (Figure 2–8). This feeding will provide a stimulus for the yeast to multiply.

7. Allow the starter to sit at cool room temperature for 4 more days (Figure 2–9). Do not stir. Continue to make observations.

8. On the ninth day, begin a regular maintenance schedule of three feedings per day, approximately every 4 to 6 hours. After the third feeding, the starter is covered and left overnight. Continue this schedule for 2 more days.

Note: Only 1 pound plus 2 ounces (2 cups; 510 g) of the starter is used at the beginning of every day (Figure 2–10). This is done to keep the amount of starter to a manageable amount. The remaining starter should be discarded and the container washed out thoroughly every morning before the first feeding. It is important to use clean containers and utensils when preparing a sourdough starter to prevent harmful bacteria and mold from developing.

FIGURE 2–9

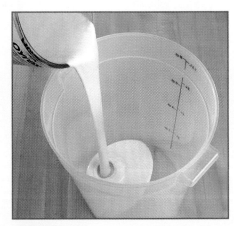

FIGURE 2–10

First Feeding: This is best done in the morning.

Ingredients: 1 pound 2 ounces (2 cups; 510 g) starter, 8 fluid ounces (1 cup; 240 mL) water at 78°F (26°C) and 6 ounces (1¼ cups; 170 g) unbleached, white bread flour.

While the starter is weighed on the scale in another clean container, thoroughly clean out the 2-gallon (7½-liter) plastic container the starter was in originally. Place the starter back into the container and add the flour and water. Sir the mixture with a wooden spoon and cover with plastic wrap secured with a large rubber band. Leave the starter at room temperature.

9. *Second Feeding:* 4 to 6 hours later. The second feeding will be double what the first one was. Add 2 cups (480 mL) water at 78°F (26°C) and 12 ounces (2⅓ cups; 340 g) bread flour. Stir mixture with a wooden spoon and cover with plastic wrap secured with a rubber band. Leave the starter at room temperature.

10. *Third Feeding:* 4 to 6 hours later. Add 32 fluid ounces (4 cups; 960 mL) water at 78°F (26°C) and 1 pound plus 8 ounces (4¾ cups; 680 g) bread flour. Stir the mixture with a wooden spoon and cover with plastic wrap secured with a large rubber band. Allow the starter to sit at room temperature overnight for approximately 12 to 15 hours.

11. *The Next Day:* Stir the starter down and discard all but 1 pound 2 ounces (2 cups; 510 g). Repeat the feedings using the same schedule as described.

12. Do this for 1 more day, for a total of 3 days of regular feedings to get the starter ready for baking. The starter should smell yeasty and be bubbly (Figure 2–11).

FIGURE 2–11

Observations:

- Day 1—Starter is made.
- Day 2—After stirring, starter turns a light brown with some bubbles and some activity. Starter smells very mild, a little yeasty.
- Day 3—Mixture separates a bit and smells acidic. It is very active. After being stirred down, starter separates into a yellowish liquid and foam. Natural bacteria multiplies and vies for living conditions with the yeast. (*Note:* It is in this "battle" so to speak between yeasts and bacteria that makes the starter seem like something has gone horribly wrong. Do not give up. This is a normal phase and passes within a few days.)
- Day 4—Mixture deflates and separates completely into a yellowish liquid on top and more solid, batter-like material on the bottom. It smells very sour and acidic. The sourness is from the bacteria that are beginning to overrun the yeast. A feeding of flour and water helps encourage the yeast to multiply.
- Days 5–8—The starter is left alone for 4 days at room temperature. A yellowish liquid remains on the top and culture separates. Bubbles begin to appear.
- Days 9–11—Regular feedings are given. The starter is fed three times a day, every 4 to 6 hours. The starter is extremely active after each feeding, producing lots of bubbles with a yeasty smell and thick batter-like consistency.
- Day 12—The starter is ready for baking. (*Note:* The starter should be used between 12 and 15 hours after its last feeding.)

Maintaining the Starter

To maintain the starter without growing any more of it, match the volume of starter with the same volume of flour and water. A little extra flour is usually given to make sure that the yeast has enough food, if a feeding is ever delayed. For every 1 pound 2 ounces (2 cups; 510 g) of starter, feed it 8 fluid ounces (1 cup; 240 mL) water at 78° F (25° C) and 6 ounces (1¼ cups; 185 g) white bread flour three times a day.

The starter can also be stored in the refrigerator in an airtight container for up to several weeks without any feedings. Whenever storing the starter in the refrigerator, it is important to label it with the amount of the starter and the date when it was placed in the refrigerator. To bring the starter back to a healthier state for baking, bring the starter to room temperature for approximately 2 hours and begin 2 to 3 days of feedings (3 per day) before using it again to bake bread.

BUTTERMILK, DRIED PLUM, AND HAZELNUT BREAD

Makes two round loaves, each weighing approximately 1 pound 14 ounces (850 g)

Additional Ideas That Use the Recipes in This Chapter

Lessons demonstrated in this recipe:

- How to prepare a yeast bread using a sponge.
- The sponge helps lighten a dough containing whole grains, producing a less dense full-flavored bread.
- Buttermilk is added for moistness and gives the bread a pleasant, sour taste.
- Wheat bran, raw wheat germ, and whole wheat flour are added for a nutty flavor and texture.
- The high moisture content of the dough softens the dried plums.
- The sponge is completely used up in the recipe.

STEP A: 2-HOUR POOLISH OR SPONGE

MEASUREMENTS				INGREDIENTS
U.S.		METRIC	BAKER'S %	
6 fluid ounces	¾ cup	180 mL	20%	water at 78°F (25°C)
	½ teaspoon	1½ g	0.2%	instant active dry yeast
¼ ounces	⅛ cup	5 g	0.5%	wheat bran
½ ounce	⅛ cup	15 g	1.6%	raw wheat germ
2½ ounces	½ cup	70 g	8%	bread flour
2½ ounces	½ cup	70 g	8%	whole wheat flour
			38.3%	Total Step A, percentage

1. In the bowl of an electric mixer using the paddle attachment, mix the water, the yeast, wheat bran, wheat germ, and flours on low speed. Increase the speed to medium and mix for 2 to 3 minutes to develop the gluten. Scrape down the bowl with a rubber spatula. Cover the bowl with plastic wrap. Leave at room temperature for approximately 2 hours or until the mixture doubles in volume and looks puffed up. If the room temperature is below 70°F (21°C), place the mixture in a proof box.

STEP B: DOUGH

MEASUREMENTS				INGREDIENTS
U.S.		METRIC	BAKER'S %	
12 fluid ounces	1½ cups	360 mL	40%	warm water at 110°F (43°C)
½ teaspoon		2½ g	0.3%	instant active dry yeast
12 fluid ounces	1½ cups	360 mL	40%	buttermilk
1 ounce	2 tablespoons	30 g	3.3%	light brown sugar
3½ ounces	¾ cup	100 g	11%	whole wheat flour
1 ounce	½ cup	30 g	3.3%	wheat bran
2 ounces	½ cup	55 g	6%	raw wheat germ
¾ ounce	1 tablespoon	20 g	2.2%	salt
1 pound + 4 ounces	4 cups	565 g	62%	bread flour, plus more if needed
4½ ounces	1 cup	130 g	14%	pitted dried plums, cut into ¼-inch (6-mm) pieces
4 ounces	1 cup	115 g	13%	chopped hazelnuts, toasted
			195.1%	Total Step B, percentage
			233.4	Total Buttermilk, Dried Plum, Hazelnut Bread percentage

1. Place the bowl with the poolish onto the base of the electric mixer. Add the water and yeast. Using the paddle attachment, blend the mixture on low speed.

2. Add the buttermilk, brown sugar, whole wheat flour, wheat bran, and wheat germ. Blend using the paddle attachment.

3. Add the salt and enough of the bread flour to make a thick, sticky mass of dough.

4. Change from the paddle to the dough hook and mix well. Add more bread flour, if necessary. The dough should be slightly wet and sticky.

5. Using a bowl scraper, scrape the dough onto a lightly floured work surface. Knead it lightly using the bowl scraper to lift it over onto itself for 2 to 3 minutes.

6. Knead in the dried plums and hazelnuts until well distributed (Figure 2–12).

7. Place the dough into a large bowl that has been sprayed with nonstick cooking spray. Flip the dough over so the greased side is up. Cover the bowl with plastic wrap and allow it to rise for approximately $1\frac{1}{2}$ to 2 hours or until it has doubled in volume. This can be done at room temperature or in a proof box.

8. Preheat a deck oven or an oven with a pizza stone or baking tiles to 450°F (230°C).

9. Punch the dough down by pulling the outer edges into the middle. Divide the dough into two equal pieces. Shape each piece into a round free-form loaf. Place the loaves on a wooden board sprinkled generously with cornmeal. Cover the loaves with a clean kitchen towel or greased plastic wrap and allow them to proof at room temperature for approximately 1 hour or until they have doubled in volume.

10. Using a razor blade, make decorative slashes on the surface of each loaf and slide them into the oven (Figure 2–13). Bake for 3 to 4 minutes, spritzing the loaves every minute or so with a water bottle. Continue baking, without opening the oven, for another 20 minutes.

FIGURE 2–12

FIGURE 2–13

11. Reduce the oven temperature to 400°F (205°C) and rotate the loaves. Continue baking for another 15 minutes or until the loaves are a dark golden brown. The crust should feel hard and the bread should sound hollow when thumped with a finger. Cool on a rack.

Buttermilk, Dried Plum, and Hazelnut Bread

OVERNIGHT POOLISH BREAD

Makes two round loaves, each weighing approximately 1 pound 13¾ ounces (845 g)

Lessons demonstrated in this recipe:

- How to prepare a yeast bread using a sponge.
- A sponge is prepared and allowed to ferment overnight in the refrigerator and is completely used up in the recipe.
- A combination of flours is used for flavor and texture.
- A long, slow fermentation for the poolish starter helps increase flavor while decreasing the dough's overall fermentation time.

STEP A: OVERNIGHT POOLISH (SPONGE)

1. Make one recipe of Overnight Poolish sponge that has been brought back to room temperature.

STEP B: DOUGH

MEASUREMENTS				INGREDIENTS
U.S.		METRIC	BAKER'S %	
1 pound + 11½ ounces	5½ cups	780 g	85%	unbleached bread flour
3 ounces	½ cup	85 g	9%	semolina flour
2 ounces	½ cup	55 g	6%	raw wheat germ
20 fluid ounces	2½ cups	600 mL	65%	water at 78°F (26°C)
	1½ teaspoon	4½ g	0.5%	instant active dry yeast
1 tablespoon		18 g	2%	kosher salt
				extra bread flour for dusting
8 ounces		234 g	25%	Overnight Poolish Sponge starter (from Step A)
			192.5%	Total Overnight Poolish Bread percentage

1. In a bowl, whisk to blend the two flours and the wheat germ. Set aside.

2. Add the water and the yeast to the starter already in the electric mixing bowl and blend using the paddle attachment until it breaks up and becomes loose.

3. Add two thirds of the flour mixture and blend on low speed (Figure 2–14).

FIGURE 2–14

FIGURE 2–15

FIGURE 2–16

4. Change to the dough hook and continue adding the remaining flour until a slightly sticky and elastic dough forms (Figure 2–15). Continue mixing on medium speed for 30 to 60 seconds.

5. Turn the machine off, cover the bowl with a clean kitchen towel, and allow the dough to rest for 10 minutes.

6. Add the salt and blend until it is well incorporated, about 60 seconds (Figure 2–16).

7. Remove the dough from the bowl and knead it on a work surface, using little or no flour, until smooth and elastic (Figure 2–17).

8. Place the dough into a large mixing bowl that has been sprayed with nonstick cooking spray. Flip the dough over once so the greased side is up. Cover with plastic wrap and allow it to rise at room temperature (approximately 75° to 80°F; 24° to 27°C) or until doubled in size, about 2 hours.

9. Punch down the dough and divide the dough in half using a dough cutter. Cover each half and allow the dough to rest for 10 minutes.

FIGURE 2–17

10. Shape each half into a smooth ball (also known as a *boule*) (Figure 2–18) and place the dough into two heavily floured round baskets or bannetons (9 to 10 inches in diameter; 22.5 to 25 cm) (Figure 2–19). Flour the tops lightly, cover with a clean kitchen towel, and allow them to proof for 45 minutes to 1 hour. Bannetons are baskets made of willow in which the dough is allowed to proof. After unmolding, the pattern of the basket is imprinted on the dough, forming an attractive loaf. If a banneton is not available, a greased stainless steel or glass bowl can be used. Be sure that the bowl is twice the size of the dough to accommodate its rising.

11. While the boules are proofing, preheat a bread oven or a conventional oven with baking tiles or a pizza stone on the bottom to 450°F (230°C).

12. Gently flip one basket of dough over onto a floured peel. Using a razor, make shallow cuts in the dough (Figure 2–20). Slide the dough into the oven. Repeat with the remaining basket of dough.

13. Bake the loaves for 20 minutes, opening the oven briefly and spritzing the bread with a spray bottle filled with water once or twice every minute for the first 5 minutes of baking.

14. Using a peel, rotate the loaves 180 degrees and continue baking the breads for 10 to 15 minutes more or until they are golden brown and sound hollow when tapped with a finger. The internal temperature of each bread should read between 195° and 200°F (91° and 94°C). Cool on racks.

FIGURE 2–18

FIGURE 2–19

FIGURE 2–20

Overnight Poolish Bread

GARLIC-INFUSED OLIVE, ONION, AND SWEET PEPPER FOCACCIA

Lessons demonstrated in this recipe:

- How to prepare a yeast bread using a natural preferment.
- The starter should be used in the recipe within 12 to 15 hours after its last feeding.
- A small amount of commercial yeast is added to the dough for a lighter texture.
- The moisture level in the dough is high to create an open hole structure.

STEP A: SOURDOUGH STARTER

1. **Make Potato Sourdough Starter. Be sure to use it within 12 to 15 hours after its last feeding.**

STEP B: GARLIC-INFUSED OIL

MEASUREMENTS			INGREDIENTS
U.S.		METRIC	
8 each		8 each	large garlic cloves, skins left on, smashed with the flat of a knife
8 fluid ounces	1 cup	240 mL	extra virgin olive oil

1. **Preheat the oven to 300°F (149°C). In a small heatproof container, place the smashed garlic and oil.**

2. **Bake for 1 hour. Remove the garlic and oil from the oven and allow the mixture to cool for 30 minutes.**

3. **Pour the mixture through a sieve and discard the garlic pieces. Use this oil in the recipe.**

STEP C: CARAMELIZED ONIONS, OLIVES, AND ROASTED SWEET PEPPERS

MEASUREMENTS			INGREDIENTS
U.S.		METRIC	
½ fluid ounce	1 tablespoon	15 mL	olive oil
2 each		2 each	large yellow onions, peeled, halved, and thinly sliced
	½ teaspoon	2½ g	granulated sugar
	2 teaspoons	10 mL	balsamic vinegar
2 each		2 each	large sweet bell peppers (whole)
4¾ ounces	1 cup	135 g	Kalamata olives, pitted and coarsely chopped
1½ fluid ounces	3 tablespoons	45 mL	garlic-infused olive oil (Step B)

1. **In a large sauté pan, heat the olive oil on medium heat. Add the sliced onions, sugar, and vinegar. Mix together until well combined.**

FIGURE 2–21

FIGURE 2–22

FIGURE 2–23

2. Cover the pan and cook for 15 minutes, stirring frequently. The onions should be brown and caramelized. Remove them from the heat and allow to cool (Figure 2–21).

3. Roast whole bell peppers over an open flame or under a broiler, rotating them until the skins are completely blackened. Place them in a heatproof bowl covered with plastic wrap. The steam from the hot peppers will loosen their skins (Figure 2–22). Remove the skins, the stems, and the seeds; coarsely chop the peppers. Set aside.

4. Combine the onions, bell peppers, olives, and garlic-infused olive oil in a mixing bowl (Figure 2–23). Cover the mixture and chill until ready to use.

Makes two half sheet pans

STEP D: FOCACCIA DOUGH

MEASUREMENTS				INGREDIENTS
U.S.		METRIC	BAKER'S %	
24 fluid ounces	3 cups	710 mL	69%	water at 78°F (25°C)
1 pound + 2 ounces	2 cups	510 g	50%	potato sourdough starter
	½ teaspoon	2½ g	0.2%	instant active dry yeast
1½ ounces	3 tablespoons	45 mL	4.4%	garlic-infused olive oil
2 ounces	½ cup	55 g	5%	raw wheat germ
14¼ ounces	3 cups	405 g	40%	all-purpose flour
1 pound + 4 ounces	4 cups	565 g	55%	bread flour, unbleached, plus more for dusting
	1 tablespoon	18 g	1.8%	kosher salt
				olive oil for brushing
				freshly ground coarse black pepper
				kosher salt for sprinkling
			225.4%	Total Step D percentage

1. In the bowl of an electric mixer, place the water, starter, yeast, garlic-infused olive oil, wheat germ, and the flours.

2. Using the dough hook, mix the ingredients on medium speed for 3 to 4 minutes until a wet, sticky dough forms.

3. Stop the machine, cover the bowl with a clean kitchen towel, and allow the dough to rest for 10 minutes.

4. On medium speed, add the salt and mix well for 6 more minutes to develop the gluten.

5. Remove the bowl from the mixer, cover with plastic wrap, and allow the dough to ferment for 1 to 1½ hours at room temperature between 70° and 80°F (21° and 27°C) or until doubled in volume. The dough will be very sticky but do not add more flour at this point.

6. Pour the dough out onto a lightly floured surface. Using a dough cutter, cut the dough in half. Each half should weigh approximately 2 pounds and 6 ounces (1.1 kg). Brush two half sheet pans with olive oil and place one half of the dough in each pan. Pat the dough gently and evenly into the pans (Figure 2–24). Try not to pop any bubbles that formed during fermentation. Do not be concerned if the dough does not go into the edges of the pan. As the dough proofs, it will spread.

7. Gently cover the dough with plastic wrap sprayed with nonstick cooking spray. Allow the dough to proof for 1 hour or until the dough looks puffy.

8. Preheat the oven to 500°F (260°C). Gently peel off the plastic wrap from the dough. Brush the top of each focaccia with olive oil. Divide the olives, onions, and peppers evenly over each of the two focaccia. Sprinkle each with fresh cracked pepper and kosher salt (Figure 2–25A and B).

9. Place the focaccia in the oven and immediately lower the temperature to 450°F (230°C). Bake for 15 minutes.

10. Rotate the focaccia and continue baking for another 15 minutes or until they are brown and crusty.

FIGURE 2–24

FIGURE 2–25A

FIGURE 2–25B

Garlic-Infused Olive, Onion, and Sweet Pepper Focaccia

BASIL AND ROSEMARY PESTO SOURDOUGH BREAD

Makes two approximately 1 pound 12½-ounce (810-g) rounds

Lessons demonstrated in this recipe:

- How to prepare a full-flavored bread using a natural sourdough starter that allows for a longer, slower fermentation time.
- No commercial yeast is added. The bread relies solely on the natural starter for leavening.
- A pesto of basil, rosemary, and oil melts into the dough, infusing the bread with great flavor.
- A small amount of flax seed meal provides texture.

STEP A: BASIL AND ROSEMARY PESTO   MAKES 4¾ OUNCES (½ CUP; 135G)

U.S.		METRIC	INGREDIENTS
1 each		1 each	large garlic clove
2 ounces	2 cups	50 g	packed fresh basil leaves
	2 tablespoons	12 g	fresh rosemary, finely chopped
½ ounce	2 tablespoons	10 g	shredded Parmesan cheese
½ ounce	2 tablespoons	20 g	pine nuts
⅓ fluid ounce	2 teaspoons	10 mL	lemon juice
			2 pinches kosher salt
2 fluid ounces	4 tablespoons	60 mL	extra virgin olive oil

1. Through the opening at the top of a food processor with the motor running, add the garlic and pulse the machine until it is finely chopped.

2. Open the top and add the basil, rosemary, Parmesan, pine nuts, lemon juice, and salt. Pulse until finely chopped.

3. Through the opening at the top, slowly pour in the oil while the machine is running (Figure 2–26).

FIGURE 2–26

STEP B: BASIL AND ROSEMARY PESTO SOURDOUGH BREAD
MEASUREMENTS   INGREDIENTS

U.S.		METRIC	BAKER'S %	
18 fluid ounces	2¼ cups	540 mL	64%	cool water at 70° to 75°F (21° to 24°C)
12 ounces	1⅓ cups	345 g	41%	potato starter
1½ ounces	½ cup	45 g	5%	flax seed meal
1 pound + 14 ounces	6 cups	850 g	100%	white bread flour
½ ounce	3 teaspoons	18 g	2.1%	kosher salt
4¾ ounces	½ cup	135 g	16%	Pesto from Step A
			228.1%	Total Basil and Rosemary Pesto Sourdough Bread percentage

1. In the bowl of an electric mixer using the paddle attachment, place the water, starter, flax seed meal, and half the flour. Blend together on low speed for 30 seconds to thoroughly combine.

2. Change to the dough hook attachment and add the remaining flour, reserving 2 ounces (½ cup; 60 g). Continue to knead until a soft, slightly sticky dough forms. Add some of the reserved flour only if the dough is sticky but you may not need all of it.

3. Stop the mixer and flip the dough over to make sure it is being mixed completely. Mix again on low speed until the dough is well blended.

4. Stop the machine and cover the bowl with a clean kitchen towel for 15 minutes to allow the dough to rest.

5. On low speed, add the salt and blend for 2 or 3 minutes until the salt is completely incorporated.

6. Remove the dough from the bowl and knead it on a work surface until it is smooth and elastic, about 3 minutes. Resist the urge to add more flour. If the dough sticks to the work surface, use a dough scraper to scrape it up or a few drops of oil onto the work surface to reduce friction.

7. Place the dough in a large bowl that has been sprayed with a nonstick cooking spray. Flip the dough over so the greased side is up. Cover the bowl with plastic wrap and allow it to rise at room temperature (about 70° to 78°F; 21° to 25°C) for about 3 to 3½ hours or until it is doubled in volume and a finger mark pushed into the dough does not spring all the way back.

8. Punch the dough down by pulling the edges up and over into the center. Place the dough onto a work surface and knead it for 1 minute.

9. Divide the dough into two equal pieces, each weighing approximately 1 pound 14 ounces (855 g) and cover each piece with a clean kitchen towel or plastic wrap. Allow the dough to rest for 10 to 15 minutes.

10. Shape each piece of dough into a small ball. Slightly flatten by docking each ball with your fingers to make several small indentations (Figure 2–27A). Spread one half of the pesto mixture over the indentations of each piece of dough (Figure 2–27B). Working with each piece of dough separately, gather the edges of the dough like a drawstring purse to enclose the pesto (Figure 2–27C). Flip it over so that the smooth side is up and shape it into a smooth ball (Figure 2–27D).

11. Place each ball seam side facing up into two well-floured bannetons. Cover the bannetons with a clean kitchen towel and allow the dough to rise for about 1 hour or until they appear puffed.

12. Sprinkle some flour on the top of each round and wrap well in plastic wrap. Place the bannetons with the dough into the refrigerator for 12 to 22 hours to develop flavors (Figure 2–28).

13. Remove the bannetons from the refrigerator and remove the plastic wrap. Place a clean kitchen towel on top of each banneton and allow them to warm up to room temperature for approximately 2 hours.

14. Preheat an oven with baking tiles or a pizza stone to 500°F (260°C).

15. Invert the bannetons onto a well-floured baker's peel. Make decorative cuts with a razor blade.

FIGURE 2–27A

FIGURE 2–27B

FIGURE 2–27C

FIGURE 2–27D

FIGURE 2–28

16. Slide each round into the oven. Immediately turn the oven down to 450°F (230°C). Spritz the rounds with a water bottle. Bake for 5 minutes, spritzing once every minute for 5 minutes.

17. Bake for 20 minutes without opening the oven. Turn the rounds to rotate them and continue baking for another 10 to 12 minutes or until they are golden brown and sound hollow when thumped with a fist. A thermometer placed in the center of the dough should read approximately 210°F (99°C). Cool on racks.

Variation: White Sourdough Bread
Follow the directions except omit the pesto filling. Bake as directed.

Basil and Rosemary Pesto Sourdough Bread

Professional Profile

BIOGRAPHICAL INFORMATION

Joseph George
Executive Chef
Midland Country Club
Midland, MI

1. Question: *When did you realize that you wanted to pursue a career in baking and pastry?*

Answer: *When I was 19 I was exposed to my first professional kitchen and I knew that I wanted to be in this field.*

2. Question: *Was there a person or event that influenced you to go into this line of work?*

Answer: *James Krutcher was the first chef I worked under and he helped me go in this direction.*

3. Question: *What did you find most challenging when you first began working in baking and pastry?*

Answer: *I came from a culinary background, so crossing over to baking and pastry forced me to be a lot more exact and scientific.*

4. Question: *Where and when was your first practical experience in a professional baking setting?*

Answer: *I was doing a stint at a restaurant, Tapawingo, and the baking and pastry staff walked out. The chef told me to get in there and do it.*

5. Question: *How did this experience affect your later professional development?*

Answer: *I didn't know much about baking and pastry before that experience. As I developed my skills though, I gained confidence. That confidence allowed me to explore new areas and continually learn.*

6. Question: *Who were your mentors when you were starting out?*

Answer: *Steve Stallard has been a major influence on much of my career.*

7. Question: *What would you list as the greatest reward of your professional life?*

Answer: *By continually gaining knowledge I can pass more on to those who come after me. In that way, I can do for others what my mentors did for me.*

8. Question: *What traits do you consider essential for anyone entering the field?*

Answer: *Anyone going into baking and pastry must have patience and dedication.*

9. Question: *If there was one message you would impart to all students in this field what would that be?*

Answer: *Anyone making a career in this area needs to understand how much of a self-sacrifice it is.*

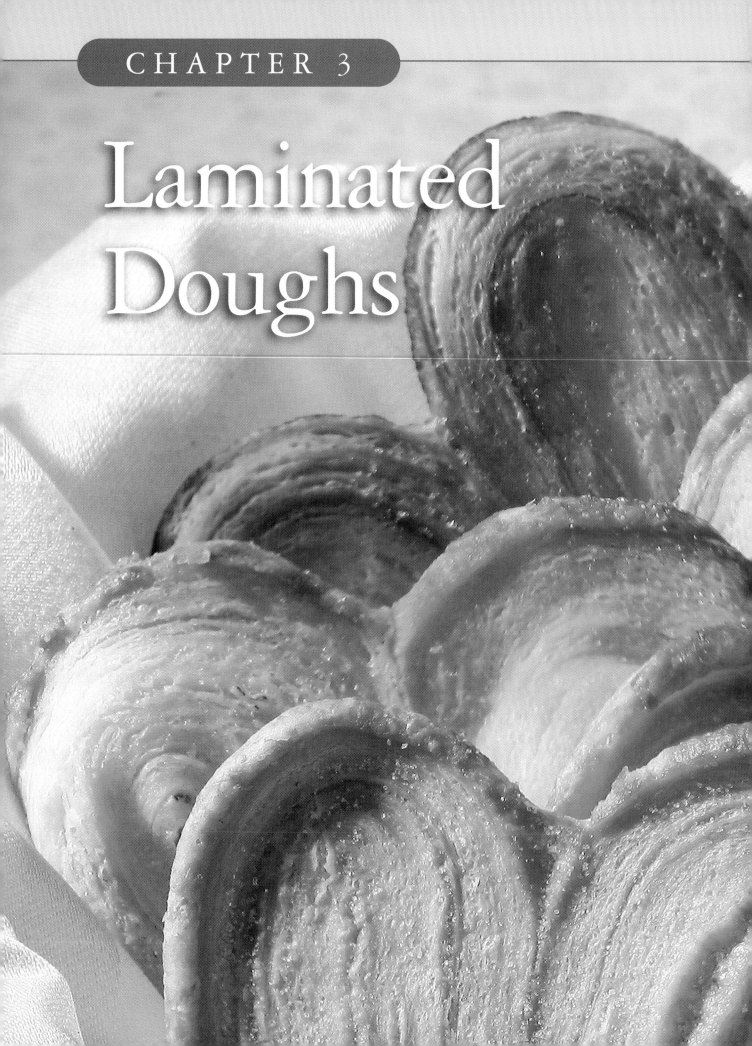

CHAPTER 3

Laminated Doughs

After reading this chapter, you should be able to:

- Define a laminated dough.

- List three different types of laminated doughs.

- Demonstrate the three-fold and four-fold folding techniques.

- Work with laminated doughs and prepare the recipes at the end of this chapter.

In Chapter 1, you learned about yeast and its role in the preparation of

straight yeast breads. In this chapter, rich sweet doughs known as *laminated doughs* are discussed.

Laminated dough is a category of very rich, fat-filled doughs used to make three main types of rich pastry: Danish pastry, croissants, and puff pastry. Only Danish and croissant doughs use yeast in their preparation. Other names used for laminated doughs are *rolled-in doughs* or *layered doughs.* Fats are incorporated into a base dough through a series of folds and turns that produce hundreds of flaky layers when baked. Pastries prepared using laminated doughs can be filled before or after baking using sweet or savory fillings.

This chapter explores the three different types of laminated dough, the differences between them, how to prepare them step-by-step, and their applications in various recipes.

Laminated Doughs Defined

All laminated doughs involve three steps: (1) preparing a base dough, (2) enclosing fat inside the dough, and (3) making a series of folds to produce thin layers of dough with fat in between them. Laminated doughs start out with a yeasted base dough for Danish and croissant doughs or a non-yeasted base dough for puff pastry that is rolled out to a rectangle. The fat (butter or solid shortening) is shaped into a rectangle and placed in the middle of the dough. In puff pastry doughs, the fat is mixed with some flour to give it more structure before it is folded into the dough.

> All laminated doughs involve three steps:
> 1. Preparing a base dough
> 2. Enclosing fat inside the dough
> 3. Folding and layering

The dough is then folded around the fat. The "fat sandwich" is then rolled out to a rectangle and then folded again. This is referred to as a turn. The pastry is then wrapped and chilled for approximately 1 hour to relax any gluten that may have developed. The steps of rolling, folding, and chilling are repeated until very thin layers of "dough, fat, dough" are made. The number of turns the dough has received in between each rest should be recorded. This is traditionally done by pressing as many fingers into the dough as the number of turns completed. The dough can also be marked after wrapping it in plastic using a marker or pen.

After an overnight rest, the dough is ready to shape. The dough cannot be kneaded at this stage because any mixing would incorporate the butter into the dough, destroying any layering. Traditionally, a laminated dough is cut and rolled out to a specific length and width. Fillings can be added and the dough can be cut and rolled into various shapes to make such pastries as snails, bear claws, envelopes, tarts, twists, coffee cakes, and croissants. Because laminated doughs can be sweet or savory, they are very versatile. When a laminated dough has undergone all of its turns and it is cut in half, it resembles a book with thin "pages." These pages are actually very thin sheets of dough with sheets of fat evenly distributed between each one.

How many turns the dough receives is determined by the final product desired. Danish and croissant doughs generally receive three turns, whereas puff pastry dough receives four turns. Why the extra turn for puff pastry? Puff pastry contains no yeast or leavening ingredients; steam is its only leavening agent. The more turns that the dough undergoes, the more layers that are formed. The greater the number of layers, the greater the rise in the pastry.

Solid Fats and Plasticity

Before you learn how to enclose fat into the base dough, you need to learn about solid fats and which ones are appropriate for laminated doughs. The type of fat chosen to be enclosed into the base dough will affect the flakiness and flavor of the finished product. Ideally, the fat should be neither too soft nor too hard. If the fat is too soft, it can ooze out of the dough or the dough will absorb it and prevent any layering from occurring in the final product. If the fat is too hard, it will be broken into small pieces during the rolling out process and create tears in the dough with uneven layering of fat.

Solid fats are traditionally used in laminated doughs. The fats that are traditionally used for laminated doughs are butter, solid vegetable shortening, margarine, and special shortenings made especially for laminated doughs.

Solid vegetable shortening, margarine, and other special shortenings tend to be used for three reasons. First, they are inexpensive. Second, they have a higher melting point than butter. Lastly, they do not get as hard when refrigerated. The last two reasons refer to the fat's plastic quality or plasticity.

Solid fats have a specific degree of plasticity. Plasticity is a physical property of fats and refers to how easily a particular fat can be molded or shaped. The degree of plasticity or how plastic a fat is depends upon the temperature surrounding it.

Fats with the highest plasticity are shortenings. This means that shortenings remain flexible, soft, and workable over a wider range of temperatures. They are used most often for laminated doughs in large-scale commercial bakeries. This is because they will not soften as each turn is being made and they remain closest in consistency to the base dough, even if kitchen temperatures vary.

Butter, on the other hand, becomes too soft to work with at warm room temperature because of its low melting point, and it becomes too hard to work with at refrigerated temperatures. However, butter contributes the best taste to laminated doughs, because it has a lower melting point and does not leave a greasy film on the tongue as do other fats. It is still the fat of choice in laminated doughs because of the rich flavor it provides.

Preparing a Base Dough

The first step in preparing a laminated dough is to make the base dough. The base dough is also known as the détrempe. The base dough will vary depending upon which of the three pastries is being made.

When preparing the base dough, some bakers add a small amount of an acid such as lemon juice or vinegar to the ingredients. The acid relaxes gluten, thereby decreasing elasticity and making it easier to roll out the dough.

Danish Base Dough

Of the three types of laminated doughs, Danish doughs tend to have the richest base dough because of the addition of eggs, milk, and sugar. Yeast is also added.

Croissant Base Dough

Croissant base dough is less rich than a Danish dough in that it contains some milk, little sugar, and no eggs. It, too, contains yeast like a Danish dough.

Puff Pastry Base Dough

Puff pastry base dough traditionally does not contain sugar. Water is added to the dough, which ultimately creates steam in the oven. Because yeast is not added to puff pastry dough, the steam is its only source of leavening.

Enclosing Fat Inside the Dough

The next step in preparing the laminated dough is to enclose fat inside the base dough. It is important to note the temperature and consistency of the fat. If the fat is too cold, it will be too hard to be evenly distributed throughout the layers, possibly tearing the dough. If the fat is too soft, it can ooze out the edges of the dough. It is crucial that the dough and fat are of the same temperature and consistency. Ideally, the temperature of the dough and the fat should be approximately 60°F (16°C). However, the important lesson to remember is that neither the dough nor the fat should be too soft or too hard.

In order to work with the fat more easily, some bakers blend it with a small amount of flour so it has more structure. The flour absorbs some of the moisture in the fat, making it easier to work with when rolling out the dough. The added flour also prevents the fat from leaking out during the baking of the finished product.

There are several ways to enclose the fat inside the dough.

1. One technique shapes the base dough into one large ball. A large "X" is cut into it with a knife. Each of the four cut edges is pulled out, forming a four-leaf clover shape. One block of fat is placed directly in the middle and each "leaf" is gently pulled up over the butter, enclosing it. This technique can be cumbersome because the butter is in one large lump.

2. Another way to enclose the fat involves rolling out the base dough to a large square and then placing a square of butter diagonally in the center of it. Each side of the dough is folded in over the butter to form a square. It is then chilled for 30 minutes before any turns are started.

3. The technique used in this book involves placing the butter between two pieces of plastic wrap. Using a rolling pin, the butter is pounded to soften it and then rolled and shaped into a rectangle. It can be chilled if it gets too soft. The base dough is then rolled into a rectangle with the rectangle of butter placed across two thirds the length of the dough (Figure 3–1). The short end (with no butter) is brought over the center, sealing the ends closed. The buttered end of the dough is pulled up over the dough like a letter and sealed forming a "fat sandwich." This technique makes the most even layers.

At this point, the dough is rotated 90 degrees and rolled out again into another rectangle. This time the dough can be folded in a letterfold for Danish and croissant dough or in a bookfold for puff pastry. Make sure the short ends of the dough are the ends being folded inward.

> It is important to keep the fat the same temperature and consistency as the base dough.

Folding and Layering

To create hundreds of layers, the dough is repeatedly rolled out and folded in a specific way. The sequence of rolling out the dough and folding it is known as a *turn.* Initially, enclosing the fat into the base dough is not counted as a turn. The more turns that the dough undergoes, the more layers that are created. The dough is then rested in between making each turn. There are two types of folding techniques used on a laminated dough to complete a turn. One technique is

FIGURE 3–1

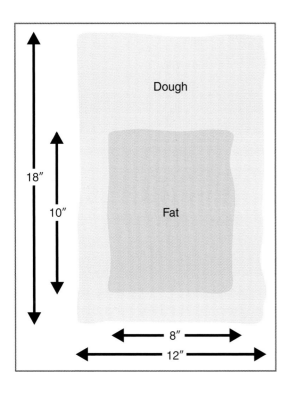

known as the three-fold or letterfold, in which the dough is rolled out into a rectangle and then folded like a letter into thirds (Figure 3–2).

The second technique is known as the four-fold or bookfold in which the dough is rolled into a rectangle with the shorter two ends folded in to meet each other in the middle; then the two halves are brought together in the middle to close like a book (Figure 3–3).

Danish and Croissant Doughs

Danish and croissant doughs are traditionally folded using the letterfold method. The dough is given three letterfold (or three-fold) turns for a total of three times. It is often referred to as a 3 × 3. After each turn, the dough is chilled and allowed to rest.

Puff Pastry Dough

Puff pastry dough is traditionally given a bookfold (or four-fold) turn for a total of four times. It is referred to as a 4 × 4. Because no yeast or other leavening agent is used, puff pastry is dependent on the increased number of layers formed within the dough. One bookfold turn creates more layers than one letterfold turn. The more layers created, the greater the height in the final pastry. Puff pastry is also given a rest in between each turn.

To compare the three types of laminated dough, see Table 3-1, Three Types of Laminated Dough.

Why Resting the Dough Is Necessary

After each turn, laminated doughs are placed in the refrigerator to rest. Resting accomplishes two goals. First, resting allows any gluten that has formed to relax. Second, during rolling and folding, heat is generated. Often, the fat warms up and becomes soft (especially when using butter). Resting the dough allows the fat time to firm up just enough before the next turn. The resting period can last anywhere from 20 minutes to 1 hour between turns. The fat should not be allowed to get too hard. Some chefs place the dough in the freezer to speed this process.

(*Note:* No matter which fold is used, the open ends are always folded in toward the center to ensure that fat stays enclosed.)

FIGURE 3–2

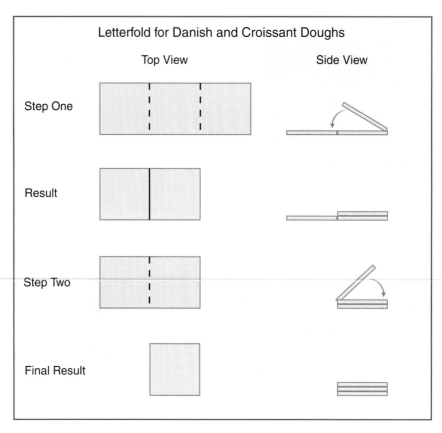

FIGURE 3–3

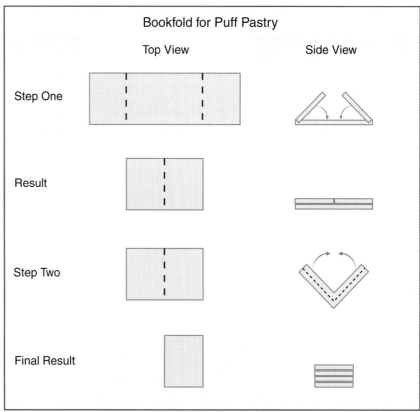

Table 3–1 Three Types of Laminated Dough

DANISH DOUGH	CROISSANT DOUGH	PUFF PASTRY DOUGH
1. Contains yeast, eggs, milk and sugar	1. Contains yeast, milk, little sugar, and no eggs	1. Contains no yeast and a lot more water
2. The dough is given 3 three-fold or letterfold turns, also known as a 3 × 3	2. The dough is given 3 three-fold or letterfold turns, also known as a 3 × 3	3. The dough is given 4 four-fold or book-fold turns, also known as a 4 × 4
3. Yeast leavened	3. Yeast leavened	3. Steam leavened
4. The base dough is the richest of the three types types		

How Laminated Doughs Puff Up and Rise

The rising power of Danish and croissant doughs is due to the yeast in the dough as well as steam. Puff pastry is steam leavened only, with no added advantage of yeast.

Similar to a flaky pie pastry, the many thin layers of fat within a laminated dough act as spacers. Once in the oven, each layer of fat melts, leaving a space between the dough above and below. These spaces fill with hot air and steam from moisture within the dough and fat layers, which expand, causing a pushing up against the dough layers, separating and lifting them to form many flaky layers. Laminated doughs are typically baked at 400°F (205°C) or higher to produce enough steam to help the layers rise.

Choosing a Flour for the Base Dough

Even though there are standard recipes for puff pastry and for Danish and croissant doughs, opinions differ in the type of flour to be used in the base dough. Some bakers choose all-purpose flour to decrease gluten and toughness. Some choose bread flour to give added structure to the risen layers. Using bread flour or other high-protein flour may not seem like the correct type of flour to use considering a tender, flaky product is desired. Yet laminated doughs need to have enough strength so that as they puff up their ultrafine layers do not break. The gluten strands should have enough resiliency to stretch, yielding paper-thin layers. The higher protein content of bread flour gives a certain amount of "give" or elasticity to the dough. Sometimes a combination of high-protein and low-protein flours is the best compromise for making tender, yet resilient, pastry.

Dimensions of the Base Dough

The exact dimensions of the rectangle made from the base dough are not important. There are two points to keep in mind, however. First, make sure the dough is not rolled too thin or the height of the baked pastry will be compromised. Second, a bookfold or four-fold turn requires the dough to be a longer length than does a letterfold or three-fold turn.

Tips for Successful Laminated Pastries
For the Base Dough

■ Use a small amount of an acid such as lemon juice or vinegar to denature some of the proteins in the flour, thereby relaxing the gluten just enough to allow better rolling out of the dough, reducing shrinkage.

■ Avoid overworking the dough to prevent too much gluten from forming.

For Enclosing the Fat

- Keep the fat and the dough the same temperature and consistency.
- For small batches, place the butter between two pieces of plastic and hit it with a rolling pin. The friction from hitting the fat will produce heat, softening it to the consistency of the dough. For large-scale operations, the fat can be softened in an electric mixer using the paddle attachment.

For Folding and Layering

- Always keep the work surface lightly floured.
- Roll the dough in one direction only for each turn to maintain the rectangular shape.
- Keep the edges of the dough straight by squaring off the dough with the rolling pin. This is to maintain the rectangular shape of the dough.
- When rolling the dough, if any fat becomes exposed, sprinkle flour on top of it and continue rolling.
- Before folding the dough to complete a turn, be sure to brush excess flour off the dough using a pastry brush.
- If at any time butter oozes out or is exposed before folding the dough, place the exposed side up so the next fold will encase the exposed fat back into the dough.
- Before each turn, rotate the dough 90 degrees so that when the short, open ends are folded in, the open sides exposing the fat are folded into the center, keeping the fat inside.
- After each turn, be sure the edges of the dough being folded over match the edges of the dough underneath to keep the layers intact.
- After a turn is completed, press a finger into the dough to designate how many turns have been completed. Once wrapped in plastic, a marker or pen can also be used.
- After all the required turns, chill the dough overnight to relax the gluten and facilitate the rolling out process before shaping.

For Shaping, Proofing, and Baking

(Note: Because puff pastry dough does not contain yeast, proofing is not necessary.)

- Brush off any excess flour before cutting and shaping the dough.
- Never proof the formed Danish pastry or croissants above 85°F (29°C) or the fat will melt and ooze out, destroying layers.

Freezing Laminated Doughs

After all turns have been completed, unbaked laminated doughs can be wrapped airtight and frozen for 2 to 3 months. If a Danish or croissant dough is frozen, it is recommended to use one fourth more yeast in the base dough because some yeast may die during freezing. The dough can be thawed for several hours in the refrigerator and then shaped, proofed, and baked.

The dough can also be rolled, cut, shaped, and then frozen as individual pieces. The pieces are thawed in the refrigerator, proofed, and then baked.

Fully baked Danish and croissants can be frozen by wrapping them airtight in plastic and placing them in a plastic bag before freezing. The baked pieces that freeze best are those that do not contain pastry creams, fruit fillings, or whipped cream.

RECIPES

Croissant Dough (This Chapter, page 83)
Danish Dough (This Chapter, page 79)
Puff Pastry Dough (This Chapter, page 85)

DANISH DOUGH

Makes approximately
2 pounds 9 ounces
(1.16 kg) dough

Lessons demonstrated in this recipe:

- How to prepare a typical Danish dough using yeast as a leavening agent.
- Cardamom, a spice from a pod used in Middle Eastern and Indian cuisine, is related to the ginger family and is often used in Danish doughs because it gives off a pleasant, spicy, almost floral aroma.
- Using some bread flour allows better absorption of liquids and better gluten development to ensure thin, strong, flaky layers.
- Danish dough is given 3 three-fold turns or a 3 × 3.
- Allowing the dough to rest in between turns relaxes any gluten that develops.

MEASUREMENTS				INGREDIENTS
U.S.		METRIC	BAKER'S %	
4 fluid ounces	½ cup	120 mL	22%	lukewarm water (110°F; 43°C)
4 fluid ounces	½ cup	120 mL	22%	whole milk scalded and cooled to lukewarm (110°F; 43°C) in a small saucepan
½ ounce	3½ teaspoons	10.5 g	1.9%	instant active dry yeast
10 ounces	2 cups	285 g	53%	bread flour, plus more if needed
9 ounces	2 cups	255 g	47%	all-purpose flour
3½ ounces	½ cup	100 g	19%	granulated sugar
¼ ounce	1 teaspoon	6 g	1.1%	salt
	¾ teaspoon	1.5 g	0.3%	ground cardamom
½ ounce	1 tablespoon	15 g	2.8%	unsalted butter, softened
2 each		94 g	17%	large eggs, room temperature
1 pound	2 cups	455 g	84%	unsalted butter, cold but not hard (leave at room temperature for 30 minutes to soften slightly)
			270.1%	Total Danish Dough percentage

1. *Danish Base Dough*

   Add the water to the saucepan containing the scalded milk. Add the yeast and stir to dissolve.

2. In the bowl of an electric mixer using the paddle attachment, combine the flours, sugar, salt, and cardamom. Blend the dry ingredients on low speed until they are well combined.

3. On low speed, gradually add the warm yeast and milk mixture, softened butter, and eggs until a soft dough forms (Figure 3–4). Stop the machine and feel the dough. If it feels too sticky, sprinkle in a small amount of bread flour.

4. Remove the dough from the bowl and knead it until it feels smooth and elastic but still soft. Do not overwork the dough. Shape the dough into a rectangular disk. Wrap the dough in plastic wrap and chill it for 30 minutes (Figure 3–5).

5. *Preparing the Butter to Be Enclosed*
Place the butter on a piece of plastic wrap that is at least two or three times its size on a clean, dry work surface. Cover the butter with another piece of plastic wrap the same size as the first. Using a rolling pin, hit the butter to soften it slightly and then begin to roll and push it into an even 6- by 12-inch (15- by 30-cm) rectangle (Figure 3–6). Do not overhandle the butter. It should, ideally, be the same consistency as the dough. Set it aside; do not chill.

6. *Enclosing the Butter*
On a lightly floured work surface, roll the chilled dough into a 9- by 18-inch (22.5- by 45-cm) rectangle. Place the rectangle of butter across two thirds of the rectangle and fold like a letter, bringing the unbuttered end up over to the middle of the butter, sealing the dough all long the edges (Figure 3–7A). Fold the remaining buttered portion of the dough on top, completely sealing the butter inside (Figure 3–7B). Be sure to stretch the corners of the dough to square it off, pressing it down to seal the edges together (Figure 3–7C). Using a pastry brush, dust off any excess flour (Figure 3–8).

7. *Completing One Three-fold or Letterfold turn*
Rotate the dough 90 degrees so that when the dough is rolled out, the open ends will become the short sides of the rectangle (Figure 3–9). It will be these short sides that are folded into the center to make sure the fat stays enclosed. Roll out the dough to a 9- by 18-inch (22.5- by 45-cm) rectangle (Figure 3–10). Using a pastry brush, dust off any excess flour.

FIGURE 3–4

FIGURE 3–5

FIGURE 3–6

FIGURE 3–7A

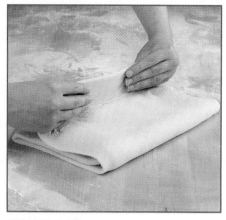

FIGURE 3–7B

FIGURE 3–7C

FIGURE 3–8

FIGURE 3–9

FIGURE 3–10

8. Take one short end of the dough and fold it one third of the way toward the middle, like a letter, pressing it down slightly. Take the other short end and pull it over the dough to meet the other side as if you are folding a letter. Press the dough down slightly so the edges remain sealed. This step completes one letterfold turn (Figure 3–11). Press one finger into the dough to show one turn is complete. Wrap the dough in plastic wrap and chill for 1 hour. A marker can also be used on the plastic to mark one completed turn (Figure 3–12).

9. Repeat rolling, folding, chilling, and marking the number of completed turns two more times for a total of three turns. After the third and final turn is complete, wrap the dough tightly in plastic wrap two times and chill it overnight in the refrigerator. The dough will expand in the refrigerator because of the yeast, so be sure it is swaddled securely within the plastic.

FIGURE 3–11

FIGURE 3–12

CROISSANT DOUGH

Makes 2 pounds 4 ounces (1 kg) croissant dough

Lessons demonstrated in this recipe:

- How to prepare a typical croissant dough giving the dough 3 three-fold turns or a 3 × 3.
- Adding an acid such as vinegar to the base dough helps relax gluten, making it easier to roll out the dough.
- Using some high-protein flour creates stronger gluten sheets, which are able to expand without falling apart, to ensure thin, strong, flaky layers.
- Allowing the dough to rest in between turns relaxes any gluten that develops.

MEASUREMENTS				INGREDIENTS
U.S.		METRIC	BAKER'S%	
1½ ounces	2 tablespoons	40 g	7%	honey
1 ounce	2 tablespoons	30 g	6%	light brown sugar, packed
12 fluid ounces	1½ cups	360 mL	67%	milk, scalded and cooled to lukewarm (110°F; 43°C)
½ ounce	3½ teaspoons	10.5 g	1.9%	instant active dry yeast
10 ounces	2 cups	285 g	53%	bread flour
9 ounces	2 cups	255 g	47%	all-purpose flour, plus more for dusting
	1 teaspoon	5 mL	0.9%	apple cider vinegar
	2½ teaspoons	15 g	2.8%	salt
12 ounces	1½ cups	340 g	63%	unsalted butter, cold but not hard (leave at room temperature for 30 minutes to soften slightly)
			248.6%	Total Croissant Dough percentage

1. ***Croissant Base Dough***

   In the bowl of an electric mixer using the paddle attachment on low speed, blend the honey, brown sugar, and lukewarm milk until well combined.

2. Sprinkle the yeast over the milk mixture and blend on low speed (Figure 3–13).

3. In another mixing bowl, whisk together both flours. On low speed, add the flours, the vinegar, and the salt to the milk and yeast mixture (Figure 3–14). Blend until a soft dough forms. If the dough feels too sticky, add another ½ to 1 ounce (15 to 30 g) all-purpose flour.

FIGURE 3–13

FIGURE 3–14

FIGURE 3–15

4. Knead the dough on a lightly floured surface until smooth. Do not overknead. Using your hands, shape the dough into a rough rectangle about 1 to 2 inches (2.5 to 5 cm) thick. Wrap the dough in plastic wrap and chill it for about 30 minutes (Figure 3–15).

5. *Preparing the Butter to Be Enclosed*
Place a sheet of plastic wrap on a work surface and lay the butter in the center of it. Place another piece of plastic wrap on top.

6. Hit the butter with a rolling pin until it softens. Once it has softened a bit, roll the butter into a 6- by 12-inch (15- by 30-cm) rectangle. Chill the butter while the dough is being rolled out.

7. *Enclosing the Butter*
On a lightly floured surface, roll the dough into a 10- by 15-inch (25- by 37.5-cm) rectangle. Peel one sheet of plastic off the butter and flip the butter onto the middle of the dough, lining up the butter with the dough so that the butter covers approximately two thirds the length of the dough.

8. Fold the unbuttered third of the dough up over to the center. Then fold the remaining buttered third over the top, just like a letter is folded. Be sure to pull the edges of the dough being folded over so they match the edges of the dough underneath. Press to make sure the edges are sealed. If the edges do not seal, brush a small amount of water on the bottom layer to act as glue.

9. *Completing One Three-fold or Letterfold Turn*
Rotate the dough 90 degrees so that when the dough is rolled out, the open ends become the short sides of the rectangle. Again, roll out the dough to a 10- by 15-inch (25- by 37.5-cm) rectangle and fold it in thirds like a letter. This completes one three-fold turn. Using a pastry brush, brush off any excess flour. Press one finger into the dough to show one turn has been completed. Wrap the dough in plastic wrap and chill it for 1 hour. A marker can also be used to mark the plastic wrap to show the number of turns completed.

10. Repeat rolling, folding, and marking the dough with a finger to show the number of turns completed two more times for a total of 3 three-fold turns, chilling the dough in between each completed turn. After three turns have been completed, wrap the dough twice in plastic wrap and chill it overnight in the refrigerator. Because the yeast in the dough will cause it to expand overnight, be sure to wrap the dough securely.

PUFF PASTRY DOUGH

Makes approximately
2 pounds 8 ounces
(1.13 kg) puff pastry dough

Lessons demonstrated in this recipe:

■ How to prepare a typical puff pastry dough giving the dough 4 four-fold turns or a 4 × 4.

■ Preparation of puff pastry relies on steam as the sole leavening agent.

■ Adding an acid to the dough allows it to roll out more easily, relaxing gluten strands.

■ Mixing flour into the butter to be folded into the dough helps prevent it from leaking out between the layers during baking.

■ Using some high-protein flour allows the gluten sheets to remain pliable so they stretch and do not break during baking.

MEASUREMENTS				INGREDIENTS
U.S.		**METRIC**	**BAKER'S %**	
8 fluid ounces	1 cup	240 mL	53%	cold water
	1 teaspoon	6 g	1.3%	salt
	1 teaspoon	5 mL	1.1%	lemon juice or apple cider vinegar
2 ounces	4 tablespoons	55 g	12%	unsalted butter, melted
4 ounces	1 cup	115 g	25%	cake flour
10 ounces	2 cups	285 g	62%	bread flour
1 pound	2 cups	455 g	100%	unsalted butter, softened slightly
2 ounces	½ cup	60 g	13%	bread flour
			267.4%	Total Puff Pastry Dough percentage

Puff Pastry

1. *Base Dough*

   In the bowl of an electric mixer using the paddle attachment, blend the water, salt, lemon juice or vinegar, melted butter, cake flour, and enough of the bread flour to make a soft dough. Do not overmix.

2. Shape the dough into a rough rectangle, wrap it in plastic wrap, and chill it for 30 minutes.

3. *Preparing the Butter to Be Enclosed*

   In the bowl of an electric mixer using the paddle attachment, blend the 1 pound (2 cups; 455 g) of butter and the bread flour until the mixture is approximately the consistency of the base dough.

FIGURE 3–16

FIGURE 3–17

4. Lay a long piece of plastic wrap on a work surface. Using a rubber spatula, scrape the butter mixture onto the center of the plastic (Figure 3–16). Cover with another piece of plastic and, using a rolling pin, gently hit and roll the butter to spread it into an 8- by 12-inch (20- by 30-cm) rectangle (Figure 3–17). Chill the butter until the base dough has been rolled out.

5. *Enclosing the Butter*
   Roll out the chilled dough to a 12- by 18-inch (30- by 46-cm) rectangle, and place the butter on the dough so that it covers the bottom two thirds of the rectangle.

6. Fold the top third of the unbuttered dough down to the middle, partially covering the butter. Now fold the bottom buttered portion of the dough over the center to meet the other side, so that a rectangle forms. Press down slightly, making sure the edges of the dough meet. The butter is now enclosed (Figure 3–18).

FIGURE 3–18

7. *Completing One Four-Fold Turn or Bookturn*

Rotate the dough 90 degrees so that when the dough is rolled out, the open ends become the short sides of the rectangle. Roll the dough to a 9- by 18-inch (23- by 46-cm) rectangle. Fold the top edge of the dough to the center and the bottom edge up to the center (Figure 3–19A). The two edges should meet but not overlap. Now bring the two halves together as if you are closing a book (Figure 3–19B). This is one completed four-fold turn. Press one finger into the dough to show one turn is complete, wrap in plastic wrap, and chill for 30 minutes to 1 hour. The plastic wrap can also be marked with a dot to reduce any confusion as to how many turns have been completed.

8. Repeat rolling, folding, and marking the number of turns three more times for a total of 4 four-fold or bookturns, chilling the dough in between each completed turn. After the final turn is complete, wrap the dough twice in plastic wrap and chill it overnight in the refrigerator.

FIGURE 3–19A

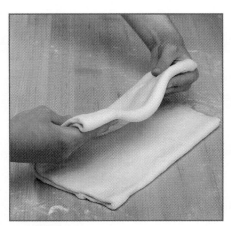

FIGURE 3–19B

RASPBERRY DANISH SPIRALS

Makes approximately eight 3½-inch (9-cm) Danishes

(*Note:* for the following recipes it is important to make a full recipe of the laminated dough or else sufficient layering will not be achieved. Any of the fillings in this section of the chapter can be used interchangeably in any of the other recipes in this section.)

Additional Ideas That Use the Recipes in This Chapter

STEP A

1. Prepare one recipe of Danish dough, using one half for this recipe and reserving the other half for another use.

STEP B: FORMING THE SPIRALS

MEASUREMENTS			INGREDIENTS
U.S.		METRIC	
3½ ounces	½ cup	100 g	granulated sugar
	1 teaspoon	1.5 g	ground cinnamon
3½ ounces	½ cup	100 g	chopped walnuts
			1 large egg, lightly beaten (to be used as glue)
12 ounces	¾ cup	340 g	seedless raspberry jam, preferably an "oven-proof" jam containing a gelling agent that will not melt out during baking

1. Roll out the dough on a lightly floured surface into a rectangle ¼-inch (6-mm) thick. Using a pizza cutter, trim the edges of the dough so they are straight (Figure 3–20). In a separate bowl, mix the sugar, cinnamon, and walnuts together.

2. Using a pastry brush, lightly brush the dough with the beaten egg and evenly sprinkle the sugar, cinnamon, and nut mixture over half of the dough. Using hands, spread the mixture evenly (Figure 3–21).

FIGURE 3–20

FIGURE 3–21

3. Fold the unfilled half of the dough over the sugar, cinnamon, and nut mixture and, using a rolling pin, lightly roll over the dough so both sides adhere (Figure 3–22A and B).

4. Using a pizza cutter, cut the dough crosswise into ½-inch (1.25-cm) thick strips.

5. Working with each strip separately, gently pull a strip lengthwise to stretch it slightly and twist it over and over until it is tightly wound (Figure 3–23). Coil the twisted dough around itself to form a spiral and place the coil on a parchment-lined sheet pan (Figure 3–24). Repeat with the remaining strips.

6. Allow the spirals to proof in a proof box at no higher than 85°F (30°C) for approximately 30 minutes to 1 hour or until they appear puffy.

FIGURE 3–22A

FIGURE 3–22B

FIGURE 3–23

FIGURE 3–24

7. Make an indentation in the center of each coil and fill with some jam (Figure 3–25).

8. Preheat the oven to 400°F (205°C).

9. Bake the spirals for 10 to 12 minutes, or until they are golden brown.

STEP C: CONFECTIONERS' SUGAR ICING

MEASUREMENTS			INGREDIENTS
U.S.		METRIC	
4 ounces	1 cup	115 g	confectioners' sugar, sifted
	1 to 1½ tablespoons	15 to 22.5 mL	milk, plus more if needed
	1 teaspoon	5 mL	vanilla extract

1. Whisk together the ingredients until a smooth, thick icing forms. If the mixture is too thick, add more milk.

2. While the spirals are still hot, drizzle each with the sugar icing (Figure 3–26).

FIGURE 3–25

FIGURE 3–26

Raspberry Danish Spirals

PALM LEAVES OR PALMIERS

Makes approximately two dozen palm leaves

STEP A:

1. Prepare one recipe of puff pastry dough, using one half for this recipe and reserving the other half for another use.

STEP B: FORMING THE PALM LEAVES

MEASUREMENTS			INGREDIENTS
U.S.		**METRIC**	
3½ ounces	½ cup	100 g	granulated sugar, plus more if needed

1. On a sugared work surface, roll out the dough to a rectangle measuring 10 by 12 inches (25 by 30 cm).

2. With the long side facing you, sprinkle some of the sugar over the dough. Fold the two long sides in to meet in the center. Press down gently to adhere (Figure 3–27).

3. Sprinkle some more sugar on top and fold the shorter sides in to meet at the center. Press down gently to adhere.

4. Fold the dough in half as if you are closing a book (Figure 3–28). Sprinkle more sugar over the dough on both sides.

Note: Do not leave the wrapped dough in the refrigerator overnight. The sugar will absorb moisture, making the dough wet and sticky.

5. Wrap the dough tightly in plastic wrap and chill it in the refrigerator for 1 hour or until it is firm enough to slice. Alternatively, the wrapped dough can be placed in the freezer for 20 to 30 minutes to firm up more quickly.

6. Preheat the oven to 400°F (205°C). Using a sharp knife, slice the dough into ⅛-inch (3-mm) thick slices and place them cut side up on a parchment-lined sheet pan (Figure 3–29). Be sure to allow at least 2 inches (5 cm) in between each slice. Sprinkle the top of each palm leaf with granulated sugar.

7. Bake the palm leaves for approximately 12 to 14 minutes or until they are golden brown and crisp. They will crisp up even more as they cool.

FIGURE 3–27

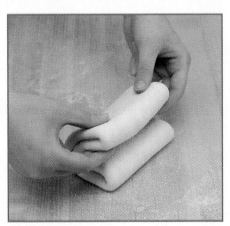

FIGURE 3–28

FIGURE 3–29

Palm Leaves or Palmiers

Almond Danish Braid

Makes one loaf 15 by
5 inch (37.5- by 12.7-cm);
approximate weight
2 pounds 2 ounces (970 g)

STEP A

1. Make one recipe of Danish dough, using one half in this recipe and reserving the other half for another use.

STEP B: ALMOND FILLING

MEASUREMENTS			INGREDIENTS
U.S.		METRIC	
5 ounces	1¼ cups	145 g	yellow or white cake broken into pieces*
4 ounces		115 g	almond paste
1 ounce	2 tablespoons	30 g	granulated sugar
2 ounces	4 tablespoons	55 g	butter, softened
¾ fluid ounce	1½ tablespoons	20 mL	water

*Cake crumbs can be made from the Two Stage Golden Cake, Chapter 7.

1. In a food processor, process cake pieces until finely ground.

2. Add the almond paste, sugar, and butter and process the mixture until well combined. With the motor running, add the water and process the mixture until a paste forms.

STEP C: ASSEMBLY

MEASUREMENTS			INGREDIENTS
U.S.		METRIC	
1¾ ounces	½ cup	50 g	sliced almonds, lightly toasted
			1 large egg white beaten with 1 teaspoon (5 mL) water (to be used as a glue and a wash)

1. Roll the Danish dough out to a 10- by 15-inch (25- by 37.5-cm) rectangle on a lightly floured surface. Brush off the excess flour from the dough. Pick up the dough and place it onto a parchment-lined sheet pan.

2. With a pizza cutter, lightly mark a line parallel to the long side approximately 3 inches (7.5 cm) in from the edge, going completely down the length of the dough without cutting through to the other side. Do the same on the opposite long side (Figure 3–30).

FIGURE 3–30

FIGURE 3–31

FIGURE 3–32

3. Spread the almond filling over the narrow rectangle that has formed. Scatter half the almonds over the filling (Figure 3–31). Using a pizza cutter, make 3-inch (7.5-cm) long cuts on an angle that are approximately $\frac{3}{4}$ inch (2 cm) apart all the way down the length of the dough on both sides. These cuts should resemble fringe on both sides of the dough (Figure 3–32).

4. Lightly egg wash the fringed strips. Begin at the end of the dough where you started to cut the fringed pieces and fold the piece of dough on top of the narrow rectangle to come over the filling by 1 inch (2.5 cm) (Figure 3–33). Fold in the strips alternating left and right sides and continuing down the full length of the dough so that it resembles a braid (Figure 3–34). Reserve the egg wash by covering it with plastic wrap and place it in the refrigerator.

5. Place the braid in a proof box at no higher than 85°F (29°C) for 30 minutes to 1 hour or until the braid appears puffy.

6. Preheat the oven to 375°F (190°C).

7. Using a pastry brush, brush the entire surface of the braid with the reserved egg wash.

8. Bake the braid for 25 to 30 minutes or until the braid is golden brown.

FIGURE 3–33

FIGURE 3–34

STEP D: CONFECTIONERS' SUGAR ICING

U.S.		METRIC	INGREDIENTS
4 ounces	1 cup	115 g	confectioners' sugar, sifted
	1 to 1½ tablespoons	15 to 22.5 mL	milk, plus more if needed
	1 teaspoon	5 mL	vanilla extract

1. Whisk the ingredients together until a smooth, thick icing forms. If the mixture is too thick, add more milk.

2. Cool the braid to lukewarm and drizzle with confectioners' sugar icing. Immediately scatter the remaining almonds over the top before the icing becomes firm.

Almond Danish Bread

CROISSANTS

Makes approximately 12
4-inch (10-cm) croissants.

STEP A

1. Make one recipe of croissant dough, using one half in this recipe and reserving the other half for another use.

STEP B: ROLLING AND CUTTING CROISSANTS

1. On a lightly floured work surface, roll the dough into a 12- by 16-inch (30- by 40-cm) rectangle. The dough should be approximately ⅛ inch (3 mm) thick. Brush off any excess flour and square off the edges using a ruler and a pizza cutter.

2. Review plan on page 98, top. Using a pizza cutter and a ruler, remeasure the length of the dough and cut it in half crosswise. Do not separate the two halves. Cut down the length of the entire rectangle, dividing it into thirds beginning at one of the short sides (Figure 3–35). There should be a total of six rectangles.

3. Review plan on page 98, bottom. Separate the six rectangles and cut each one diagonally to form two triangles (Figure 3–36). There should be a total of 12 triangles. If at any point in rolling the dough it becomes too soft, gently place it on a sheet pan and chill it for 10 to 15 minutes.

FIGURE 3–35

FIGURE 3–36

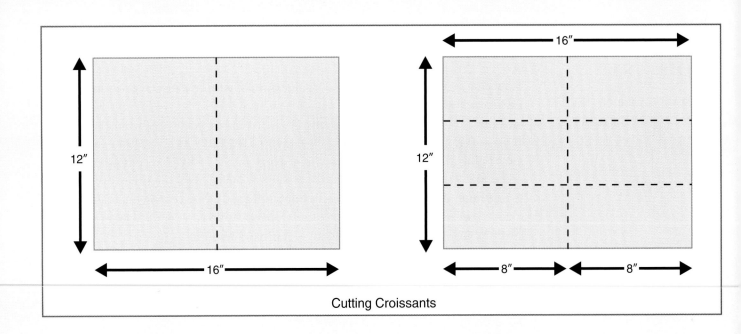

Cutting Croissants

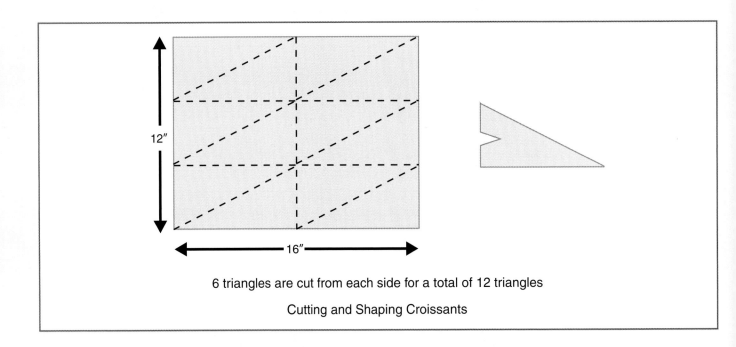

6 triangles are cut from each side for a total of 12 triangles

Cutting and Shaping Croissants

4. Using the pizza cutter, make a small ½-inch (1.2-cm) slit at the base of each triangle. Taking one triangle, gently roll over it with a rolling pin so its length is stretched about 50% longer than its original length (Figure 3–37). Do not press down hard with the rolling pin or the layers will be flattened. Using both hands, gently pull the base so that it widens out even more and begin to roll the triangle from the base, pulling the tip of the triangle to elongate it. Roll it into a tight crescent shape, making sure that the tip of the triangle is tucked underneath the crescent so it will not unroll during baking. Place the croissant on a parchment-lined sheet pan, curving the ends to resemble a crescent as it is placed on the baking sheet. Repeat slitting, rolling, and shaping the remaining chilled dough to make a total of 12 croissants, maintaining a space between each one. Place the croissants in a proof box set at no higher than 85°F (29°C) for 1 hour or until they appear spongy and puffed.

5. Preheat the oven to 425°F (219°C).

6. Spritz the croissants lightly with water using a spray bottle before placing them in the oven. Once the croissants are in the oven, reduce the temperature to 400°F (205°C). Bake for 10 minutes and then rotate the pan and continue baking the croissants for an additional 5 to 10 minutes more or until they are golden brown.

TIP An egg wash made by blending 1 large egg with 1 teaspoon (5 mL) water can be brushed onto the croissants instead of spraying them with water before baking.

TIP Croissants can be baked and frozen in plastic bags and stored for 2 months. Place the frozen croissants on a sheet pan and bake for approximately 10 minutes in a preheated 375°F (190°C) oven.

FIGURE 3–37

The previous method is just one way croissant dough can be cut and shaped on a small scale. Another way to cut a small batch of croissant dough is to use a metal tool known as a rolling cutter (Figure 3–38). A rolling cutter resembles a row of small pizza cutters joined together. It can expand and contract like an accordion to make different sized cuts. The rolling cutter is opened to the width desired for the croissant and rolled across the dough diagonally. The cutter is then rolled from the other direction to form triangles. Bakers who make croissants on a large-scale use a tool known as a croissant cutter (Figure 3–39). The dough is rolled out and the croissant cutter is rolled down the length of the dough cutting triangles and slits at the same time.

FIGURE 3–38

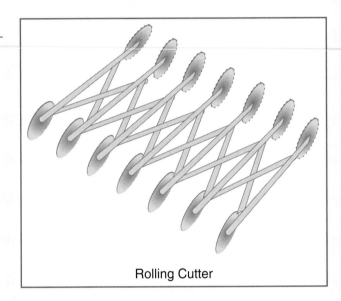

Rolling Cutter

FIGURE 3–39

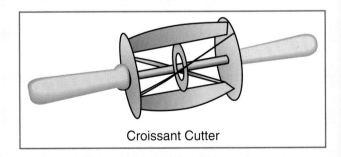

Croissant Cutter

Croissants

PAIN AU CHOCOLAT COFFEE CAKE

Makes one 9-inch (23-cm) round cake or 8 servings

STEP A

1. Make one recipe of croissant dough, using half in this recipe and reserving the other half for another use.

STEP B: CHOCOLATE FILLING

MEASUREMENTS			INGREDIENTS
U.S.		METRIC	
13½ ounces	3 cups	385 g	chocolate cake crumbs*
11 ounces	½ cup	310 g	warm chocolate ganache (Chapter 8, page 366)

*Cake crumbs can be made from the Fudgy Chocolate Cake, Chapter 7.

1. Mix the ingredients in a bowl until a thick and spreadable paste forms. Set aside.

STEP C: ASSEMBLY

MEASUREMENTS			INGREDIENTS
U.S.		METRIC	
			1 egg white mixed with 1 teaspoon (5 mL) water to be used as glue
1¾ ounces	½ cup	50 g	chopped walnuts
6 ounces	1 cup	170 g	miniature semisweet chocolate chips
			1 whole egg mixed with 1 teaspoon (5 mL) water to be used as a wash

1. Using a dough cutter, cut off one third of the croissant dough and keep the remaining two thirds of the dough chilled until needed.

2. Roll the dough into an 11-inch (27.5-cm) circle using the bottom of a 9-inch (22.5-cm) false bottom pan as a guide. Spray nonstick cooking spray into a 9-inch (22.5-cm) round false bottom pan and fit the circle of dough into the bottom. Press to fit (Figure 3–40). Place the pan onto a half sheet pan to catch any butter that might ooze out as it bakes. Set aside.

FIGURE 3–40

3. Roll the remaining two thirds of the dough into a 10- by 14-inch (25- by 35-cm) rectangle and, using a pastry brush, brush it with the egg white and water mixture. Spread the chocolate filling evenly over the dough right up to the edges. Sprinkle the walnuts and the chips evenly over the filling (Figure 3–41).

4. Starting at the long end, roll up the dough into a tight coil. Cut into eight equal pieces approximately 2 inches (5 cm) wide (Figure 3–42).

5. Brush the reserved circle of dough with the egg mixture and place seven of the rolled pieces evenly around the outside edge of the pan with the cut side facing up. Place the last roll in the center (Figure 3–43). Cover the egg wash and reserve it in the refrigerator. Cover and allow the coffee cake to proof at no higher than 85°F (29°C) for approximately 1 hour or until the dough looks puffy.

6. Preheat the oven to 400°F (205°C).

7. Using a pastry brush, brush the reserved egg wash over the top and sides of the coffee cake (which is still on the half sheet pan) and bake for 40 minutes or until golden brown.

STEP D: CONFECTIONERS' SUGAR ICING

MEASUREMENTS			INGREDIENTS
U.S.		METRIC	
4 ounces	1 cup	115 g	confectioners' sugar, sifted
	1 to 1½ tablespoons	15 to 22.5 mL	milk, plus more if needed
	1 teaspoon	5 mL	vanilla extract

1. Whisk the ingredients together until a smooth, thick icing forms. If the mixture is too thick, add more milk.

2. Remove the coffee cake from the oven and allow it to cool for 5 to 10 minutes.

3. Remove the sides of the pan and drizzle the icing over the top while it is still warm. Cool completely.

FIGURE 3–41

FIGURE 3–42

FIGURE 3–43

Pain au Chocolat Coffee Cake

APPLE CRANBERRY TURNOVERS

Makes approximately nine 7-inch (17.5-cm) long turnovers

STEP A: PUFF PASTRY

1. Make one recipe of puff pastry, using half in this recipe and reserving the other half for another use.

STEP B: APPLE CRANBERRY FILLING

MEASUREMENTS			INGREDIENTS
U.S.		METRIC	
1½ pounds		750 g	Granny Smith apples
3½ ounces	1 cup	100 g	fresh cranberries
3½ ounces	½ cup	100 g	granulated sugar
1 ounce	⅛ cup	30 g	light brown sugar (packed if measuring by volume)
1 fluid ounce	⅛ cup	30 mL	apple cider or apple juice
1 tablespoon		15 mL	lemon juice
¾ teaspoon		1.25 g	ground cinnamon

1. Peel, core, and chop two thirds of the apples (1 pound; 455 g) into 1-inch (2.5-cm) pieces. Reserve the remaining apples.

2. In a large, heavy saucepan, place the apples, cranberries, both sugars, apple cider or juice, and lemon juice. Cook over medium-high heat for approximately 10 to 20 minutes or until the apples soften and the cranberries have burst open. Remove the pan from the heat and set aside.

3. Peel, core, and chop the remaining apples into ½-inch (12-mm) pieces. Add the apples and the cinnamon to the cooked apple-cranberry mixture.

4. Cool completely before using.

STEP C: ASSEMBLY

MEASUREMENTS			INGREDIENTS
U.S.		METRIC	
			1 large egg mixed with 1 teaspoon (5 mL) water to be used as a glue and wash
As desired		As desired	granulated sugar for sprinkling

1. Preheat the oven to 400°F (205°C).

2. Roll out the puff pastry to a square 18 by 18 inches (46 by 46 cm). Trim any rough edges to make sure they are even.

FIGURE 3–44

FIGURE 3–45

3. Cut nine 6-inch (15.25-cm) squares using a pizza cutter (Figure 3–44).

4. Using a pastry brush, egg wash around the edges of each square and place 1.5 ounces (1½ heaping tablespoons; 45 g) of apple-cranberry filling into the center of each one. Fold the squares diagonally to form triangles, sealing the edges well (Figure 3–45). Using a sharp knife, make one or two slits through the top of each triangle to allow steam to escape. Transfer the filled turnovers onto a parchment-covered sheet pan. Using a pastry brush, egg wash the tops of each turnover and sprinkle each one generously with sugar.

5. Bake for 20 to 22 minutes or until the turnovers are golden brown and the fruit is bubbling through the slits.

Apple Cranberry Turnovers

PUFF PASTRY TRIANGLES WITH LIME CREAM

Makes six to eight triangles, 7 inches (18 cm)

STEP A: PUFF PASTRY

1. Make one recipe of puff pastry, using half in this recipe and reserving the other half for another use.

STEP B: PUFF PASTRY TRIANGLES

MEASUREMENTS			INGREDIENTS
U.S.		METRIC	
2 fluid ounces	¼ cup	60 mL	heavy cream
			granulated sugar for sprinkling

1. Roll out one half of the puff pastry dough into a large square ¼-inch (6-mm) thick. Trim the edges using a pizza cutter (squaring the edges off) and cut 5- by 5-inch (12.5- by 12.5-cm) squares. Cut each square in half diagonally and lay the triangles 1 inch (2.5 cm) apart on a sheet pan lined with parchment paper. Using a size 12 or 14 plain round pastry tip, poke out random holes in the triangles (Figure 3–46A). Chill the triangles for 30 minutes. While the triangles are in the refrigerator, start the lime curd for the lime cream filling (Step C).

2. Preheat the oven to 450°F (230°C).

3. Brush the triangles with heavy cream and sprinkle each one generously with granulated sugar. Bake the triangles for approximately 7 to 9 minutes or until they are puffed and light brown. Cool and set aside (Figure 3–46B).

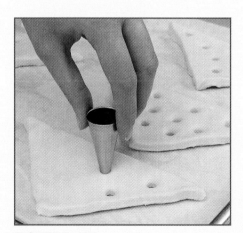

FIGURE 3–46A

FIGURE 3–46B

Makes 4 cups
(32 ounces; 907 g)

STEP C: LIME CREAM FILLING

U.S.		METRIC	INGREDIENTS
			Lime Curd
7 ounces	1 cup	200 g	granulated sugar
			1 pinch salt
6 fluid ounces	¾ cup	180 mL	fresh lime juice
3 each		140 g	large eggs
3 each		60 g	large egg yolks
2 ounces	4 tablespoons	55 g	unsalted cold butter, cut into small cubes
			finely grated zest from 3 limes
			1 to 3 drops green food coloring, optional
			Whipped Cream
	¾ teaspoon	2 g	unflavored gelatin, softened in 1½ tablespoons (22.5 mL) cold water
12 fluid ounces	1½ cups	360 mL	heavy cream, chilled

1. Place a medium heatproof bowl with a strainer over it into an ice water bath and set it aside.

2. In a heavy medium saucepan, place the sugar, salt, lime juice, eggs, and yolks.

3. Set the saucepan over medium-low heat and bring it to a simmer, stirring constantly, with a whisk. Do not allow the mixture to boil or the eggs will curdle. When a thermometer inserted into the mixture reads 160°F (71°C) and the mixture has thickened, remove it from the heat and stir in the cold butter (Figure 3–47).

4. Strain the mixture immediately into the prepared bowl set over the ice water bath (Figure 3–48). Stir in the zest and food coloring (optional). Place a piece of plastic wrap directly on top of the curd and chill it until cold.

5. To prepare the whipped cream, place the bowl of softened gelatin over a pan of hot water to melt it. It should feel warm to the touch. Do not allow it to get too hot or it will warm up the heavy cream and prevent it from beating up and thickening.

FIGURE 3–47

FIGURE 3–48

6. In the bowl of an electric mixer using the whip attachment, beat the heavy cream on high speed until soft peaks form (Figure 3–49). Slowly add the melted gelatin and beat until stiff peaks form (Figure 3–50).

7. Using a rubber spatula, fold the whipped cream into the cold lime curd (Figure 3–51). Chill the mixture until ready to use.

STEP D: ASSEMBLY

MEASUREMENTS		INGREDIENTS
U.S.	METRIC	
		confectioners' sugar for dusting

1. On a dessert plate, arrange one puff pastry triangle. Place a spoonful of lime cream on top. Arrange another puff pastry triangle on top. Dust with confectioners' sugar. Serve at once.

FIGURE 3–49

FIGURE 3–50

FIGURE 3–51

Puff Pastry Triangles with Lime Cream

Professional Profile

Hilary DeMane, CEPC, CCE
Chef Instructor
Indiana University of Pennsylvania
Indiana, PA

1. Question: *When did you realize that you wanted to pursue a career in baking and pastry?*

Answer: *Baking and cooking was always part of my family life growing up. We made everything from scratch and did all the special dishes for holidays. Then, when I was 16, I got my first job in a small bakery in my home town of Greenwich, CT.*

2. Question: *Was there a person or event that influenced you to go into this line of work?*

Answer: *The owner of that small bakery was an inspiration. She taught me a lot about baking and let me work on a lot of the types of things you would expect to find in such a place—muffins, breads. I wasn't ready for anything too elaborate but it was a good foundation.*

3. Question: *What did you find most challenging when you first began working in baking and pastry?*

Answer: *For me the challenge was understanding the science behind the process. When I was still at home I often made bread. One time I made it and the dough wouldn't rise. I kept watching it hoping for a miracle. I baked it and it came out like a brick. Later I realized I put in too much salt and the yeast couldn't work. Baking isn't about miracles; it is about balancing ingredients and knowing the science behind what happens.*

4. Question: *Where and when was your first practical experience in a professional baking setting?*

Answer: *After my training I worked at a resort in Hawaii and then I got a job on a cruise ship. I found that I was the first woman to hold this pastry position on the Holland America Line. There were 1,600 passengers and the bakeshop consisted of four people and one helper. It was a lot of responsibility and a lot of work. Seventy-hour weeks were standard and 80-hour weeks were not uncommon.*

5. Question: *How did this experience affect your later professional development?*

Answer: *On the ship I was working with people from 10 different countries and I came in contact with people from all over the world. The big lesson I learned was the ability to work with people from all different backgrounds. The other important lesson was that organization was key. You can't handle that large an operation without good organizational skills.*

6. Question: *Who were your mentors when you were starting out?*

Answer: *One of the greatest influences has been Ewald Notter. I must have taken at least a dozen classes from him and I am sure I'll be taking more. He is just a wonderful teacher. He really is an inspiration.*

7. Question: *What would you list as your greatest rewards?*

Answer: *Participating and winning in professional competitions are very rewarding. When you finally win a gold medal in a highly competitive event it is so satisfying. If you are not involved in competitions I don't think you can understand the time and commitment that are necessary. And I wouldn't be in education if I didn't find teaching so rewarding.*

8. Question: *What traits do you consider essential for anyone entering the field?*

Answer: *To be successful you must have discipline, drive, and maybe most of all, patience.*

9. Question: *If there was one message you would impart to all students in this field what would that be?*

Answer: *You have to love it and give it everything. If you love it then it's not work. You also have to care about every aspect of your job. You have to care that each step is done well, even if you are just cleaning up.*

CHAPTER 4

Working with Fats in Pies and Tarts

After reading this chapter,
you should be able to:

- Explain the difference between the three types of pastry crusts: pâte brisée, pâte sucrée, and pâte sablée.

- Explain the role fat plays in making pies and tarts.

- List the six steps to reduce gluten formation.

- Demonstrate how to work with fats in pies and tarts by preparing the recipes in this chapter.

Pie and tart crusts traditionally contain a great deal of fat. This is one of the reasons we love them so much. Fat makes the crust tender, light, and flaky. A recurrent theme in this book is the role fats play as tenderizers in baked goods; this role is no less important for pies and tarts.

Pies are defined as a crust topped with a sweet or savory filling. Some pies have a top and bottom crust, a bottom crust only, or a top crust only. Pies are traditionally baked in a pie pan, a shallow, slope-sided pan made of metal, tempered glass, or other decorative materials such as ceramic.

Tarts are, in actuality, just pies without a top crust, although there may be some exceptions. Tarts tend to be a showcase for a spectacular arrangement of fruits, mousses, pastry creams, and chocolate. The shape of a tart has more variations than for a pie. Tart pans may be geometrically shaped (e.g., round, square, rectangular) or shaped like hearts or flowers, for example. The pans may or may not be fluted. Tarts may also be prepared freeform without a pan.

In general, tarts have a "dressy" look. They lend a degree of sophistication to the end of a formal meal. Both pies and tarts can be prepared in individual portion sizes.

This chapter discusses the three types of pastry crusts and how fats are distributed in each one. The topic of gluten is also revisited, primarily how to decrease its formation in the preparation of tender, flaky pies and tarts.

Recipes using the different types of pastry crusts appear at the end of the chapter.

The Difference between Tenderness and Flakiness in a Pastry Crust

Many bakers seem to have a special knack for creating the most tender, flaky pastry crusts. The first step in learning how to work with fats in preparing pies and tarts is to understand the difference between tenderness and flakiness.

Certain fats are referred to as shortenings because tenderness results when the fat shortens strands of gluten, preventing them from joining together and producing a tough pastry crust.

Flakiness results when pieces of fat, acting as spacers within the dough, melt in the oven and leave spaces of air in their place. These spaces of air expand and any moisture in the dough turns to steam, pushing up against each layer of dough. This expansion separates the layers of dough, producing flakiness in a similar manner to laminated doughs.

Choosing the Right Fat

As you learned in Chapter 3, while most fats tenderize and reduce gluten, different types of solid fat have different levels of plasticity and produce differences in flakiness.

Solid fat, such as vegetable shortening, is ideal for preparing crusts for pies and tarts. It produces the flakiest pastry crust because of its high degree of plasticity. Lard, a fat rendered from

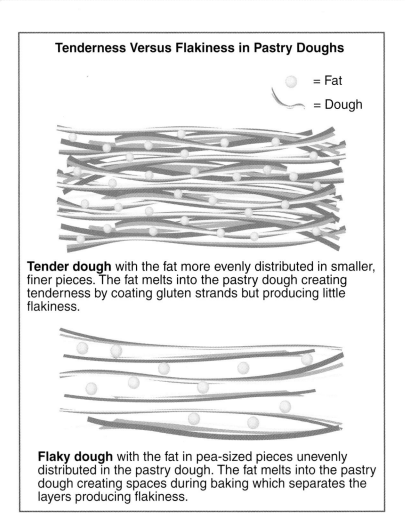

Tenderness Versus Flakiness in Pastry Doughs

= Fat

= Dough

Tender dough with the fat more evenly distributed in smaller, finer pieces. The fat melts into the pastry dough creating tenderness by coating gluten strands but producing little flakiness.

Flaky dough with the fat in pea-sized pieces unevenly distributed in the pastry dough. The fat melts into the pastry dough creating spaces during baking which separates the layers producing flakiness.

pigs, also produces a flaky crust, but because of its high level of saturated fat and health concerns, it is not used as often.

Butter, on the other hand, provides the best flavor despite its lower plasticity. As the temperature of the kitchen goes up, doughs containing butter become difficult to handle. The higher the melting point of the fat used, the longer it will maintain its shape before melting in a hot oven, thereby increasing the pastry's flakiness. This is the main reason why liquid fats such as oils are not used. Although they do tenderize, they produce a crust that crumbles easily and contributes little to flakiness.

Some chefs substitute a portion of the fat for a different one to enhance the flavor of the crust. Some substitutions include high-fat dairy products such as cream cheese or sour cream. For a savory crust, cheeses such as Cheddar can be used. Although these fatty dairy products will produce a very tender crust, they may not be as flaky.

Three Different Types of Pastry Doughs

There are three basic types of pastry doughs: pâte brisée, pâte sucrée, and pâte sablée. Pâte brisée is the flakiest of the three types. Pâte sucrée and pâte sablée are referred to as "short doughs" and produce the most tender crusts. This is because the fat is softened and creamed with sugar. Because the fat is blended in so thoroughly (unlike in pâte brisée), gluten strands are shortened so that they are not able to form a strong structure. The addition of a larger amount of sugar also decreases gluten formation, producing a very tender, but not a flaky crust. Some short dough recipes may contain a small quantity of baking powder. Because pâte sucrée and pâte sablée are quite soft after mixing, they need to be chilled for several hours or overnight until they are firm enough to work with. Pâte sucrée holds its shape when rolled out, whereas pâte sablée is too tender.

In general, pâte sucrée and pâte sablée are baked with no filling and then filled once they have cooled. Typically, they are baked at lower temperatures than pâte brisée because of their high sugar content which causes burning at temperatures greater than 375°F (190°C). Tarts made from pâte sucrée and pâte sablée generally are filled while still in the pan and unmolded when served. (See Table 4-1, The Three Pastry Mixing Methods; and Table 4-2, Comparing the Three Types of Pastry Doughs.)

Pâte Brisée

Pâte brisée is a rich, flaky dough containing flour, salt, butter (or other fat), and ice water. In French, pâte brisée means "broken pastry." Broken pastry refers to the tender flakes that break off as one cuts into this rich crust.

There are two types of pâte brisée: flaky pie dough and mealy pie dough. They are prepared using the same ingredients. The difference between them is in how thoroughly the fat pieces are blended into the dry ingredients. In a typical flaky pie dough, the fat is blended in only until it breaks up into pea-sized pieces; in a mealy pie dough, the fat is blended in more thoroughly until it breaks up into finer pieces resembling corn meal.

Flaky pie dough tends to produce a flaky crust because the larger fat pieces act as spacers separating the layers of dough during baking. As flaky pie dough is rolled out, pieces of fat can be easily seen interspersed within the dough. Flaky pie doughs are ideal for top crusts.

Mealy pie doughs produce a more tender crust. This is because the fat is almost completely blended into the flour. The word *mealy* is used because the dough resembles meal. Each grain of flour becomes entirely coated with fat, greatly reducing gluten formation. Less water is needed in a mealy pie dough because the fat-covered flour particles become less able to absorb it. This creates a very tender, water-resistant crust that is ideal for the bottom crust of a fruit or custard pie. Mealy pie doughs are less likely to get soggy because of their resistance to allowing moisture in.

This dough usually has little or no sugar in it. Traditionally, a pastry blender is used to combine the dry ingredients with the fat for small batches of dough. However, a food processor may be used when the fat is too firm or frozen, taking care not to overprocess the ingredients.

RECIPES
Pâte Brisée (This chapter, page 121)
Pâte Brisée with Cream Chesse (This chapter, page 123)

Pâte Sucrée

Pâte sucrée is a French term for "sweet" or "sugar dough." It resembles a sugar cookie dough more than it does a flaky pie dough or pâte brisée. In pâte sucrée, the fat is creamed with the sugar. These two ingredients act as tenderizers, interfering with the network of gluten and creating a tender, not flaky pastry. When it is baked, pâte sucrée has a rich, buttery flavor with a crisp texture. This is a rich dough containing flour, a small amount of sugar, butter, and eggs. Eggs and often rich dairy products (e.g., milk, cream, or sour cream) replace the water that is in pâte brisée. Pâte sucrée can be rolled out just like a pâte brisée.

The Pâte Brisée Method

1. Place the dry ingredients in a medium-sized mixing bowl.

2. Cut in the cold fat using a pastry blender until it becomes the size of peas for flaky pie dough and the size of small particles resembling meal for mealy pie dough.

3. Drizzle in ice water.

4. Gently gather the ingredients into a ball of dough. Flatten into a disk and wrap in plastic wrap and chill.

The Pâte Sucrée Method

1. Cream together the butter and the sugar in the bowl of an electric mixer.

2. Gradually add the eggs and dairy products.

3. Add the flour and the salt.

4. Gently gather the ingredients into a ball of dough. Flatten into a disk and wrap the dough in plastic wrap and chill.

RECIPE

Pâte Sucrée (This chapter, page 124)

Pâte Sablée

Pâte sablée is a French word meaning "sandy dough" and is a very sweet, crumbly short dough used for sweet tarts and cookies. Of the three types of pastry dough, this dough contains the most sugar and is called "sandy dough" for a reason. The fat and sugar, acting as tenderizers, prevent gluten from forming, so the dough crumbles easily. This dough, when baked, can even be eaten and enjoyed as a rich shortbread cookie all by itself!

Because pâte sablée is so tender and crumbly, it is difficult to roll out. Instead, it is simply crumbled and sprinkled into the tart pan and pressed into place by hand. It is best used in tarts, small individual tarts, and cookies. When baked, pâte sablée has a rich, tender texture that easily breaks apart.

RECIPE

Chocolate Pâte Sablée (This chapter, page 127)

> The Pâte Sablée Method
>
> 1. Cream together the butter and sugar in the bowl of an electric mixer.
> 2. Gradually add the eggs.
> 3. Add the flour, other dry ingredients, baking powder, and salt.
> 4. Gently gather the ingredients into a ball of dough. Flatten into a disk and wrap the dough in plastic wrap and chill.

Table 4–1 **The Three Pastry Mixing Methods**

PÂTE BRISÉE METHOD	PÂTE SUCRÉE METHOD	PÂTE SABLÉE METHOD
(Broken dough)	(Sugar dough)	(Sandy dough)
1. Place dry ingredients in bowl.	1. Cream butter and sugar.	1. Cream butter and sugar.
2. Cut in cold fat.	2. Gradually add eggs and/or milk or cream.	2. Gradually add eggs.
3. Drizzle in ice water.	3. Add flour and salt.	3. Add flour, salt, and occasionally a leavening agent like baking powder.
4. Wrap in plastic wrap and chill.	4. Wrap in plastic wrap and chill.	4. Wrap in plastic wrap and chill.

Table 4–2 **Comparing the Three Types of Pastry Doughs**

	PÂTE BRISÉE	PÂTE SUCRÉE	PÂTE SABLÉE
Translated from French	Broken pastry	Sweet or sugar dough	Sandy dough
Texture after baking	Flaky pie dough: unsweetened, tender, very flaky Mealy pie dough: unsweetened, tender, not as flaky	Sweet, rich, crisp, not flaky	Sweet, rich, crumbly, not flaky
Best used in	Flaky pie dough: top crusts of pies Mealy pie dough: bottom crusts of pies, especially those with wet fillings such as fruit and custards	Tarts and small, individual tarts	Tarts and small, individual tarts; cookies

Six Ways to Ensure a Tender, Flaky Pastry Crust

There are six ways to ensure a tender, flaky pastry crust. The first step contributes to flakiness. The remaining five steps are directed at decreasing gluten from forming to create the most tender crust. They include:

1. Use a solid, cold fat.
2. Use a low-protein flour.
3. Add an acid.
4. Avoid using too much water.
5. Do not overmix.
6. Allow the dough to rest.

Use a Solid, Cold Fat

Fat Fact: The colder and harder the fat, the flakier the crust.

Even though butter has low plasticity, it is the fat of choice in pastry crusts. Some bakers use a combination of fats—vegetable shortening and butter—to achieve both flakiness and flavor.

If using butter, cut it into small cubes, wrap it, and freeze it for a short period of time. This prevents it from melting into the dough too quickly in the oven and increases flakiness. (Freeze cubed butter for approximately 20 minutes if using a pastry blender; freeze the butter cubes for approximately 1 hour if a food processor is used.) The butter should not be too hard to cut into the dry ingredients. The general rule is the colder and harder the fat, the flakier the crust.

Use Low-Protein Flour

Flours containing fewer proteins develop less gluten. Most pastry crusts use pastry or all-purpose flour.

Add an Acid

Adding a very small amount of an acid like orange juice, lemon juice, or vinegar helps break down and denature the proteins in the flour, preventing gluten from forming. The acid also allows the dough to be rolled out more easily with less shrinkage. Many pastry chefs substitute cold orange or lemon juice for some of the ice water, not only to decrease gluten but also to add flavor.

Avoid Using Too Much Water

It can be tempting to add too much water when making a pastry crust, especially when the pastry seems dry. Remember that gluten strands do not develop until the proteins in the flour come into contact with water.

One of the most common mistakes that bakers make is adding too much water when making a pastry crust. Sprinkle in the water gradually instead of pouring it in all at once. The entire amount may not be needed. The less water added, the more tender the crust. The water should be very cold, so add ice to a bowl of cold water, then measure from there. Using ice water prevents the pieces of fat from warming up and melting into the dough before baking.

Do Not Overmix

Any type of mixing will encourage some gluten to develop. Never knead a pastry dough; instead, push the dough gently against the sides of the bowl until it comes together and then gather it into a ball. Overhandling the dough with hands will warm up the fat, reducing flakiness.

Allow the Dough to Rest

Resting the dough relaxes gluten strands, allowing the dough to be more easily rolled out and shaped with less shrinkage. When gluten has been overdeveloped, rolling out the dough becomes an impossible task. It is similar to pulling on a too tightly wound rubber band. The more you try to stretch it, the more it resists and shrinks back. Some gluten is desirable, though, to give a certain amount of structure to the crust.

Resting also allows the fat in the dough to firm up, preventing it from melting into the dough. Generally, the dough is placed in the refrigerator to rest but it can be placed in the freezer briefly just until the dough is firm enough to roll out.

If the dough does become overworked, wrap it in plastic wrap and allow it to rest in the refrigerator for at least 30 minutes, or even longer.

Blind Baking

There are times when a pastry shell must be baked before it can be filled. After the tart shell is cooled, it can be filled with pastry cream, mousse, whipped cream, ice cream, or fruit. Tarts such as these contain fillings that do not need cooking or that have been cooked separately and, once in the tart shell, are ready to be served. When a tart shell is baked with nothing in it, it is called blind baking.

When blind baking a pastry crust it is important to prevent the sides of the crust from shrinking down and the bottom of the crust from puffing up.

Three Ways to Blind Bake a Pie Shell

- The unbaked tart shell can be covered with a piece of parchment paper and filled with dried beans or pie weights. This prevents the dough from puffing up during baking. The shell is baked halfway, and then the parchment paper with beans or weights is removed. The shell is placed back in the oven to finish baking and ensure that the crust browns evenly.
- A second way to blind bake a pastry shell is to prick holes all over the bottom and sides of the unbaked pie shell. This is referred to as stippling or docking. A fork, paring knife, or a special tool made just for this purpose can be used. The holes prevent the crust from puffing up and shrinking in the oven.
- A third way is to set an empty pie pan on top of the unbaked pie shell and invert it onto a sheet pan. It is baked halfway, then the crust is flipped over so it is right side up. The extra pan is then removed and the pie shell is placed back in the oven to finish baking. Baking the crust upside down prevents the sides from shrinking.

Helpful Tips to Roll Out a Pastry Crust

There are many ways to roll out a pastry crust:

- Before you begin to roll out the dough, use your hands to form the dough into the shape it will ultimately be in when it is rolled out. For example, if a round pie pan will be used, form the disk of dough into a rough circle before you begin to roll it out. If a rectangular tart pan will be used, form the dough into a rough rectangle.
- Be sure the work surface and the rolling pin are well floured. Before transferring the dough to the pan, brush off excess flour using a pastry brush.
- After rolling out the dough in one direction, rotate the dough to make sure it is not sticking to the work surface. Roll out the dough evenly so the crust will be uniform in thickness.

Tips for a High-Quality Pastry Crust

1. Make sure the fat is cold.

2. Make sure the water or liquid is ice cold.

3. Chill the dough before rolling.

4. When rolling out a pastry dough, keep the work surface well dusted with flour to prevent the dough from sticking and then brush off excess flour with a pastry brush after rolling is completed.

5. Place the unbaked pie or tart shell on a sheet pan to make it easier to bring it to and from the oven.

TIP Chilling or freezing the raw pastry shell briefly before baking allows the fat to harden, ensuring a flaky crust, and relaxes gluten to prevent shrinkage.

- If the dough becomes too sticky or soft to work with, rewrap it and chill it in the refrigerator for 20 to 30 minutes.
- Stickier doughs may be rolled out between two pieces of plastic wrap and chilled for a short time to firm up. When you are ready to transfer the dough to the pan, peel off the top sheet of plastic. The pan can be placed on top of the dough (upside down) and the dough flipped over into the pan. The remaining piece of plastic wrap can then be peeled off.
- Place the pan on top of the rolled-out dough that is 1 to 2 inches (2.5 to 5 cm) larger than the pan to ensure you have enough dough to go up the sides.
- Roll the dough over a rolling pin or fold it into quarters and transfer it to the pan.
- Do not stretch the dough as you fit it into the pie or tart pan. This will cause the pastry to shrink as it bakes.
- Decoratively crimp the edges of the dough if using a pie pan.
- Roll over the top edge of a tart pan with a rolling pin to remove the excess dough.

Preventing a Soggy Bottom Crust

The following are tips to avoid a soggy bottom crust:

- Use a mealy pie dough for the bottom crust, especially for fruit or custard pies.
- Bake the crust for a long enough time. It should not look doughy.
- Bake the crust on the lowest rack of the oven at least for part of the time to ensure it bakes through completely.

For crusts to be blind baked, the following tips apply:

- Just before baking is complete, brush the bottom of the crust with egg white or egg wash and return it to the oven just long enough to cook and set the egg. This forms a barrier or protective coating against the filling.
- After baking, brush the bottom of the crust with melted chocolate or scatter chopped chocolate over the hot crust. Once melted, spread the chocolate evenly over the bottom. Chill the crust just until the chocolate has hardened before filling.

PÂTE BRISÉE DOUGH OR FLAKY PASTRY DOUGH

Makes enough pastry for two 9- to 10-inch (22.5- to 25-cm) single pie crusts or one 9- to 10-inch (22.5- to 25-cm) double pie crust

Lessons demonstrated in this recipe:

- How to prepare a pâte brisée or flaky pie crust.
- The butter is frozen briefly so that it will not melt too quickly in the oven, thus ensuring a flaky crust.
- An acid such as lemon juice is added to help the dough roll out more easily with less shrinkage.
- Keeping the water icy cold ensures the fat will not melt during mixing.
- Resting the dough allows any gluten that developed during mixing to relax so the crust will roll out easily and not shrink in the oven.

MEASUREMENTS				INGREDIENTS
U.S.		**METRIC**	**BAKER'S %**	
12¾ ounces	3 cups	360 g	100%	pastry flour
	2 tablespoons	30 g	8%	granulated sugar
	½ teaspoon	3 g	0.7%	salt
10 ounces	2 sticks + 4 tablespoons	285 g	79%	unsalted butter cut into small cubes, wrapped in plastic wrap, and frozen for approximately 1 hour
	1 teaspoon	5 mL	1.4%	fresh lemon juice
4 to 4¼ fluid ounces	8 to 8½ tablespoons	120 to 130 mL	33% to 36%	ice water
				additional ice water, if needed
			225%	Total Pâte Brisée percentage

1. Combine the flour, the sugar, and the salt in the work bowl of a food processor. Pulse the mixture until the dry ingredients are well combined. *Pulsing* means to turn the food processor on and off in spurts. This prevents overmixing.

2. Add the frozen butter cubes, scattering them on top of the flour (Figure 4–1). Pulse the mixture about 8 times, or until the butter becomes the size of peas (Figure 4–2). Stop the machine and feel the butter pieces with your fingers to make sure they are not too small.

FIGURE 4–1

FIGURE 4–2

FIGURE 4–3

FIGURE 4–4

TIP The dough can also be made by hand using a pastry blender. For larger batches, use an electric mixer with a pastry blender attachment.

TIP If preparing a mealy pie dough, process the mixture in step 2 until the butter and dry ingredients are very fine and resemble meal.

3. Gradually add the lemon juice and half the water, and pulse the mixture until just combined. Add more water until the dough holds together. All of the water may not be needed, so stop the machine and feel the dough to see if it holds together (Figure 4–3). At this point the mixture can be poured into a bowl, if desired, to gather it into a ball. Do not overmix or knead the dough.

4. Form the dough into a disk. Wrap the dough in plastic wrap and chill for 1 hour (Figure 4–4). How to shape and bake this dough depends on what is being made with it. For specific recipes on how to use this dough, see Additional Ideas That Use the Recipes in This Chapter.

PÂTE BRISÉE WITH CREAM CHEESE

Makes enough pastry for a single 9- to 10-inch (22.5- to 25-cm) pie crust

Lessons demonstrated in this recipe:

- How to prepare a pâte brisée substituting cream cheese for a portion of the fat.
- The butter is frozen for a short period of time and the cream cheese is chilled to ensure that they will not soften or melt during mixing.
- Resting the dough allows gluten to relax.
- An acid such as cold orange juice is added to make rolling out easier with less shrinkage while providing flavor to the crust.

MEASUREMENTS				INGREDIENTS
U.S.		**METRIC**	**BAKER'S %**	
6½ ounces	1½ cups	185 g	100%	pastry flour
	⅛ teaspoon	0.5 g	0.3%	salt
	1 teaspoon	6 g	3.2%	grated orange zest
3½ ounces		100 g	54%	cold cream cheese, cut into cubes
4 ounces		115 g	62%	unsalted butter, cut into small cubes, wrapped in plastic wrap, and frozen for 20 to 30 minutes
1½ fluid ounces	3 tablespoons	45 mL	24%	ice cold orange juice
			243.5%	Total Pâte Brisée with Cream Cheese percentage

1. In the bowl of a food processor, add the flour, the salt, and the orange peel. Pulse the mixture to blend.

2. Add the cream cheese to the dry ingredients and pulse the mixture a few times to distribute it until the cream cheese resembles small pea-sized pieces (Figure 4–5). Stop the machine and feel the pieces to make sure they are the right size.

3. Add the frozen butter and pulse the mixture 8 to 10 times to reduce the butter to pea-sized pieces.

4. Add half of the orange juice, pulsing the machine only until the mixture is just combined. Pour the mixture into a bowl and with your hands gather the dough together to form a ball (Figure 4–6). If the dough feels too dry, add the remaining orange juice. Shape the dough into a disk. Wrap it in plastic wrap and allow it to rest in the refrigerator for at least 1 hour or overnight. How to shape and bake this dough depends on what is being made with it. For specific recipes on how to use this dough, see Additional Ideas That Use the Recipes in This Chapter.

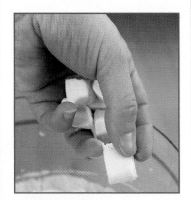

FIGURE 4–5

FIGURE 4–6

TIP The dough can also be made by hand using a pastry blender. For larger batches, use an electric mixer with a pastry blender attachment.

PÂTE SUCRÉE DOUGH

Makes enough pastry for one 10-inch (25-cm) tart shell

Lessons demonstrated in this recipe:

- How to prepare a pâte sucrée dough.
- A low-protein flour is used to prevent gluten formation.
- Flour particles will become completely coated with fat, causing less gluten to form.
- The fat in the whole milk or heavy cream coats gluten strands, making a more tender crust.
- The egg helps enrich and bind the dough while providing tenderness.
- Resting the dough in the refrigerator allows any gluten that develops to relax and firms up the dough enough to be rolled out.

MEASUREMENTS				INGREDIENTS
U.S.		METRIC	BAKER'S %	
8½ ounces	2 cups	240 g	100%	pastry flour
	⅛ teaspoon	0.5 g	0.2%	salt
5 ounces		140 g	58%	unsalted butter, softened
3½ ounces	½ cup	100 g	42%	confectioners' sugar, sifted if lumpy
1 each		47 g	20%	large egg
	1 tablespoon	15 mL	6%	whole milk or heavy cream
			226.2%	Total Pâte Sucrée percentage

1. In a small bowl, blend the flour and salt and set aside.

2. In the bowl of an electric mixer using the paddle attachment, cream the butter with the sugar until the mixture is light (Figure 4–7).

3. Add the egg and the milk or cream. Blend well (Figure 4–8).

4. On low speed, add the flour and salt, and mix the ingredients until they are just combined (Figure 4–9).

FIGURE 4–7

FIGURE 4–8

FIGURE 4–9

5. Form the dough into a disk (Figure 4–10). Wrap the dough in plastic and chill for at least 20 minutes in the freezer, or at least 1 hour in the refrigerator, until it firms up enough to be rolled out.

Blind Baking the Tart Shell

6. Roll out the dough onto a lightly floured surface $\frac{1}{4}$-inch (6-mm) thick to an 11- to 12-inch (27$\frac{1}{2}$- to 30-cm) circle. Fit the dough snugly into a 10-inch (25-cm) false-bottom tart pan with fluted edges (Figure 4–11). Trim the excess dough and chill the unbaked tart shell for at least 30 minutes.

7. Preheat the oven to 375°F (190°C). Using a fork, dock the tart shell all over the bottom and sides to prevent any puffing up of the dough (Figure 4–12). Fill the shell with parchment paper and dried beans (Figure 4–13). Place the tart shell on a half sheet pan and bake for 20 to 25 minutes. Remove the tart shell from the oven and gently lift off and remove the parchment paper with the beans. Return the tart shell to the oven and bake for another 5 to 10 minutes, until the bottom of the tart shell is light brown and there are no areas of crust that look doughy and underbaked.

8. Cool the tart shell completely, leaving it in the pan.

FIGURE 4–10

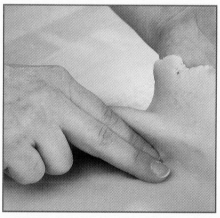

FIGURE 4–11

FIGURE 4–12

FIGURE 4–13

Pâte Sucrée Tart Shell

CHOCOLATE PÂTE SABLÉE DOUGH

Makes enough pastry for one 10-inch (25-cm) tart shell

Lessons demonstrated in this recipe:

- How to prepare a pâte sablée dough.
- Ground nuts and cake crumbs are added for their flavor and tenderizing ability.
- A low-gluten flour is used to help tenderize the dough.
- Baking powder is added to leaven, resulting in a lighter, more tender pastry.
- Pâte sablée dough is too tender to be rolled out and should be crumbled into the tart pan.

MEASUREMENTS				INGREDIENTS
U.S.		METRIC	BAKER'S %	
4 ounces	1 cup	115 g	48%	walnuts
4 ounces	½ cup + 2 tablespoons	130 g	54%	granulated sugar
6 ounces		170 g	71%	unsalted butter, softened
1 each		47 g	20%	large egg
8½ ounces	2 cups	240 g	100%	pastry flour
	1 tablespoon	8 g	3.3%	Dutch processed cocoa powder
	1 teaspoon	4 g	1.7%	baking powder
	½ teaspoon	0.75 g	0.3%	cinnamon
2 ounces	½ cup [heaping]	60 g	25%	chocolate cake crumbs*
			323.3%	Total Chocolate Pâte Sablée Dough percentage

Chocolate cake crumbs can be made from the Fudgy Chocolate Cake, Chapter 7.

1. In a food processor, process the walnuts and the sugar until they are finely ground. Transfer the mixture to the bowl of an electric mixer.

2. In the bowl of an electric mixer, cream the butter with the ground nuts and sugar until the mixture is light. Add the egg and blend the mixture well (Figure 4–14).

3. In a small bowl, blend the flour, cocoa powder, baking powder, cinnamon, and the cake crumbs.

4. Slowly add the dry ingredients into the butter and nut mixture (Figure 4–15). Blend the mixture well.

FIGURE 4–14

FIGURE 4–15

TIP Excess pâte sablée dough can be rolled out and cut into a variety of shapes, baked at 325°F (165°C), and eaten as a shortbread cookie.

5. Wrap the dough in plastic wrap and chill for at least 3 to 4 hours or overnight (Figure 4–16). The dough must be firm in order to use it.

Blind Baking the Tart Shell

6. Breaking up the firm dough with your hands, crumble it onto the bottom of a 10-inch (25-cm) false-bottom tart pan (Figure 4–17). Pat the dough evenly over the bottom and sides of the tart pan to a ¼ inch (6 mm) thickness (Figure 4–18). Reserve the excess dough if there is any for another use. Chill the tart shell for 30 minutes.

7. Preheat the oven to 325°F (165°C). Place a piece of parchment paper into the tart shell and fill it with dried beans. Place the bean-filled tart pan on a sheet pan. This makes it easier to take it in and out of the oven. Bake it for 25 to 35 minutes or until it is lightly browned. Do not allow the crust to get too brown. Crusts made with cocoa can taste bitter if overbaked.

8. Gently lift off and remove the beans and parchment paper, and return the tart shell to the oven for 5 to 10 extra minutes or until the bottom is lightly browned and there are no areas of crust that look doughy and underbaked. Allow it to cool completely, leaving it in the pan.

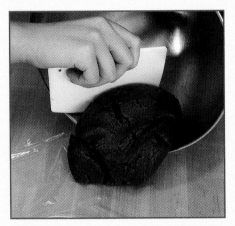

FIGURE 4–16

FIGURE 4–17

FIGURE 4–18

Chocolate Pâte Sablée Tart Shell

INDIVIDUAL NECTARINE ALMOND GALETTES

Makes 8 to 10 individual galettes (4 to 5 inches or 10 to 12.5 cm in diameter)

Additional Ideas That Use the Recipes in This Chapter

Galette is a French word that refers to a free-form tart that is filled with different types of fruit and baked. They are easy to make and can be made as individual tarts or as one large galette to feed several people.

STEP A: PÂTE BRISÉE

Make one recipe of pâte brisée, adding 2½ ounces (½ cup; 70 g) natural almonds with skins (toasted) to step 1 of the Pâte Brisée recipe, pulsing until the almonds are finely chopped.

STEP B: NECTARINE FILLING

MEASUREMENTS			INGREDIENTS
U.S.		METRIC	
1¾ ounces	¼ cup	50 g	granulated sugar
	½ teaspoon	0.75 g	ground cinnamon
	1 tablespoon	8 g	cornstarch
2¾ pounds		1.25 kg	8 nectarines pitted and thinly sliced
	1 teaspoon	5 mL	almond extract

1. In a large bowl, whisk the sugar, cinnamon, and cornstarch until well combined. Add the sliced nectarines and the almond extract to the dry ingredients and mix gently until the nectarines are coated (Figure 4–19). Set the filling aside.

STEP C: ASSEMBLY

MEASUREMENTS			INGREDIENTS
U.S.		METRIC	
As needed		As needed	all-purpose flour for dusting
			1 large egg, lightly beaten, to be used as a wash
6 ounces	½ cup	170 g	peach preserves
1⅓ ounces	½ cup	40 g	sliced almonds, toasted

1. Preheat oven to 400°F (205°C). Line two sheet pans with parchment paper. Set aside.

2. Divide the dough into 8 to 10 equal pieces. With a rolling pin, roll each piece of dough into a rough circle between ⅛ and ¼ of an inch (3 to 6 mm) thick and about 5 to 6 inches (12.5 to 15 cm) in diameter (Figure 4–20). Using an offset spatula, transfer the pastry rounds so they fit on the prepared baking sheets. Be sure to space them so they are not touching. Using a pastry brush, lightly egg wash each round (Figure 4–21).

FIGURE 4–19

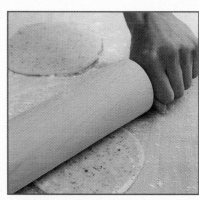

FIGURE 4–20

FIGURE 4–21

TIP After cooling, galettes can be individually wrapped in plastic wrap, placed in plastic bags, and frozen for 3 to 4 months. They can be unwrapped, placed on a parchment-lined sheet pan while still frozen, and placed in a preheated 375°F (190°C) oven for 15 to 20 minutes or until the crust becomes crisp and flaky.

3. With a slotted spoon, place about 2 tablespoons (30 mL) of filling (draining out any liquid you can through the holes in the spoon) into the center of the dough (Figure 4–22).

4. Using your fingers, fold the outer edges of the dough in toward the nectarine filling, forming a rim over the nectarines (Figure 4–23).

5. Repeat filling the remaining rounds of dough. Brush the dough around each galette with egg wash.

6. Bake the galettes for 30 to 35 minutes or until brown. Remove them from the oven.

7. In a small saucepan, melt the peach preserves over low heat. With a pastry brush, brush the edges of the crust and the nectarines with preserves (Figure 4–24). Sprinkle sliced toasted almonds all around the edges of each galette (Figure 4–25).

FIGURE 4–22

FIGURE 4–23

FIGURE 4–24

FIGURE 4–25

Individual Nectarine Almond Galettes

PIZZA DESSERT TART WITH THE WORKS

Recipe makes one
12-inch (30.5-cm) pizza

STEP A

Make one recipe of Pâte Brisée with Cream Cheese.

STEP B: MIXED BERRY PIZZA SAUCE

MEASUREMENTS			INGREDIENTS
U.S.		**METRIC**	
4¼ ounces	1 cup	120 g	frozen raspberries, thawed and drained
3 ounces	½ cup	85 g	fresh strawberries, sliced or frozen strawberries, thawed and drained (it is not necessary to slice them, if they were frozen)
2½ ounces	¼ cup	70 g	strawberry or raspberry jam or preserves
	1 teaspoon	5 mL	fresh orange juice
	1 teaspoon	5 mL	balsamic vinegar

In a food processor, purée all the ingredients and pour the mixture into a bowl. Refrigerate the sauce until needed.

STEP C: TOPPINGS

MEASUREMENTS			INGREDIENTS
U.S.		**METRIC**	
	1 tablespoon	15 mL	milk
	2 teaspoons	10 g	coarse sugar
6 ounces	1 cup	170 g	diced fresh or drained canned pineapple
1 ounce	¼ cup	30 g	dried cranberries or cherries
2 ounces	½ cup	60 g	high-quality white chocolate warmed in a microwave for 20 to 30 seconds on low power and shredded using a grater.
2 ounces	⅓ cup	60 g	mini semisweet chocolate chips or coarsely chopped semisweet chocolate
1¼ ounces	⅓ cup	35 g	coarsely chopped walnuts, toasted in a 400°F (205°C) oven for 5 to 10 minutes
½ ounce	⅓ cup	15 g	shredded coconut, toasted
	1 tablespoon	3 g	confectioners' sugar put through a sieve for dusting over the top of the pizza

STEP D: ASSEMBLY

1. Set the oven rack on the lowest position and preheat the oven to 400°F (205°C). Spray a 12-inch (30-cm) pizza pan with nonstick cooking spray.

2. Roll out the pâte brisée dough onto a lightly floured surface into a 12-inch (30-cm) circle, using the pizza pan as a guide, about ¼ inch (6 mm) thick. Fit the dough into the pizza pan (Figure 4–26). Roll the edges over like a cuff on a pant leg, if desired to create a higher edge. With a pastry brush, brush the edges with milk and sprinkle them with coarse sugar (Figure 4–27).

3. Spread 4 ounces (½ cup; 119 mL) of the mixed berry pizza sauce evenly over the crust. Reserve the remaining sauce for another use.

4. Scatter the crust with the pineapple and the dried cranberries or cherries (Figure 4–28). Bake for about 40 minutes or until the crust is golden brown.

5. Remove the pizza tart from the oven and immediately sprinkle the shredded white chocolate "cheese," mini chocolate chips, walnuts, and coconut evenly over the top (Figure 4–29). Allow the pizza to cool. Dust the top with confectioners' sugar and serve the pizza at room temperature cut into slices.

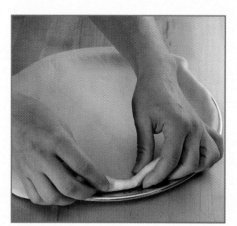

FIGURE 4–26

FIGURE 4–27

FIGURE 4–28

FIGURE 4–29

Pizza Dessert Tart with the Works

FRESH BERRY TART FILLED WITH COCONUT PASTRY CREAM

Makes one 10-inch (25-cm) tart, serving 8 to 10

STEP A: COCONUT PASTRY CREAM

MEASUREMENTS			INGREDIENTS
U.S.		METRIC	
4 fluid ounces	½ cup	120 mL	whole milk
9½ fluid ounces	1 cup + 3 tablespoons	285 mL	coconut milk
2 ounces	4 tablespoons	60 g	cream of coconut, well stirred
3½ ounces	½ cup	100 g	granulated sugar
1¼ ounce	¼ cup	35 g	cornstarch
4 each		76 g	large egg yolks
	1 teaspoon	5 mL	coconut extract
	1 teaspoon	5 mL	vanilla extract

1. Place the milk, coconut milk, and cream of coconut into a saucepan. Whisk the mixture and bring it to a boil. Remove it from the heat and set it aside.

2. In a heatproof mixing bowl, whisk the sugar, cornstarch, and egg yolks until there are no lumps. The mixture will be thick.

3. Temper the egg yolk mixture by slowly adding dribbles of the hot milk into it, whisking constantly until all the milk has been added (Figure 4–30). Pour the mixture back into the saucepan and place it over medium-high heat. Bring the mixture to a boil whisking constantly. The mixture will thicken (Figure 4–31). Continue to boil and whisk the custard for 20 more seconds.

4. Remove the mixture from the heat and whisk in both extracts. Pour the pastry cream into a bowl and place a piece of plastic wrap directly onto it to prevent a skin from forming (Figure 4–32). Place the pastry cream in the refrigerator for approximately 3 to 4 hours or until it is cold. To hasten the cooling process, place the bowl of pastry cream over an ice water bath, stirring frequently.

FIGURE 4–30

FIGURE 4–31

FIGURE 4–32

STEP B: PÂTE SUCRÉE

1. Make one recipe Pâte Sucrée and bake it blind.

STEP C: CHOCOLATE-COATED PASTRY CRUST

U.S.		METRIC	INGREDIENTS
3 ounces	½ cup	90 g	semisweet chocolate, chopped

1. Bake the pâte sucrée crust blind as directed in the recipe for pâte sucrée dough. Immediately after removing the baked tart shell from the oven, sprinkle the chopped semisweet chocolate evenly over the bottom of the hot crust. Wait 1 to 2 minutes and then spread it evenly with an offset spatula (Figure 4–33). Allow the tart to cool completely and chill it for approximately 15 minutes or until the chocolate hardens. The chocolate will protect the bottom crust from becoming soggy after it is filled.

STEP D: ASSEMBLY

U.S.		METRIC	INGREDIENTS
			Clear Glaze
	2 teaspoons	10 mL	cold water
	1 teaspoon	3 g	unflavored powdered gelatin
4 fluid ounces	½ cup	120 mL	water
3½ ounces	½ cup	100 g	granulated sugar
			fresh fruit such as strawberries, raspberries, blackberries, and blueberries
½ ounce	¼ cup	15 g	shredded coconut, ligthly toasted, finely chopped

1. Pour 2 teaspoons (10 mL) cold water into a small bowl. Sprinkle the gelatin over the water and allow it to bloom for 5 minutes.

2. In a small saucepan, make a simple syrup by bringing the 4 fluid ounces (½ cup; 120 mL) water and the sugar to a boil on high heat. Boil the mixture for approximately 1 minute or until the sugar is completely dissolved. Remove the pan from the heat and whisk the bloomed gelatin into the hot simple syrup. Whisk the glaze until the gelatin has melted. Set the glaze aside until the tart has been filled.

FIGURE 4–33

3. Pour the cooled coconut pastry cream into the chocolate-lined crust and spread it evenly over the bottom. Slice the strawberries, if desired. Place the berries decoratively over the pastry cream, laying the fruit down in an attractive pattern (Figure 4–34).

4. Stir the glaze over an ice water bath until it begins to thicken slightly or until it has the consistency of raw egg white. Using a pastry brush, glaze each piece of fruit and the rim of the tart (Figure 4–35).

5. Immediately sprinkle the toasted coconut around the outer edge of the tart. Chill the tart until ready to serve. The tart should be served the same day it is made and removed from the pan right before it is served.

Note: Tart shown removed from pan for demonstration purposes.

FIGURE 4–34

FIGURE 4–35

Fresh Berry Tart Filled with Coconut Pastry Cream

CHOCOLATE RASPBERRY TART

Makes one 10-inch (25-cm) tart, serving 8 to 10

STEP A; CHOCOLATE PÂTE SABLÉE CRUST

1. Make one recipe of Chocolate Pâte Sablée Dough.

2. Bake the chocolate pâte sablée crust blind in a 10-inch (25-cm) false bottom tart pan as directed in the recipe for Chocolate Pâte Sablée Dough. Allow it to cool.

STEP B: GANACHE (CHOCOLATE CREAM FILLING)

MEASUREMENTS			INGREDIENTS
U.S.		METRIC	
12 fluid ounces	1½ cups	355 mL	heavy cream
12 ounces		340 g	semisweet chocolate, coarsely chopped
	2 tablespoons	30 mL	seedless raspberry preserves
	10 tablespoons	296 mL	seedless raspberry preserves
	1 tablespoon	15 mL	raspberry liqueur
8½ to 13 ounces	2 to 3 cups	240 to 370 g	fresh raspberries

1. In a saucepan, bring the cream to a boil. Remove the pan from the heat and add the chopped chocolate. Whisk the ganache gently until the chocolate is melted and the mixture is smooth. Stir in 2 tablespoons (30 mL) of the preserves.

2. Let the ganache cool for 1 hour at room temperature. Pour it into the cooled tart shell (Figure 4–36). Chill the tart for approximately 1 hour or until the ganache feels slightly firm.

3. In a small saucepan, blend the remaining preserves and liqueur. Heat the mixture on low heat until it becomes thin and fluid.

4. Arrange the raspberries in concentric rows on top of the ganache.

5. Using a pastry brush, brush the raspberries with the warmed preserve mixture (Figure 4–37). Chill the tart. Remove the tart from the pan right before it is served.

FIGURE 4–36

FIGURE 4–37

Chocolate Raspberry Tart

INDIVIDUAL RHUBARB BERRY "COBBLERS"

Makes 12 4-ounce (120-mL) ramekins

Lessons demonstrated in this recipe:

- How to prepare a starch-thickened fruit filling made with cornstarch.
- The filling is brought to a boil to reach the full thickening strength of the starch.
- Crisp pastry circles are placed on top of the cooked hot filling instead of the traditional biscuit topping of a cobbler.

STEP A: PASTRY DOUGH FOR CIRCLES

INGREDIENTS

| 1 recipe Pâte Brisée or Pâte Brisée with Cream Cheese (this chapter) |
| 1 large whole egg whisked with 1 teaspoon (5 mL) water (to be used in egg wash) |
| Granulated sugar for sprinkling |

1. Make 1 recipe **Pâte Brisée** or **Pâte Brisée with Cream Cheese** through step 4.

2. Preheat the oven to 425°F (219°C). Divide the dough in half. Rewrap and place one half of the dough back into the refrigerator. Roll out the remaining half onto a floured work surface to a rough rectangle 1/8 inch (3 mm) thick.

3. Cut at least 12 3-inch (7½-cm) diameter plain or fluted circles using a cookie cutter. Place circles on a sheet pan covered with parchment paper. Repeat with the other half of the dough. There will be extra circles.

4. Using a pastry brush, egg wash each circle and sprinkle generously with sugar. Bake the circles for 8 to 12 minutes or until lightly browned. Remove from the oven and allow to cool. Set aside.

> **TIP** The pastry circles can be baked, cooled, and stored in an airtight container for 1 to 2 days.

Makes 6 cups (48 fluid ounces; 1.4 L)

STEP B: RHUBARB BERRY FILLING

MEASUREMENTS			INGREDIENTS
U.S.		**METRIC**	
1½ to 2 pounds		685 to 910 g	fresh rhubarb, cut crosswise into ½-inch (1¼-cm) thick pieces
7 ounces	1 cup	200 g	granulated sugar
	½ teaspoon	¾ g	cinnamon
	½ teaspoon	1 g	ginger
¾ ounce	3 tablespoons	30 g	cornstarch
1½ fluid ounces	3 tablespoons	45 mL	orange juice
7½ ounces	1 cup	215 g	fresh whole blackberries or raspberries
7½ ounces	1 cup	215 g	fresh strawberries, cut into ½-inch (1¼-cm) thick pieces
½ ounce	1 tablespoon	15 mL	orange-flavored liqueur
			zest from 1 orange

1. Place the rhubarb, sugar, cinnamon, and ginger in a large pot (Figure 4–38). Cook over medium-high heat until sugar is melted, about 5 to 6 minutes.

FIGURE 4–38

FIGURE 4–39A

FIGURE 4–39B

FIGURE 4–40

2. In a small bowl, whisk the cornstarch and the orange juice together until smooth. Mix into the hot rhubarb mixture and bring the mixture to a boil until it has thickened, stirring constantly, about 4 to 5 minutes (Figure 4–39A and 4–39B). Pour the hot mixture into a clean bowl. At this stage the filling can be cooled, covered with plastic, and chilled for 1 to 2 days.

3. Stir the berries, the orange-flavored liqueur, and the zest into the rhubarb and set the mixture aside (Figure 4–40).

STEP C: ASSEMBLY

1. Preheat oven to 425°F (219°C). Spoon 4 ounces ($\frac{1}{2}$ cup; 115 g) of the filling into 12 4-ounce (120-mL) ramekins. Place the filled ramekins on a sheet pan.

2. Bake for 10 to 15 minutes or until fruit is hot and bubbling.

3. Top each ramekin with a sugared pastry circle and serve immediately (Figure 4–41).

FIGURE 4–41

Professional Profile

BIOGRAPHICAL INFORMATION

James P. Sinopoli, CEPC
Chef Instructor
Stratford University
School of Culinary Arts
Arlington, VA

1. Question: *When did you realize that you wanted to pursue a career in baking and pastry?*

Answer: *I was a cook at a Hyatt in New Jersey when some shift changes resulted in my slot being taken over. The chef said, "Go to the bakeshop," but I knew nothing about baking. He sent me to The Culinary Institute of America in Hyde Park to take a class on tortes and gateau. The moment I stepped through those doors I knew I had to come back there to get my education. I was fascinated by the artistry side of pastry.*

2. Question: *Was there a person or event that influenced you to go into this line of work?*

Answer: *The "event" was basic survivorship—I needed a job! But the pastry chef at that Hyatt in New Jersey, Phyllis Levy, really encouraged me.*

3. Question: *What did you find most challenging when you first began working in baking and pastry?*

Answer: *Having cooked for close to 15 years prior to going off to school I found the hardest challenge was the science part of baking.*

4. Question: *Where and when was the first practical experience in a professional baking setting?*

Answer: *After graduating from the CIA, I went to work as a chocolatier for Birnn Chocolates in Highland Park, NJ.*

5. Question: *How did this experience affect your later professional development?*

Answer: *The chocolate shop taught me about chocolate, its uses and abuses.*

6. Question: *Who were your mentors when you were starting out?*

Answer: *I would be remiss if I didn't say Julia Child. I loved everything she did. I also learned so much from Jan Bandula, CMPC.*

7. Question: *What would you list as your greatest rewards of your professional life?*

Answer: *Some of the best rewards are seeing people just go crazy when they consume something that you have poured your heart into. I also had the privilege to work for the past 2 years as the contract pastry chef at Blair House. Having your food appreciated by world leaders and having the opportunity to occasionally meet them was very exciting.*

8. Question: *What traits do you consider essential for anyone entering the field?*

Answer: *Passion. You must really have it for the food and for hospitality. This is a meet and greet business. You must be a people person to be in it.*

9. Question: *If there was one message you would impart to all students in this field what would that be?*

Answer: *There are easier ways to make money, but if you really love what you are doing then other things in your life will be much easier for you. Think about it; you love what you do so at least 8 to 12 hours a day you are guaranteed to be happy!*

CHAPTER 5

Using Chemical and Steam Leaveners

After reading this chapter, you should be able to:

- Describe what a chemical leavener is.

- Explain the role leaveners play in baked goods.

- Differentiate between baking powder, baking soda, and ammonium carbonate.

- Describe the roles of air and steam in the leavening of baked goods.

- Define a quick bread.

- Describe the role carbon dioxide plays in the leavening of quick breads, cakes, and cookies.

- Demonstrate how to work with chemical and steam leavening agents by preparing the recipes at the end of this chapter.

In Chapters 1, 2, and 3, you have seen just how important yeast is as a leavening agent in preparing such baked goods as breads, pizza, cinnamon rolls, croissants, and rich coffee cakes.

In this chapter, other leavening agents are discussed. Quick breads, cakes, and cookies use chemical leaveners such as baking powder, baking soda, and ammonium carbonate to help them rise.

Chemical leaveners react much more quickly than yeast in leavening baked goods, and there is no waiting period needed to allow batters or doughs to ferment.

Quick breads refer to a category of breads that are leavened with carbon dioxide gas from chemical leaveners. They include such baked goods as banana breads, short cakes, biscuits, and muffins. They are, in effect, quick to prepare.

Besides chemical leaveners, air and steam are also important leavening agents. Air is always present whenever mixing occurs. Steam, produced during the baking process, is an important leavening agent for specialty baked goods such as éclair paste, popovers, and puff pastry.

Specific mixing methods for quick breads and cakes are discussed in Chapters 6 and 7. This chapter is an introduction to various leavening agents other than yeast and their applications in specific baked goods.

Chemical Leaveners

Quick breads may include banana bread, lemon bread, date nut bread, scones, biscuits, shortcakes, and blueberry muffins.

A leavener is a substance that helps the mixed ingredients rise in the oven. Leaveners are added as ingredients in both quick breads and yeast breads. There are different types of leaveners. Chemical leaveners are, in fact, chemicals that react with the liquid ingredients within the recipe and the heat of the oven to produce carbon dioxide gas. The gas bubbles expand in the oven and push against batters and doughs to help the product to rise. Chemical leaveners are used in many baked goods like cakes, cookies, and quick breads. There is no rising time needed. Chemical leaveners include baking soda, baking powder, and ammonium carbonate.

Acids, Bases, and Neutralization Reactions

Before discussing acids or bases, the concept of pH must be defined. pH is a measurement used to describe how basic or acidic a substance is. The range is between 0 and 14. Water has a pH of 7 and is considered neutral; anything less than 7 is referred to as acidic and anything greater than 7 is referred to as basic.

An easy way to think of an acidic ingredient is to think of foods that are sour like citrus juices (e.g., lemon or orange juice), vinegar, sour cream, and yogurt. Other acidic ingredients include buttermilk, brown sugar, honey, molasses, chocolate, and cocoa powder (not Dutch processed).

The most common bases used in baking include sodium bicarbonate (baking soda) and ammonium carbonate.

When an equal amount of an acid is combined with an equal amount of a base, a neutralization reaction occurs. This means that, as the two substances are combined, they form a mixture containing salt and water (Figure 5–1). This mixture has a pH of 7, which is neutral (neither acidic nor basic). Neutralization reactions occur in batters that contain chemical leaveners and they ultimately help baked goods to rise.

Baking Soda

Baking soda is a chemical known as *sodium bicarbonate* or *bicarbonate of soda.* When baking soda comes in contact with moisture and an acid, a neutralization reaction takes place. This reaction causes not only salt and water to form but also carbon dioxide bubbles (Figure 5–2). This last by-product is most important to the baker.

The carbon dioxide gas bubbles form throughout the batter or dough, expanding as they are heated in the oven. This expansion of gases pushes up against the batter or dough, helping it to rise.

It is helpful to find which ingredient acts as the acid in a recipe for any quick bread or cake. For example, if a batter is acidic because of one of its ingredients like cocoa powder or buttermilk, adding baking soda helps the batter to rise and neutralizes the batter's acidity at the same time. Sometimes, a slight tangy taste is desired. In that case, a smaller amount of baking soda can be used so that some of the batter's acidity can be maintained.

When an acid combines with baking soda, the reaction that forms carbon dioxide is immediate; therefore it is important to bake batters and doughs containing baking soda as quickly as possible or risk losing the leavening power.

If too much baking soda is used, an off flavor from the salt residue becomes noticeable.

As a general rule, for every cup of flour (approximately 4 ounces; 115 g) in the recipe, approximately $\frac{1}{4}$ teaspoon (1 g) of baking soda is added.

SHELF LIFE

Over time, the shelf life of baking soda is shortened because it absorbs moisture from the air, which weakens it. A quick test can be done to determine whether it is still active. Add a small amount of baking soda to some vinegar. If the baking soda is active, the mixture should foam. If kept unopened, it should last approximately 2 years.

Baking Powder

Baking powder is a mixture of different leaveners that include baking soda and one or more acids. Because the acid is already in the baking powder, acidic ingredients in a dough or batter are not necessary for carbon dioxide to be produced as they are in baking soda. Cornstarch is

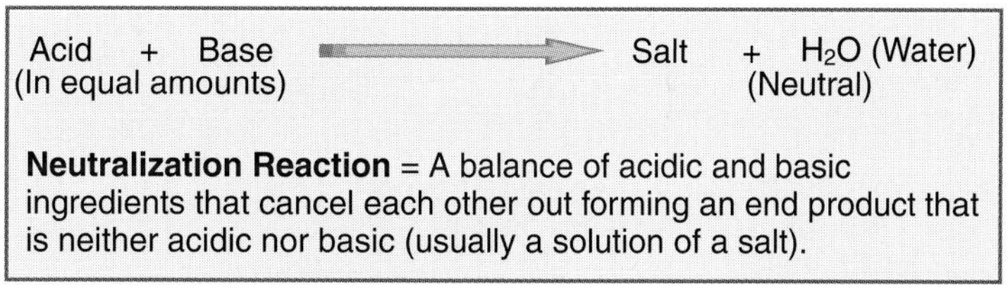

FIGURE 5–1

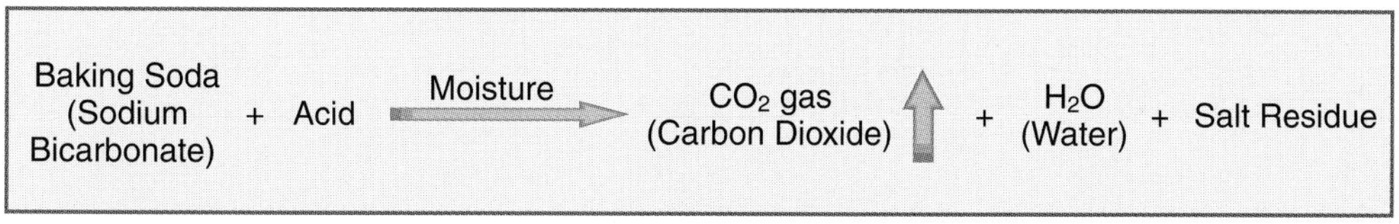

Acid   +   Base $\longrightarrow$ Salt   + H_2O (Water)
(In equal amounts)   (Neutral)

Neutralization Reaction = A balance of acidic and basic ingredients that cancel each other out forming an end product that is neither acidic nor basic (usually a solution of a salt).

FIGURE 5–2

Baking Soda
   (Sodium   +   Acid $\xrightarrow{\text{Moisture}}$ CO_2 gas ↑ + H_2O + Salt Residue
Bicarbonate)   (Carbon Dioxide)   (Water)

usually added to baking powder to prevent clumping (by absorbing excess moisture) and to keep the active ingredients separated to prevent their reacting and reducing the leavening power. There are two types of baking powders: single-acting baking powder and double-acting baking powder (Figure 5–3).

Single-acting baking powder is actually baking soda and cream of tartar (an acid). It needs only to come into contact with moisture from liquid ingredients to react. This leaves little time to get the batter into pans and into the oven before all the carbon dioxide gas is lost into the air. For this reason, single-acting baking powder is known as *fast-acting* and is rarely used.

Double-acting baking powder reacts twice. It requires both moisture and heat to react and is referred to as *slow-acting.* Once double-acting baking powder comes into contact with liquid ingredients, it produces an initial group of carbon dioxide gas bubbles. It reacts again once the batter or dough is heated in the oven. Double-acting baking powder contains two or more acids that react twice. The first reaction allows one acid to react with the baking soda before it goes into the oven. The second reaction uses another acid, which requires heat from the oven to react with the sodium bicarbonate (baking soda) to form carbon dioxide gas. This makes double-acting baking powder more versatile than baking soda.

As a guide, for every 1 cup (4 ounces; 115 g) of flour, between 1 and $1\frac{1}{4}$ teaspoons (4 to 5 g) of baking powder is added.

SHELF LIFE

To test the activity of baking powder, stir a small amount into hot water. If the baking powder is active, the mixture should foam and fizz. Baking powder has a shelf life of approximately 18 months if kept covered at room temperature.

Ammonium Carbonate

The third chemical leavener is ammonium carbonate. Ammonium carbonate, also known as ammonium bicarbonate or baking ammonia, is not as widely used as baking soda or baking powder.

Ammonium carbonate, in the presence of moisture and heat, reacts to produce ammonia, carbon dioxide gas, and water (Figure 5–4). All three of these by-products can act as leaveners. Ammonia and carbon dioxide are already gases, and water, when heated, becomes steam, which is also a gas. Because gases expand in the oven leavening takes place.

FIGURE 5–3

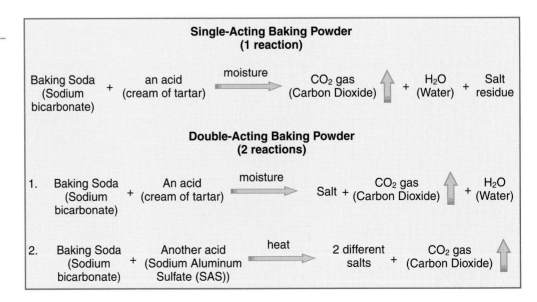

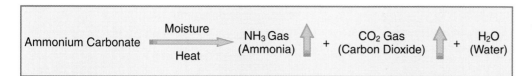

FIGURE 5–4

Ammonium carbonate is best used in small baked goods such as dry cookies, crackers, or éclair or choux paste in which the baked good will be baked at a high temperature until it is dry and crisp. The smaller the baked good with more surface area exposed and the longer the baking process, the more ammonia gas will evaporate off. This decreases the chance of the residual ammonia leaving an off smell.

Ammonium carbonate is used widely in Europe. It should not be used in moister baked goods such as cakes, muffins, biscuits, or soft cookies.

The Important Role of Air in Leavening

It is important to recognize the critical role that air plays in leavening. Air is incorporated into most mixtures just through the act of mixing the ingredients together. Mixing can take many forms such as stirring, creaming, or beating.

Air becomes an extremely important team player when using chemical leaveners. As ingredients are mixed, air is incorporated in the form of small cells or bubbles. These bubbles of air are evenly distributed throughout the batter or dough, along with other ingredients and chemical leaveners.

In the oven, carbon dioxide bubbles form and seek out these small air bubbles, enlarging them and leavening the baked good. The presence of chemical leaveners does not create new air bubbles; they just help enlarge the ones that already exist.

If the smaller air bubbles were not present, any carbon dioxide gas that formed would escape from the batter or dough because it would have nowhere to go but into the air.

Steam Leavening

Almost all baked goods use steam as a leavening agent, either as the sole leavener or in conjunction with other leaveners like yeast or chemical leaveners. That is because many ingredients within baked goods contain some form of water. Citrus juices, eggs, and dairy products such as milk, yogurt, and cream all contain a certain amount of water.

When water is heated in the oven to 212°F (100°C), it becomes a gas. This gas is known as *steam*. Steam is a powerful force that can expand over 1,000 times its original volume. The result of the steam expanding under pressure within batters and doughs causes them to expand and rise. This leavens the baked good.

Air also works as a team player with steam much like it does with chemical leaveners. As the steam seeks out the small air bubbles in the batter, the air bubbles expand and leaven the baked good.

Steam is the exclusive leavening agent in certain specialty baked goods such as popovers (a specialty quick bread), éclair paste, and puff pastry, a laminated dough. (See Chapter 3 for more on puff pastry and steam leavening.)

Popovers

Popovers are a puffy quick bread with a brown exterior and a hollow, eggy center. They grow especially tall in the oven when the batter is poured into a special pan called a *popover pan*. This pan has spaces that are deeper and spaced farther apart than the average muffin pan. Typical popover batter is very thin. It is similar in consistency to heavy cream. It contains only a few ingredients, two of which are milk and eggs. Unlike typical quick breads that use a flour with a low-protein content, popovers use bread flour. Bread flour helps develop the much needed gluten, which stretches and expands as steam is produced. Steam is produced from the liquid ingredients in the batter. This causes the interior to puff and become hollow. Popovers are baked at a very high temperature. This high temperature (425°F; 219°C) helps the liquid within the batter to form a great deal of steam. The steam expands, causing the popovers to rise. Because the pans are deep, the batter climbs the sides quickly and continues to grow, forming a high, puffed exterior. Steam develops inside breaking apart some of the egg proteins, forming a large space within. The gluten combined with the cooked egg proteins help form structure so the popover will not collapse.

RECIPE

Popovers (Chapter 6, page 180)

Éclair Paste or Choux Paste

Éclair paste or choux paste is another steam-leavened specialty dough that is used to produce cream puffs, éclairs, and other varieties of pastries that can be filled with whipped cream, fruit, mousse, ice cream, or savory fillings. The dough is also known as "pâte à choux." Pâte à choux is a French term that means "cabbage paste," so deemed because of the cabbage-like appearance of small cream puffs. The dough may be formed into a number of various shapes. When éclair paste is formed into small rounds or cream puffs, they may be referred to as *profiteroles*.

Éclair paste is so versatile that it can also be piped out into thin shapes that can be baked until crisp and used as a garnish on top of sweet and savory foods. Or, the dough can be deep-fried and served as a type of doughnut. It can also be used to make savory cheese puffs that can be made by mixing cheese into the cream puff dough before baking. Another special way to prepare éclair paste is in the shape of swans. Éclair paste is used to form necks, which are mounted on the specially shaped cream puff "bodies" to resemble birds. The neck and wings stick into whatever fills the swan (whipped cream, mousse, or ice cream). Another special use for éclair paste is to prepare a tower of pastry cream–filled cream puffs held together and drizzled with caramel, known as croquembouche (Chapter 10).

Typical éclair paste always includes a liquid such as water or milk, a fat such as butter, flour, eggs, and a little salt. Some recipes may contain sugar. Although popover batter and éclair paste dough look very different, steam does the job of leavening for each. The liquid ingredients (water, milk, and eggs) in the éclair paste form steam in the oven and help the dough to rise. The procedure to make éclair paste is simple and can be started right in a saucepan.

A liquid like water or milk is brought to a rolling boil with butter, salt, and sugar. The flour is stirred in quickly, causing the starches in the flour to swell and absorb all of the liquid at once. The dough should be very stiff in the beginning to allow it to hold as many eggs as possible and still hold its shape.

Egg whites and sometimes ammonium carbonate may be added to help make the puffs dry and crispy. Whole eggs are added for leavening and to provide structure. The more eggs that are added to the dough, the lighter the finished puffs will be because whole eggs hold some air and also contribute to leavening. If too many eggs are added, the dough will be too thin and will not be able to hold its shape. Between the eggs and the higher protein content in the bread flour, a firmer structure will be formed. Steam expands the puffs while egg protein sets the structure. As the egg proteins denature and uncoil from the heat of the oven, steam begins to form. The steam puts pressure on the proteins to expand and eventually break apart, leaving a hollow space within the puff. The proteins at this point have cooked enough so the structure of the puff is set and does not collapse.

TIP Éclair paste must be baked until golden brown and crisp or else it will soften and collapse once it is cooled.

RECIPES

Cream Puffs (This chapter, page 154)
Cream Puff Swans (This chapter, page 157)

Two Methods to Bake Éclair Paste

There are two schools of thought when baking éclair paste.

One method is to bake the dough at an extremely high temperature, approximately 425°F (219°C) for a short period of time to cause steam to form quickly and then lower the temperature to approximately 375°F (190°C) to finish the baking process. This causes the egg protein to coagulate and form a set structure.

The other method starts the baking process at a low temperature, approximately 300°F (149°C). Once the éclair paste is in the oven, the temperature is raised immediately to a high temperature of approximately 450°F (230°C). Steam quickly forms, and the puffs rise. After approximately 10 to 15 minutes, the temperature is turned back down to 300°F (149°C) to dry the puffs and brown them. Starting at a lower temperature gives the dough time to rise higher before the egg proteins actually set and a hard outer crust forms. This second method is used in this text.

CREAM PUFFS

Makes approximately 30 to 36 small cream puffs

Lessons demonstrated in this recipe:

- How to prepare a steam-leavened specialty baked good made from éclair paste.
- A high proportion of water-based liquid ingredients produce enough steam to leaven.
- A high-protein flour is used to give structure.
- Whole eggs leaven and add structure.
- Egg whites are added for crispness and lightness.
- Great volume is obtained by starting the baking process at a lower temperature, which allows the cream puffs to rise higher before they become set and form a crust.

MEASUREMENTS				INGREDIENTS
U.S.		**METRIC**	**BAKER'S %**	
6 fluid ounces	¾ cup	180 mL	127%	water
4 ounces	½ cup	115 g	81%	unsalted butter
	¼ teaspoon	1.25 g	0.9%	salt
	2 teaspoons	10 g	7%	sugar
5 ounces	1 cup	115 g	100%	bread flour
2 each		94 g	66%	large whole eggs
2 each		56 g	40%	large egg whites
			421.9%	Total Cream Puffs percentage

1. Preheat the oven to 300°F (149°C).

2. In a medium saucepan, bring the water, butter, salt, and sugar to a vigorous boil. Remove saucepan from heat. Add the flour all at once and stir mixture quickly with a wooden spoon (Figure 5–5). This will cause a ball of dough to form immediately. Continue to cook the mixture for approximately one minute until it forms a ball and pulls away from the sides of the pan. Remove the pan from the heat and allow the mixture to cool for 15 minutes to approximately 130°F (54°C) to prevent the eggs from cooking as they are added to the mixture.

FIGURE 5–5

3. Transfer the dough into the bowl of an electric mixer. Using the paddle attachment, mix the dough on low speed. Add 2 whole eggs, one at a time, blending well after adding each egg (Figure 5–6). Then add 2 egg whites, one at a time, blending well after each addition. The dough should hold its shape somewhat and should not be too loose (Figure 5–7). The color of the dough should be yellowish from the butter and eggs, and it should look glossy.

4. Using a large round tip placed into a pastry bag, or a 1 teaspoon (5 mL) measure, pipe or spoon walnut-sized balls of dough onto 2 half sheet pans covered with parchment paper, spacing the balls about 1 inch (2.5 cm) apart (Figure 5–8). With your fingertips moistened with water, touch the tops of the cream puff dough to smooth the shape (Figure 5–9). The shape should be a small 1-inch (2.5 cm) round.

5. Place the sheet pans in the oven at 300°F (149°C). Close the oven door and raise the oven temperature to 450°F (230°C). Bake for 10 to 15 minutes until well puffed. Turn the oven down to 300°F (149°C) and bake for another 15 to 20 minutes until crisp and browned. Cut open one of the puffs to see if it is dry inside. If not, continue baking.

6. Remove from the oven and with a small, sharp knife, poke a small hole into each cream puff (Figure 5–10). This helps dry out the moist inside of each puff and allows them to cool faster. Cream puffs can be baked one day ahead and kept covered in an airtight container at room temperature.

FIGURE 5–6

FIGURE 5–7

FIGURE 5–8

FIGURE 5–9

FIGURE 5–10

Three great ways to serve pâte à choux:

■ Using ice cream and one recipe of Rich Chocolate Sauce (Chapter 11), cut each pâte à choux in half crosswise and fill with small scoops of ice cream. Place three or four filled puffs in each dessert bowl and generously pour warm ganache frosting over the top. Makes six to seven servings.

■ Use as a base for the Croquembouche recipe in Chapter 10.

■ Use as a garnish on top of a dessert by piping the dough into various designs onto a parchment-lined sheet pan and baking at 375°F (190°C) for 5 to 7 minutes or until crisp and lightly browned.

CREAM PUFF SWANS

Makes approximately 13 4-inch (10-cm) long swans

Lesson demonstrated in this recipe:

■ How to shape éclair paste to form pastries resembling swans.

STEP A: MAKE ÉCLAIR PASTE

Prepare one recipe of éclair paste from the recipe for Cream Puffs.

STEP B: SHAPING THE SWANS

Creating the swan's body

Place all but 3 ounces ($\frac{1}{3}$ cup; 105 g) of the dough into a pastry bag fitted with a large star tip (approximately 14 mm). Hold the pastry bag at 90 degrees to a parchment-lined sheet pan and pipe out a 2$\frac{1}{2}$-inch (6$\frac{1}{4}$-cm) long shell making an up and over motion (Figure 5–11A and B). Pull back so that the end is tapered.

Continue to make approximately 12 more tapered shells leaving a space between each swan body.

Creating the swan's neck and head

Place the remaining dough into a pastry bag fitted with a small round tip (approximately $\frac{1}{8}$ inch; 3 mm). Pipe a small round ball (the head) and then continue to form a thin letter **S** (the neck) on a second sheet pan covered with parchment paper (Figure 5–12). Place the tip into the round ball (the head) at the top of the **S** and, squeezing gently, quickly pull back to form the swan's beak (Figure 5–13). Repeat the procedure until there are as many swan necks as there are bodies (with extra in case of breakage).

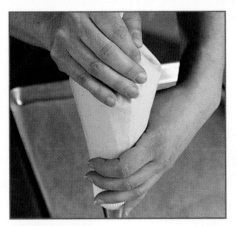

FIGURE 5–11A

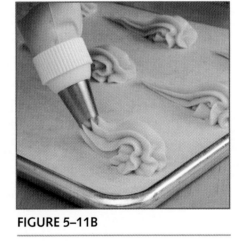

FIGURE 5–11B

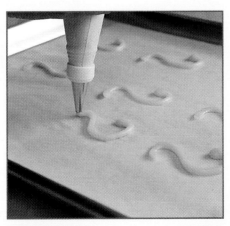

FIGURE 5–12

FIGURE 5–13

Bake the swan bodies as directed in the Cream Puff recipe. Because the swan necks are thinner, they need to be baked at a lower temperature of 375°F (190°C) for 8 to 10 minutes or until they are a pale brown and firm to the touch. Remove and allow to cool.

STEP C: STABILIZED WHIPPED CREAM

MEASUREMENTS			INGREDIENTS
U.S.		**METRIC**	
	¾ teaspoon	2.25 g	unflavored gelatin
¾ fluid ounce	1½ tablespoons	22 mL	cold water
12 fluid ounces	1½ cups	360 mL	heavy cream, cold
3 ounces	¾ cup	100 g	confectioners' sugar, sifted if lumpy plus more for dusting
	1½ teaspoons	7.5 mL	vanilla extract

1. In a small heatproof bowl, sprinkle the gelatin over the water. Stir and allow it to soften for 5 minutes.

2. Melt the gelatin over a hot water bath until it is completely liquefied. Remove the gelatin from the heat and whisk ½ fluid ounce (1 tablespoon; 15 mL) heavy cream into the gelatin to cool.

3. In the bowl of an electric mixer, use the whip attachment to beat the remaining heavy cream until it forms soft peaks.

4. Gradually add the confectioners' sugar and then the melted gelatin and cream mixture along with the vanilla extract. Beat until stiff peaks form. Do not overbeat.

STEP D: ASSEMBLY

1. Cut each cooled swan body in half crosswise. Cut the top portion of the body lengthwise to form two wings. Leave the bottom half intact.

2. Using a pastry bag fitted with a large round or star tip, pipe the whipped cream into the bottom of the swan body. Gently insert a wing into the cream on either side (Figure 5–14).

3. Insert a neck into the cream (Figure 5–15). Repeat with the remaining swans. Dust the swans with confectioners' sugar (Figure 5–16).

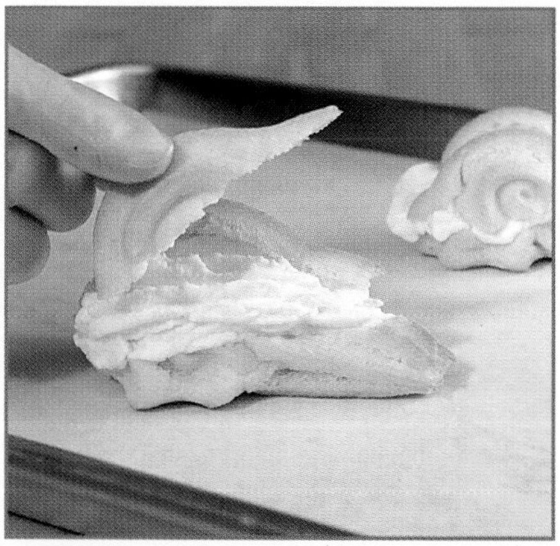

> **TIP** Cream puff swans should be refrigerated. They are best consumed the same day they are prepared. If they need to be stored overnight, do not insert the necks into the bodies until right before serving as refrigeration for over 4 hours softens them. Leave the necks at room temperature, covered in an airtight container.

FIGURE 5–14

FIGURE 5–15

FIGURE 5–16

Cream Puff Swans

Professional Profile

BIOGRAPHICAL INFORMATION

Deborah Gordon
Baking and Pastry Instructor
Florida Community College at Jacksonville
Institute of the South for Hospitality and Culinary Arts
Jacksonville, FL

1. Question: *When did you realize that you wanted to pursue a career in baking and pastry?*

Answer: *I realized I wanted to go into this field when I was 19 years old. I used to make homemade confections and give them as gifts. I also enjoyed making and decorating cakes for family and friends.*

2. Question: *Was there a person or event that influenced you to go into this line of work?*

Answer: *My mother influenced my love of cooking and baking. She was self-taught. Her mother passed when she was very young and she was an only child. She is one of the best and most creative cooks I know, even compared to those who went to culinary school.*

3. Question: *What did you find most challenging when you first began working in baking and pastry?*

Answer: *Making yeast dough was a challenge for me. It is not my strong suit.*

4. Question: *Where and when was your first practical experience in a professional baking setting?*

Answer: *My first experience was about 7 years ago at Ruth's Chris Steak House. I was the pastry chef.*

5. Question: *How did this first experience affect your later professional development?*

Answer: *I loved it. It helped me to realize my strengths and weaknesses.*

6. Question: *Who were your mentors when you were starting out?*

Answer: *My mentor was, of course, my mother. I have a lot of admiration for her natural abilities.*

7. Question: *What would you list as your greatest rewards?*

Answer: *My greatest reward has to be the satisfaction I got from opening my own retail bakery and gourmet shop and being skilled enough to run it completely on my own.*

8. Question: *What traits do you consider essential for anyone entering the field?*

Answer: *You must be responsible and be able to multitask. You have to be willing to work hard, often working long hours without a break.*

9. Question: *If there was one message you would impart to all students in this field what would that be?*

Answer: *If you don't love it, don't do it. In this field it is a lot of hard work for, sometimes, very little reward. So, if you don't love it you won't last. I would also advise anyone going into baking and pastry to set goals high and try to learn as much as possible from each place that you work.*

CHAPTER 6

Quick Bread Mixing Methods

After reading this chapter, you should be able to:

- Understand the differences between yeast breads and quick breads.

- Understand the three quick bread mixing methods.

- Demonstrate the quick bread mixing methods by preparing the recipes in this chapter.

In the previous chapter you learned how to define a quick bread.

The difference between a good quick bread and a great quick bread lies in the mixing method. Quick breads have their own mixing methods. The general goal behind each mixing method is to reduce gluten formation so the quick bread is tender. Quick bread recipes tend to use low-protein flours like all-purpose, cake, and pastry. There are exceptions for specialty quick breads that may use some higher protein flours. The presence of fat and sugar also add to a tender quality. In most quick bread recipes, the words "mix until just combined" are written over and over again. Overmixing develops gluten and toughness. Toughness is one of the pitfalls of quick bread preparations. Quick breads that are overmixed have a tendency to develop large holes or cavities throughout the inside, which is caused when a thick gluten network has formed and trapped gases (carbon dioxide) in the batter. This is referred as tunneling.

This chapter describes differences between yeast breads and quick breads and the three basic mixing methods for quick breads. They are (1) the biscuit method, (2) the muffin method, and (3) the creaming method. The popover, a specialty quick bread, has its own mixing method.

Differences between Yeast Breads and Quick Breads

Before discussing the different mixing methods for quick breads, you should know the major differences between a yeast bread and a quick bread. One similarity between yeast breads and quick breads is that they both use carbon dioxide gas to leaven baked goods. However, the two breads have more differences than similarities. (See Table 6-1, Differences between Yeast Breads and Quick Breads.)

Yeast Is Alive; Chemical Leaveners Are Not

Yeast breads are leavened with yeast, a living organism, which eats sugar and requires specific living conditions to survive. The yeast undergo fermentation in which the by-products produced are carbon dioxide gas and alcohol.

Table 6–1 Differences between Yeast Breads and Quick Breads

YEAST BREADS	QUICK BREADS
1. Leavened with carbon dioxide gas	1. Leavened with carbon dioxide gas
2. Use yeast that is alive	2. Use chemical leaveners that are not alive
3. Require time to prepare	3. Require little time to prepare
4. Most gases formed before baking	4. A small percentage of gases produced during mixing; most gases formed during baking
5. In general, gluten development encouraged, resulting in chewy texture	5. In general, gluten development discouraged, resulting in tender, more crumbly texture

Chemical leaveners are not alive. They form chemical reactions with specific ingredients within the batter or dough, causing the formation and release of carbon dioxide gas.

Preparation of Yeast Breads Requires Time; Preparation of Quick Breads Does Not

Yeast doughs must ferment, which takes time. Because the yeast are alive, they require time to multiply, eat available sugars, and give off carbon dioxide gas and alcohol. The rate of fermentation depends on the temperature of the dough, which must be sufficient to maintain the yeast in an active state.

Quick breads, on the other hand, are prepared quickly with little time spent on preparation. Quick breads must be baked relatively soon after they are mixed because of the immediate chemical reactions that occur. This ensures valuable carbon dioxide gas bubbles do not escape into the air before leavening has occurred.

Gases Are Formed before Baking in Yeast Breads; Gases Are Mostly Formed in the Oven during Baking in Quick Breads

Carbon dioxide gas produced by yeast in yeast breads is created before baking. The carbon dioxide becomes trapped within gluten sheets during fermentation and continues to form until the yeast are killed in the oven.

Quick breads prepared with chemical leaveners do not develop a strong enough of a network of gluten to trap carbon dioxide gas before baking. Therefore, batters and doughs using chemical leaveners depend upon them to release carbon dioxide gas in the oven. This release of gas should come just as the structure of the baked good forms and sets so that the gases can be retained and expand the product without it collapsing. However, it is normal for a small percentage of gas bubbles to dissipate into the air before baking.

Yeast Breads Tend to Be Chewy; Quick Breads Have a Finer, More Tender Crumb

Yeast breads and quick breads have different textures. In general, yeast breads are mixed for a longer period of time and tend to be prepared with flours containing higher protein levels. Gluten development is encouraged and the finished products are chewier in texture.

Quick breads, with few exceptions, are mixed until the ingredients are just combined. Most quick breads use flours with lower protein levels because little gluten development is desired. The result is a baked good with a finer, more tender crumb.

Different Mixing Methods

There are three mixing methods for quick breads. They include the biscuit method, the muffin method, and the creaming method. Different mixing methods may be used for the same type of quick bread. For example, one recipe for banana bread may use the muffin method and another recipe for banana bread may use the creaming method. Each banana bread will have a slightly different texture because of the different method used. (See Table 6–2, Quick Bread Mixing Methods Using Chemical Leaveners; and Table 6–3, Popovers: A Specialty Quick Bread Mixing Method Using Steam Leavening.)

Three Mixing Methods
for Quick Breads

■ Biscuit method

■ Muffin method

■ Creaming method

Table 6–2 Quick Bread Mixing Methods Using Chemical Leaveners

BISCUIT METHOD	MUFFIN METHOD	CREAMING METHOD
1. Mix dry ingredients in a bowl.	1. Mix dry ingredients in a medium bowl.	1. Cream fat and sugar until light and fluffy.
2. Cut in cold fat with a pastry blender.	2. Whisk liquid ingredients in another bowl.	2. Add eggs one at a time.
3. Add liquid ingredients and gather together until mixture forms a ball.	3. Pour liquid ingredients into dry ingredients.	3. Alternate dry and wet ingredients (beginning and ending with dry ingredients).
4. Chill to relax gluten.	4. Mix until just combined.	4. Mix until just combined.

The Biscuit Method

1. Mix the dry ingredients and the chemical leaveners in a mixing bowl.

2. Cut in the cold fat until it becomes the desired size using a pastry blender.

3. Combine all liquid ingredients in a separate bowl.

4. Add liquid ingredients to dry ingredients only until combined.

The Biscuit Method

The biscuit method is used primarily for biscuits, shortcakes, and scones. Biscuits are typically served with a meal, whereas shortcakes tend to be served for dessert. Scones are usually served as a breakfast item. All three tend to be flaky, light, and tender.

The flaky texture of baked goods using the biscuit method of mixing is produced by using a solid fat such as butter or shortening. The fat should be cold. If butter is used, it may be frozen for a short period to make sure it will not melt too quickly in the oven. Depending on the amount of flakiness desired in the finished product, the fat is mixed in with the dry ingredients until the pieces range from pea sized to smaller pieces that are the consistency of oatmeal. A pastry blender or a commercial mixer with a pastry blender attachment can be used. If the fat has been frozen, a food processor can be used by pulsing dry ingredients with the fat until the desired sized pieces are obtained. Do not overprocess the mixture or the fat will soften and blend into the dry ingredients. The technique used to combine fat into the dry ingredients is known as cutting or cutting in the fat. This cutting technique is used so that the fat is not broken down and blended in too thoroughly. Cutting also allows particles of flour to be coated with fat. The fat-coated flour particles are so slippery that as water-based ingredients are added, very little gluten is able to form. The lack of gluten produces a tender product. Depending on what is being made, the ingredients will form either a thick batter or dough.

During the baking process, the pieces of fat melt into the dough leaving air spaces which expand in the oven, forcing the dough to rise. This results in flaky layers. The longer the cold fat pieces can remain intact before melting, the flakier the finished baked good will become. The smaller the fat pieces, the more quickly they will melt into the dough, forming a more tender baked good.

Table 6–3 Popovers: A Specialty Quick Bread Mixing Method Using Steam Leavening

POPOVERS
1. Whisk egg, milk, and melted fat or oil.
2. Whisk in dry ingredients.
3. Mix until just combined.

The biscuit method of mixing resembles the method for making a flaky pie crust with the following differences:

- The ice water used in the flaky pie crust method is changed to other, richer liquid ingredients like milk, cream, or buttermilk in the biscuit method for quick breads.
- While there is no leavening used in the flaky pie crust method, chemical leaveners are typically used in the biscuit method for quick breads.

RECIPE

Nutty Chocolate Chip Shortcakes (This chapter, page 169)

The Muffin Method

The muffin method is simple and quick. That is why muffins and some loaf breads are called "quick breads." Date nut bread, banana bread, corn bread, and blueberry muffins are just four types of quick breads that may use this method. Many pancake and waffle recipes also use the muffin method of mixing. Most quick breads using the muffin method use liquid fats like oils or melted butter instead of solid fats and achieve a coarser, less cake-like texture. The muffin method uses just two bowls. In one bowl, all the dry ingredients are whisked together; in the other bowl go all the liquid ingredients. The liquid ingredients are then mixed into the dry ingredients only until combined. Overmixing is discouraged to prevent much gluten development, which leads to tough quick breads.

RECIPE

Grandma Etta's Blueberry Muffins (This chapter, page 172)

The Creaming Method

If a cake-like quality is desired in a quick bread such as a muffin or bread, the creaming method is used. The creaming method for a quick bread is the same method that is used for certain cakes. A softened solid fat is blended with sugar until light and fluffy, using an electric mixer with the paddle attachment. This blending of fat and sugar until light and fluffy is referred to as creaming. The granular, sand-like crystals of sugar push the fat against the bowl, helping to incorporate air, which lightens the color and texture. In the oven, the small air bubbles that have been incorporated into the fat enlarge due to the presence of chemical leaveners. This helps the baked good to rise.

Once the fat and sugar have been creamed, eggs are added one at a time to ensure that a proper emulsion forms (see Chapter 7 for more on emulsions). The dry ingredients are then added alternately with the liquid ingredients, mixing only until combined. As a variation to alternating the addition of liquid and dry ingredients, some recipes add all the liquid ingredients at one time and then all the dry ingredients.

RECIPES

Cake-Like Chocolate Chip Muffins (This chapter, page 177)
Nutty Banana Bread (using the creaming method) (This chapter, page 174)

A Specialty Quick Bread

A specialty quick bread discussed in Chapter 5 is the popover. Popovers use steam as their sole leavening agent.

RECIPES

Popovers (This chapter, page 180)

The Muffin Method

1. Mix dry ingredients in a medium-sized mixing bowl.

2. Blend liquid ingredients in another bowl.

3. Pour liquid ingredients into dry ingredients.

4. Mix only until combined.

The Creaming Method

1. Cream fat and sugar until light and fluffy in an electric mixer using the paddle attachment.

2. Add eggs one at a time.

3. Whisk dry ingredients in one bowl and blending liquid ingredients in another.

4. Alternately add dry ingredients and liquid ingredients, ending with the dry ingredients.

5. Mix only until combined.

Preparation of Muffin Pans

Muffin pans used for quick breads should be sprayed with nonstick cooking spray or greased to prevent sticking. Lining muffin pans with paper liners can replace greasing; however, the baked muffins will not have the same height or full appearance as the muffin without a liner. Muffins with liners are used in retail bakeries and supermarkets because the liner protects the muffins, aids in their easy removal, and prevents them from becoming dry.

A muffin without a liner looks more substantial because the batter can take up the entire muffin pan and rise up the sides unimpeded by any liner. In general, muffin pans should be filled two-thirds full.

NUTTY CHOCOLATE CHIP SHORTCAKES

Makes 10 to 16 shortcakes

A shortcake is a sweet biscuit that is baked and then split in half horizontally and filled with fruit, usually berries (e.g., strawberries, blueberries, or raspberries) and then dolloped with whipped cream. Remember that the "short" in shortcake refers to the shortening of the gluten strands by coating them with fat.

Lessons demonstrated in this recipe:

- How to prepare a quick bread using the biscuit method of mixing.
- Buttermilk is the acid that neutralizes baking soda to form carbon dioxide gas.
- The cold butter pieces distributed throughout the dough form a tender, flaky shortcake.

MEASUREMENTS				INGREDIENTS
U.S.		METRIC	BAKER'S %	
13 ounces	2¾ cups	370 g	100%	all-purpose flour
	½ teaspoon	3 g	0.8%	salt
1¾ ounces	¼ cup	50 g	14%	granulated sugar
⅔ ounce	1 tablespoon + 2 teaspoons	20 g	5%	baking powder
	½ teaspoon	2 g	0.5%	baking soda
4 ounces	½ cup	115 g	31%	unsalted butter, cold, cut into ½-inch (1.25-cm) cubes
4½ ounces	¾ cup	130 g	35%	semisweet chocolate chips
1¾ ounces	½ cup	50 g	14%	hazelnuts, pecans, or walnuts toasted in a 400°F (205°C) oven for 5 to 10 minutes and coarsely chopped
8 fluid ounces	1 cup	240 mL	65%	buttermilk, well shaken
1 each		47 g	13%	large egg
1 fluid ounce	2 tablespoons	30 mL	8%	buttermilk or milk for brushing
As needed		As needed		extra granulated sugar for sprinkling
				nonstick cooking spray
			286.3%	Total Nutty Chocolate Chip Shortcakes percentage

1. **Preheat oven to 375°F (190°C).**

2. Using a pastry blender, gently mix the flour, salt, sugar, baking powder, and baking soda in a large mixing bowl (Figure 6–1).

3. Cut in the cold, cubed butter with a pastry blender until it is the size of peas (Figure 6–2). The mixture should look like coarse oats (Figure 6–3).

4. Add the chocolate chips and nuts into the flour and butter mixture and toss together. Blend the buttermilk and the egg together in a small bowl. Pour the egg and buttermilk into the flour mixture and stir gently using a spoon until the mixture can be gathered into a ball (Figure 6–4). Do not overmix!

5. Gather the dough into a disk shape and wrap it in plastic. Chill the dough for 1 hour or up to 2 days in the refrigerator.

6. On a lightly floured surface, roll out the chilled dough to a ½ inch (1.25 cm) thickness (Figure 6–5). Cut the dough into desired shapes about 2½ to 3 inches (6 to 7.5 cm) in diameter (Figure 6–6).

Alternatively, the dough can be cut in half and each half rolled into a 7-inch (17½-cm) diameter circle that is ½ inch (1.25 cm) thick. Using a knife or pizza wheel, cut each circle into 8 pie-shaped wedges.

FIGURE 6–1

FIGURE 6–2

FIGURE 6–3

FIGURE 6–4

FIGURE 6–5

FIGURE 6–6

7. Cover a large sheet pan with parchment paper and place the shortcakes on them, leaving a 1-inch (2.5-cm) space between each one. Brush each shortcake with the remaining buttermilk and sprinkle generously with sugar (Figure 6–7). Bake the shortcakes for 16 to 20 minutes, or until they are light brown. Serve warm or at room temperature. Serve split in half with a dollop of whipped cream.

FIGURE 6–7

Nutty Chocolate Chip Shortcakes

GRANDMA ETTA'S BLUEBERRY MUFFINS

Makes approximately 12 to 14 muffins

Lessons demonstrated in this recipe:

- How to prepare a quick bread using the muffin method of mixing.
- The fat used in the muffin method is typically in liquid form.
- Two bowls are used: one for dry ingredients and one for wet ingredients.
- The wet ingredients are added to the dry ingredients.

MEASUREMENTS				INGREDIENTS
U.S.		METRIC	BAKER'S %	
9½ ounces	2 cups	270 g	100%	all-purpose flour
	1 tablespoon	12 g	4.5%	baking powder
1¾ ounces	4 tablespoons	50 g	19%	granulated sugar
	½ teaspoon	3 g	1.1%	salt
1 each		47 g	17%	large egg
8 fluid ounces	1 cup	240 mL	89%	milk
2 fluid ounces	¼ cup	60 mL	22%	canola oil
10 ounces	2 cups	285 g	106%	fresh or frozen blueberries (thawed), mixed with 1 teaspoon (2.5 g) all-purpose flour
½ to 1 ounce	1 to 2 tablespoons	15 to 30 g	9%	granulated sugar for sprinkling
			367.6%	Total Grandma Etta's Blueberry Muffins percentage

1. Preheat the oven to 400°F (205°C). Spray a 12-cup muffin pan with nonstick cooking spray. Set aside.

2. Whisk together the flour, baking powder, sugar, and salt in a large mixing bowl (Figure 6–8).

3. In another bowl, whisk together the egg, milk, and oil until blended.

4. Pour the liquid ingredients into the dry ingredients and mix only until combined (Figure 6–9A and B).

FIGURE 6–8

FIGURE 6–9A

FIGURE 6–9B

5. Toss the blueberries and 1 teaspoon of flour (2.5 g) into a bowl and fold them gently into batter (Figure 6–10). Do not overmix!

6. Fill the muffin cups two-thirds full. If there is remaining batter, spray another pan and fill in the same manner. Sprinkle each muffin with granulated sugar (Figure 6–11).

7. Bake for approximately 20 minutes, or until lightly browned and a cake tester or sharp knife inserted into the center of a muffin comes out clean.

FIGURE 6–10

FIGURE 6–11

Grandma Etta's Blueberry Muffins

NUTTY BANANA BREAD (USING THE CREAMING METHOD)

Makes one 9- by 5- by 3-inch (22.5- by 12.5- by 7.5-cm) loaf

Lessons demonstrated in this recipe:

- How to prepare a quick bread using the creaming method of mixing.
- The creaming method results in a quick bread with a cake-like texture.
- Double-acting baking powder reacts with liquid ingredients upon mixing and again in the oven to form carbon dioxide gas.
- Baking soda combined with an acidic ingredient like brown sugar reacts to form carbon dioxide gas.

MEASUREMENTS				INGREDIENTS
U.S.		METRIC	BAKER'S %	
8¼ ounces	1¾ cups	200 g	87%	all-purpose flour
1¼ ounce	¼ cup	35 g	13%	whole wheat flour
	2 teaspoons	8 g	3.5%	baking powder
	¼ teaspoon	1 g	0.4%	baking soda
	½ teaspoon	3 g	1.1%	salt
1¾ ounces	¼ cup	50 g	19%	pecans, coarsely chopped
1¾ ounces	½ cup	50 g	19%	walnuts, coarsely chopped
4 ounces	½ cup	115 g	43%	unsalted butter, softened
2½ ounces	⅓ cup	70 g	26%	granulated sugar
2½ ounces	⅓ cup	70 g	26%	light brown sugar (packed, if measuring by volume)
2 each		94 g	35%	large eggs
10 ounces	1 cup	285 g	106%	mashed bananas (from 2 or 3 very ripe bananas)
1 fluid ounce	2 tablespoons	30 mL	11%	rum
1½ teaspoons		7.5 g	3%	granulated sugar mixed with ¼ teaspoon (0.5 g) cinnamon
			392.5%	Total Nutty Banana Bread percentage

1. Preheat the oven to 350°F (175°C). Spray a loaf pan with nonstick cooking spray and set aside.

2. Whisk the flours, baking powder, baking soda, salt, and nuts in a medium bowl. Set aside.

3. In the bowl of an electric mixer, using the paddle attachment, cream the butter and the sugars on medium speed until light and fluffy (Figure 6–12). Add the eggs, one at a time, blending well and scraping down the bowl after each addition (Figure 6–13). On low speed, add mashed bananas and rum; blend well.

4. On low speed, add the flour and the nut mixture. Blend only until the ingredients are incorporated (Figure 6–14).

5. Pour the batter into the greased pan. Sprinkle the top with the granulated sugar and cinnamon (Figure 6–15). Bake for 45 to 50 minutes, or until golden brown and a cake tester or sharp knife or toothpick comes out clean. Cool in pan.

FIGURE 6–12

FIGURE 6–13

FIGURE 6–14

FIGURE 6–15

Nutty Banana Bread

CAKE-LIKE CHOCOLATE CHIP MUFFINS

Makes 12 muffins

Lessons demonstrated in this recipe:

- How to prepare a more cake-like muffin using the creaming method of mixing.
- Creaming the butter and sugar beats air into the mixture, allowing carbon dioxide from the baking powder to enlarge those air bubbles and aid the leavening process.
- When double-acting baking powder is mixed with a liquid and then heated, it reacts to form carbon dioxide gas, helping the batter to rise.

MEASUREMENTS				INGREDIENTS
U.S.		METRIC	BAKER'S %	
4 ounces	8 tablespoons	115 g	43%	unsalted butter, softened
3¾ ounces	½ cup	110 g	41%	granulated sugar
3 ounces	½ cup	85 g	32%	light brown sugar (packed, if measuring by volume)
9.5 ounces	2 cups	270 g	100%	all-purpose flour
	½ teaspoon	3 g	1.1%	salt
	2 teaspoons	8 g	3.9%	baking powder
4 fluid ounces	½ cup	120 mL	45%	milk
2 each		94 g	35%	large eggs
	1 teaspoon	5 mL	1.9%	vanilla extract
6 ounces	1 cup	170 g	63%	mini semisweet chocolate chips
	2 tablespoons	30 g	11%	granulated sugar
			376.0%	Total Cake-Like Chocolate Chip Muffins percentage

1. Preheat oven to 375°F (190°C). Spray a 12-cup muffin pan or two 6-cup muffin pans with nonstick cooking spray and set aside.

2. In the bowl of an electric mixer, using the paddle attachment, cream the butter with the sugars on medium speed until fluffy and lighter in color. Allow 2 to 3 minutes to do this.

3. In a medium bowl, whisk together all-purpose flour, salt, and baking powder; set aside.

4. Place the milk in a mixing bowl and whisk in the eggs and vanilla.

5. On low speed, gradually add the milk mixture to the creamed mixture and mix well (Figure 6–16). Stop the mixer and scrape down the sides of the bowl with a rubber spatula (Figure 6–17).

6. On low speed, slowly add dry ingredients and mix only until combined (Figure 6–18). Fill the muffin cups about two-thirds full.

7. Sprinkle the mini chocolate chips and then the sugar over each muffin (Figure 6–19).

8. Bake for 25 to 28 minutes or until a cake tester or sharp knife inserted into the center of a muffin comes out without any crumbs attached and the muffins are brown. Remove from the oven and allow to cool.

   The muffins can be placed in a large freezer bag and frozen. When ready to serve, place the frozen muffins on a sheet pan in a 350°F (175°C) oven until thawed and warm.

FIGURE 6–16

FIGURE 6–17

FIGURE 6–18

FIGURE 6–19

Cake-Like Chocolate Chip Muffins

POPOVERS

Makes 6 to 8 large popovers using a popover pan

Lessons demonstrated in this recipe:

■ How to prepare a specialty quick bread using steam as a leavening agent.

■ A high-protein flour is used to help give structure.

■ A large portion of water-based liquid ingredients produces lots of steam for leavening.

MEASUREMENTS				INGREDIENTS
U.S.		METRIC	BAKER'S %	
5 each		235 g	87%	large eggs
16 fluid ounces	2 cups	480 mL	178%	milk
1 fluid ounce	2 tablespoons	30 mL	11%	olive oil
8¼ ounces	1¾ cups	234 g	87%	bread flour
1¼ ounce	¼ cup	35 g	13%	whole-wheat flour
	1 teaspoon	5 g	1.9%	granulated sugar
	1 teaspoon	6 g	2.2%	salt
			380.1%	Total Popovers percentage

1. Preheat oven to 425°F (220°C). Spray a popover pan with nonstick cooking spray.

Note: If you don't have a popover pan, use a regular muffin pan, but fill alternating muffin cups with batter to allow room for the popovers to expand.

2. Whisk together the eggs, milk, and oil in a mixing bowl (Figure 6–20). Whisk in the flours, sugar, and salt only until combined (Figure 6–21). Do not overmix. A few lumps in the batter are fine. Fill a pitcher with batter and pour the batter into each popover tin until it is three-fourths full (Figure 6–22).

3. Bake for 35 minutes without opening the oven. (This ensures that the oven stays hot so steam will form and the popovers will rise.)

Note: Popovers can be placed in a freezer bag after cooling completely and frozen for several weeks in the freezer. Place them still frozen on a sheet pan in a 350°F (175°C) oven for 10 to 15 minutes to thaw and warm up.

4. Remove the popovers from the oven. Cool for 5 minutes and, using a knife to cut around the edge of the tin, lift each warm popover out. Fill the tins with the remaining batter and bake in the same way. Serve immediately.

FIGURE 6–20

FIGURE 6–21

FIGURE 6–22

Popovers

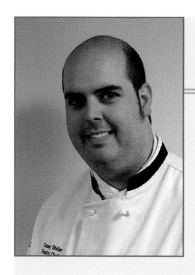

Professional Profile

BIOGRAPHICAL INFORMATION

Casey Shiller
Director of Baking and Pastry Instruction
Saint Louis Community College—Forest Park
St. Louis, MO

1. Question: *When did you realize that you wanted to pursue a career in baking and pastry?*

Answer: *In looking over college options as a high school junior in the 1990s, I had heard about Johnson & Wales University in Providence, Rhode Island. I discovered they offered a Bachelor of Science degree in Pastry Arts and Baking. I flew up to Providence that summer and took a 3-day "career-exploration" course in baking and pastry. I enjoyed myself so much I decided that I was destined for the bakeshop.*

2. Question: *Was there a person or event that influenced you to go into this line of work?*

Answer: *Attending the 3-day mini class at Johnson & Wales University sparked a life-altering realization. I could be responsible for peoples' joy everywhere; after all, who doesn't like dessert? I returned to Missouri eager to begin my newfound passion. I graduated high school 1 year early and was in Providence 32 days after retuning from the trip to "check it out."*

3. Question: *What did you find most challenging when you first began working in baking and pastry?*

Answer: *Starting out in this industry it is difficult to realize how important it is to take an extra couple of minutes to ensure your production is flawless. I learned I must get each task complete but can never cut corners or allow my quality to decrease. Someone will be purchasing and consuming my wares. If I can make every product to the best of my ability, I have succeeded, and will reap the rewards personally, professionally, and financially.*

4. Question: *Where and when was your first experience in a professional baking setting?*

Answer: *In my sophomore year in college, I had the opportunity to obtain one of the first externships at the Trump Plaza Hotel-Casino in Atlantic City, New Jersey. I became part of the staff under the direction of Executive Pastry Chef Thomas Vaccaro. There was one pastry kitchen responsible for all of the bread and dessert production for the 15 dining venues of the property. Chef Vaccaro set up a schedule that would allow me to rotate into each area of the pastry kitchen so I could gain a better understanding of how such a large operation functioned.*

5. Question: *How did this first experience affect your later professional development?*

Answer: *The culinary industry is made up of a very tightly knit group of chefs. If you work hard and strive for perfection, chefs from across the country begin to hear about you. Throughout my education, I remained a member of Chef Tom Vaccaro's team at Trump laza. Through Chef Vaccaro I became acquainted with Certified Master Chef Edward Leonard. Chef Leonard invited me to become part of the opening team of Cantare Ristorante in Chicago, Illinois, and I jumped at the opportunity.*

6. Question: *Who were your mentors when you were starting out?*

Answer: *Chef Thomas Vaccaro, Executive Pastry Chef of the Trump Plaza Casino-Hotel, shaped me into the chef that I am today. Chef Vaccaro took me under his wing and gave me insight into every aspect of the pastry field. He guided me toward a path of success, and is still a huge support and confidant. Chef Edward Leonard, one of only a few Certified Master Chefs in the nation, has also had an enormous impact on my professional life. Chef Leonard inspired me to push myself beyond the limits I thought were obtainable. He instilled the values of quality, perfection, and organization—traits that I utilize and pass on to my staff. These gentlemen opened my eyes to the wide world of competitions, both serving as coaches to the Culinary Olympic Team USA. Their guidance molded my professional success. I am lucky to have had the opportunity to gain inspiration from two of the best in the business.*

7. Question: *What would you list as the greatest rewards in your professional life?*

Answer: *As a college educator I am daily rewarded by seeing my students learn the great elements of the pastry arts. I love to see them applying the skills that I had imparted to them through lectures and hands-on lab projects. Making a direct impact in the life of an aspiring apprentice is an incomparable experience. Competition also plays a key role in my rewarding career. Through the avenues of professional competition, I am afforded the opportunity to hone my skills while being rewarded instantly for excellent performance.*

8. Question: *What traits do you consider essential for anyone entering the field?*

Answer: *The most essential trait for anyone entering the culinary industry is passion. Passion is something that cannot be taught, and will allow anyone to succeed. Direct skills may be taught: knife skills, proper cooking techniques, how to decorate, etc. Those aspiring to excel must look beyond the practical skills and see how to make their products better. If they are passionate about food, they will do everything in their power to ensure their product is far superior. If someone is eager to learn, no matter what the sacrifice, that person will have the focused drive to succeed.*

9. Question: *If there was one message you would impart to all students in this field what would that be?*

Answer: *Set goals and work vigorously to achieve those goals. Set goals that seem intangible at your current stage of thinking. Set your ambitions high, and push yourself to reach them. Persistent devotion to reach the stars will motivate you to test your limits and drive you to victory.*

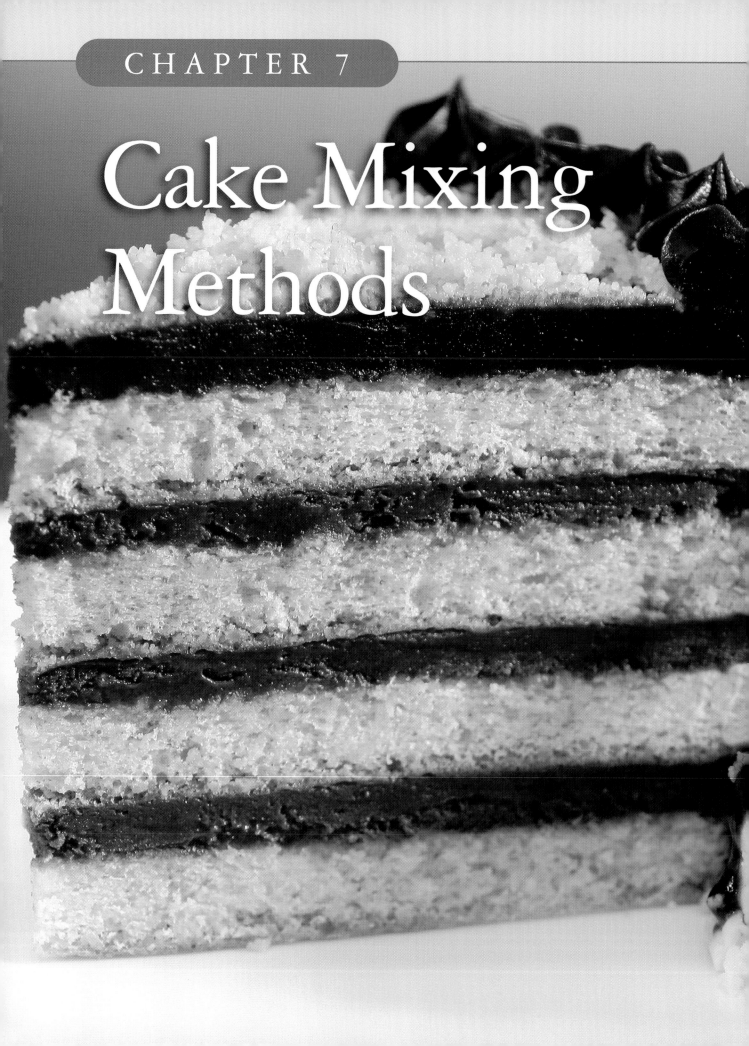

CHAPTER 7

Cake Mixing Methods

After reading this chapter,
you should be able to:

- Define the two categories of cakes.

- Describe the three cake mixing methods for cakes high in fat.

- Describe the three cake mixing methods for cakes low in fat.

- Understand the three ways to tell whether a cake is done.

- Prepare recipes using the various mixing methods in this chapter.

Cakes are a major part of almost every celebration there is. Cake varieties are endless and run the gamut from light and airy to the very heavy and rich.

Cakes are defined as a sweet, tender, moist baked pastry that is sometimes filled and frosted. Cake batters can be baked in an endless array of cake pans of various shapes and sizes. Using just a few basic cake recipes, bakers can create a wide range of cakes. By varying fillings and frostings, a basic cake can be transformed into a spectacular dessert.

Cake recipes are prepared using a few basic mixing methods. Once these mixing methods are mastered, their versatility to the baker has no bounds.

This chapter discusses the various cake mixing methods and their application in recipes. It is important to be aware that cake recipes are generally referred to as *formulas*. These formulas are just ingredients in the right proportions. Baking is a science in which exact measurements are crucial to success; following a formula exactly the way it is written will ensure that you prepare a cake to be proud of. The end of the chapter contains additional recipes using the cakes from each section and how they can be assembled with various fillings and frostings.

Two Categories of Cakes

There are really only two categories of cakes: cakes that are high in fat and cakes that are low in fat. (See Table 7–1, Cake Mixing Methods.)

Cakes High in Fat

Cakes high in fat rely on solid and liquid fats to keep gluten development low in order to produce a tender product. These types of cakes tend to have a longer shelf life because of their high fat content which slows the staling process. They tend to be moister, richer, and have a more tender crumb. There are three basic methods used to prepare these cakes:

- Creaming method
- Two-stage method
- One-stage method

Cakes Low in Fat

Cakes low in fat need some other ingredient to tenderize them. These types of cakes tend to be very high in sugar because sugar is a tenderizer. They include sponge cakes that use the air beaten into eggs to leaven them. Cakes of this type are referred to as egg-foam cakes. Egg-foam cakes tend to produce a drier, more flexible cake that does not crumble as easily as a cake high in fat. These can easily be cut crosswise into layers or rolled as for a jelly roll. Because these cakes are dry, many chefs brush them with a sugar syrup in which equal parts of sugar and water are brought to a boil and then flavored with such ingredients as liqueurs or extracts. There are three basic methods used to prepare these cakes:

- Sponge method (whole egg foam and separated egg foam)
- Chiffon method
- Angel food method

Three Mixing Methods for Cakes High in Fats

- Creaming method
- Two-stage method
- One-stage method

Three Mixing Methods for Cakes Low in Fats

- Sponge method
- Chiffon method
- Angel food method

Table 7.1 Cake Mixing Methods

CAKE METHODS HIGH IN FAT		
CREAMING METHOD	**TWO-STAGE METHOD**	**ONE-STAGE METHOD**
1. Cream fat and sugar on low to medium speed until light and fluffy.	1. Blend dry ingredients and fat on low speed.	1. Blend dry ingredients on low speed.
2. Add eggs, 1 at a time.	2. Then add liquid in two stages, scraping down the bowl after each addition.	2. Blend liquid ingredients in another bowl.
3. Alternate dry and wet ingredients (beginning and ending with dry ingredients).	3. Mix until just blended.	3. Gradually add wet ingredients to dry ingredients.
4. Mix until just blended.		4. Mix until just blended.

CAKE METHODS LOW IN FAT			
SPONGE METHOD— WHOLE EGG FOAM (GENOISE)	**SPONGE METHOD— SEPARATED EGG FOAM**	**CHIFFON METHOD**	**ANGEL FOOD METHOD**
1. Whole eggs and sugar warmed and beaten to a foam.	1. Egg yolks and a portion of the sugar are beaten until thick and light.	1. Dry ingredients with a portion of the sugar are sifted into mixing bowl.	1. Sift flour with half of sugar.
2. Sifted dry ingredients are folded in gently.	2. Egg whites are beaten with rest of sugar to form stiff peaks.	2. Oil, yolks, water, and flavorings are mixed in.	2. Beat room temperature egg whites until foamy.
3. Melted butter, sometimes added.	3. Beaten egg whites are folded into yolks and sugar alternately with sifted dry ingredients.	3. Egg whites are beaten with cream of tartar and remaining sugar until stiff.	3. Add cream of tartar and salt.
		4. Beaten egg whites are folded into batter.	4. Beat until soft peaks form.
			5. Gradually add remaining sugar, beat until stiff.
			6. Fold in flour and sugar mixture.

Cake Batters as Emulsions

With the exception of an angel food cake, all cake batters are comprised of fat and water-based ingredients that would normally separate out much like a cup of oil and water. Even if they are stirred together, eventually they will separate into two layers. This is because fats and water-based ingredients are immiscible liquids, meaning they do not stay mixed together. The natural tendency is for the two substances to separate into two layers. The fat is lighter and rises to the top and the water stays on the bottom. This natural tendency for a fat and water to separate is due to a phenomenon known as *surface tension.*

To prepare a proper cake batter in which the ingredients will not separate, another ingredient needs to be added to act as a "go-between" to help the fat and water-based ingredients stay evenly dispersed, forming a smooth batter. This ingredient is called an emulsifying agent. Emulsifying agents accomplish two tasks that contribute to the finished cake's volume and texture. First, they reduce the surface tension between the two immiscible substances, allowing them to mix together

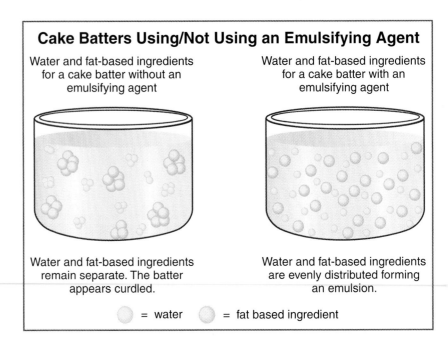

Cake Batters Using/Not Using an Emulsifying Agent

Water and fat-based ingredients
for a cake batter without an
emulsifying agent

Water and fat-based ingredients
for a cake batter with an
emulsifying agent

Water and fat-based ingredients
remain separate. The batter
appears curdled.

Water and fat-based ingredients
are evenly distributed forming
an emulsion.

= water   = fat based ingredient

The Creaming Method

1. Cream the fat and sugar on low to medium speed until light and fluffy in the bowl of an electric mixer using the paddle attachment.

2. Blend in the eggs one at a time, thoroughly blending in one before adding the next.

3. Blend the liquids and dry ingredients alternately beginning and ending with the dry ingredients. The dry ingredients are added last to ensure the liquids will be absorbed. Another advantage of adding some of the flour mixture last is that less gluten will develop.

more easily. Second, emulsifying agents help the air bubbles incorporated in the batter become more evenly distributed. This encourages more even rising of the cake, resulting in a finer crumb.

Eggs, which are typically used in cake batters, contain natural emulsifiers in the yolk and act as an emulsifying agent to bring the two immiscible ingredients, fat and water, together in an emulsion. The immiscible ingredients become evenly dispersed within the batter where they float around each other.

A properly prepared cake batter is smooth. So, when adding an emulsifying agent, it is important to gradually incorporate it into the fat mixture. If the batter appears to be curdled, it is a sign that the emulsion may not have been prepared properly. However, some cake batters look curdled until the dry ingredients are blended in.

During the creaming method, if the eggs are not added gradually into the creamed fat and sugar mixture, the batter will appear curdled. Gradually adding the eggs, little by little, allows the fat and sugar time to absorb them and form a proper emulsion.

Mixing Methods for Cakes High in Fat

There are three methods for mixing cakes high in fat: the creaming method, the two-stage method, and the one-stage method.

The Creaming Method

The first method of mixing cakes high in fat is known as the creaming method. It is the method of choice when preparing butter cakes. The creaming method starts out with softened, solid fat (such as butter) at approximately 65°F (19°C). The fat is then mixed with granulated or brown sugar in an electric mixer using the paddle attachment. The creaming comes into play as the fat is mashed against the sides of the mixer with the sand-like sugar crystals working against it, softening it even more while forcing air into it. The process of creaming is most important and cannot be rushed. Creaming should be done at a low to medium speed for between 3 and 5 minutes. A high speed may melt the fat, causing a loss of air bubbles. Creaming for too long creates a coarse texture in the finished cake. Creaming for too short a time, produces a cake with poor volume. If butter is used to be creamed with the sugar, the mixture should turn from a yellowish color (from the butter) to a lighter, whiter color (from the aerated butter). This color change is due to all the air that is mixing in with the butter and sugar. These air bubbles are held within the fat. Ultimately, these air bubbles will expand even more when mixed with the leavening ingredients (like baking powder or baking soda).

How Chemical Leavening Agents Work with Air to Leaven Cakes

Chemical leavening agents such as baking powder and baking soda work with the air incorporated from the creaming step to leaven cakes. As chemical leavening agents produce carbon dioxide gas during the baking process, those gases are attracted to the small air bubbles formed during the creaming step and enlarge them. Although gases (e.g., carbon dioxide) are formed, no new air bubbles are created by the chemical leavening agents. The existing air bubbles created during creaming just get larger. That is why the creaming step is so important to the volume and texture of a cake. These trapped air bubbles within the batter expand, pushing it upward, causing the batter to rise.

Creaming is also important to the mixing of ingredients because it is much easier to blend other ingredients into a "fluffy" fat than it would be if the fat was hard. The next step in the creaming method is for beaten eggs to be added to the creamed fat and sugar mixture. The eggs must be added slowly to allow for their absorption. If the eggs are colder than the creamed fat, the mixture may appear curdled. Next, the dry ingredients are added alternately with the remaining liquid ingredients ending with the final addition of the remaining dry ingredients. Alternating dry and liquid ingredients maintains the emulsion while minimizing gluten development. Some recipes add all of the liquid after the eggs and then add all the dry ingredients at once. Adding the dry ingredients all at once shortens the mixing process and still minimizes gluten development. Caution must be taken to prevent overmixing. Cakes that use the creaming method tend to be very tender and, if creamed properly, will attain good volume.

Recipe

Fudge Swirl Sour Cream Pound Cake (This chapter, page 193)

The Two-Stage Method

The two-stage method is used primarily for cakes in which a tender, light, moist texture is desired. The liquids are added in two stages, and the batters are usually thin. Typically, all the dry ingredients are mixed together in the bowl of an electric mixer using the paddle attachment. The fat is then added, coating the flour particles and preventing them from absorbing water and forming gluten. A portion of the liquid ingredients are then blended in at a low speed to obtain a smooth, thin batter. Once the fat and part of the liquid ingredients have been added, the remaining liquid is added. Water-based ingredients when added to flour contribute to gluten formation, but because the fat was added first, coating each particle of flour, the gluten that forms is minimal, producing a tender cake. A cake batter should never be overmixed, because this encourages gluten to form.

Traditionally, the fat used in a two-stage cake is an emulsified shortening. (See Knowing Which Fat to Use in High-Fat Cakes). This type of shortening holds a great deal of liquid and is the fat of choice for a high ratio cake because of its ability to keep fat and water-based ingredients from separating. Typically, a two-stage method cake has more sugar by weight than flour. This is known as a high ratio cake. Cakes that use the two-stage method of mixing tend to be more tender than cakes using the creaming method. These types of cakes are used most often in high volume bakeries.

> **The Two-Stage Method**
>
> 1. Mix dry ingredients with emulsified fat. The fat coats the particles of flour, preventing gluten from forming once the liquid ingredients are added.
>
> 2. Add liquids in two stages.

The One-Stage Method

1. Place the dry ingredients into the bowl of an electric mixer and, using the paddle attachment, mix until blended together.

2. Blend liquids together in another bowl.

3. Add the liquids slowly on low speed until a smooth batter is made, scraping the sides of the bowl often.

RECIPE

Two-Stage Golden Yellow Cake (This chapter, page 197)

The One-Stage Method

The one-stage method is the easiest method in that all the dry ingredients are placed in a bowl of an electric mixer and the liquid ingredients are added in one step. It is important to add the liquid gradually and scrape down the bowl frequently or else lumps of dry ingredients can form. The fat for this method is usually a liquid (either oil or melted butter). Because this method can cause a great deal of gluten formation, if overbeaten, the liquid fat may be added with the dry ingredients first to coat the flour.

Cakes using the one-stage method are known for their speed in preparation and for their fine, tender crumb.

RECIPE

Fudgy Chocolate Cake (This chapter, page 199)

Knowing Which Fat to Use in High-Fat Cakes

It is important to know the differences between fats such as oils, butter, and shortening and which one to use when preparing cakes that are high in fat.

Liquid fats such as oils, do not make very good creaming agents because oils are incapable of holding air when beaten. In one-stage method cakes, oils are the ideal fat to prevent gluten from forming.

Solid fats, such as butter and hydrogenated vegetable shortening, are capable of holding air and therefore make better creaming agents. Some types of fat hold more air than others. Butter, known for its rich flavor and appealing mouth feel, melts at a lower temperature than shortening. It is the ideal fat for butter cakes using the creaming method. However, in a cake using the two-stage method, butter would not be a good choice because it is not capable of holding the larger quantities of sugar and liquids to create an emulsion. When creaming, butter should be soft enough to be mashed and trap air bubbles, ideally between 65° and 70°F (19° and 21°C). However, it should not be melted. Because butter contains some water (up to 20 percent), it cannot hold as much air as shortening. Shortening, however, is almost 100 percent fat and contains little water.

Shortening has one advantage over butter in that it already contains small air bubbles that are evenly dispersed throughout (approximately 10 percent). This makes shortening the ideal fat for creaming in cakes other than butter cakes. A significant drawback of using shortening is its greasy feel on the tongue, which is due to its higher than body temperature melting point.

Shortening with added emulsifiers is typically used in the two-stage method because it is better able to hold immiscible ingredients like fats and water-based liquids together in a suspension where they are evenly dispersed within the batter without separating out and appearing curdled.

Hydrogenated vegetable shortenings start out as liquid oils that have undergone a chemical process known as *hydrogenation,* which changes them to a solid. This solid is less likely to go rancid than the liquid oil.

Mixing Methods for Cakes Low in Fat

There are three mixing methods for cakes that are low in fat: (1) the sponge method with two variations, (2) the chiffon method, (3) and the angel food method.

Cakes low in fat tend to incorporate some form of an egg foam using whole eggs or just the whites. Chapter 8 discusses the ability of eggs to hold air. Egg foams are made by beating eggs with sugar into a foam. Cream of tartar may be added to increase the egg foam's stability. The foam is then folded into a batter in which the air bubbles expand in the oven, leavening the cake. Egg-foam type cakes include angel food cakes, sponge cakes, chiffon cakes, and genoise. For more on the leavening qualities of egg whites, see Chapter 8.

Some cakes using the foaming method contain very little fat. The sponge and the chiffon methods use egg yolks, which add some fat to the cake. The angel food method is the only method in which only the whites are used, making this type of cake basically fat free!

Many recipes for egg-foam cakes use superfine sugar. Superfine sugar is granulated sugar that has been pulverized to a finer texture. This finer sugar dissolves instantly, creating a smoother, less gritty egg foam.

The Sponge Method

The sponge method uses both the yolk and the white of the egg. There are two main types of sponge cake: whole egg foams (genoise) and separated egg foams. In *whole egg foam cakes,* warm whole eggs and sugar are beaten into a foam using the whip attachment. The eggs are warmed to attain a greater volume and then beaten. Sifted dry ingredients are then folded into the foam. This is known as a *genoise* or French sponge cake. Some recipes call for melted butter to be blended in at the end. These cakes tend to be drier and tougher than a high fat cake, but the butter helps tenderize them. Because of their strong structure, genoise sponge cakes are easily split into layers. Soaking the cake layers in a flavored sugar syrup adds moistness.

The second type of sponge cake is known as a *separated egg-foam cake.* In this type of cake, the yolks are beaten with part of the sugar to form a foam, then the egg whites are beaten with the remaining sugar to form stiff peaks. The beaten whites are folded into the yolk and sugar mixture alternating with the sifted dry ingredients. Separated egg-foam cakes tend to be moister than whole egg-foam cakes.

RECIPES
Chocolate Sponge Cake Roll (This chapter, page 205)
Hazelnut Genoise (This chapter, page 201)

The Chiffon Method

The chiffon method differs from the other two methods in that it contains a liquid fat, usually oil, and a chemical leavener like baking powder. Therefore, chiffon cakes get their leavening from two sources—air and baking powder. They are not as fragile as the other types of egg foam cakes in that if some air bubbles are lost in mixing, the baking powder can act as a backup leavener. Chiffon cakes tend to be moister because of the oil and typically are baked in a tube pan.

RECIPE
Citrus Chiffon Cake (This chapter, page 208)

The Whole Egg-Foam Method

1. Whole eggs and sugar are warmed and beaten until thick. They are then beaten until cold and they form a ribbon when dropped from the beater that slowly disappears as it hits the batter.

2. Sifted dry ingredients and sometimes melted butter are folded in alternately.

(*Note:* To allow the egg foams to attain the greatest volume, the egg yolks or whites are beaten using the whip attachment. Notice that chiffon cakes use the paddle to prepare the main batter and the whip to beat the whites.)

The Separated Egg-Foam Method

1. Egg yolks and part of the sugar are beaten until thick and lighter in color.

2. Egg whites are beaten with the remaining sugar to form stiff peaks.

3. Beaten egg whites are folded into the egg yolks and sugar alternately with sifted, dry ingredients.

The Chiffon Method

1. The dry ingredients, with part of the sugar, are sifted into the mixing bowl.

2. The oil, egg yolks, water, and flavorings are mixed in using the paddle attachment.

3. Egg whites are beaten with cream of tartar and the remaining sugar until stiff.

4. The beaten egg whites are folded into the batter.

The Angel Food Method

1. Start with egg whites at room temperature.

2. In a mixing bowl, sift flour and cornstarch with half the sugar and set aside.

3. In the bowl of an electric mixer, using the whip attachment, beat the egg whites with cream of tartar and salt until soft peaks form.

4. Slowly beat in the remaining sugar that was not mixed with the flour. Beat until stiff peaks form.

5. Using a rubber spatula, fold in the flour and sugar mixture into the beaten egg whites.

The Angel Food Method

The angel food method uses only the egg whites, which are low in fat. Angel food cakes tend to be light and airy, and need no frosting or other adornments. They are often served with a fresh fruit compote or a sauce.

RECIPE

Coconut Angel Food Cake (This Chapter, page 211)

How to Tell When a Cake Is Done

There are three ways to tell when a cake is done. First, the cake should spring back when you gently press it with your finger. Second, the sides of the cake should pull away from the pan. Third, a cake tester, when inserted into the center of the cake, should come out clean and free of crumbs or batter.

There are a few tools that can be used as cake testers. One type of cake tester consists of a long metal wire with a ring or handle at one end. Alternatively, a thin wooden skewer or a small knife can be used.

Testing Cake Doneness

Three ways to tell when a cake is done:

■ The cake springs back when you gently press it with your finger.

■ The cake pulls away from the sides of the pan.

■ A cake tester, wooden skewer, or thin knife inserted gently into the center of the cake comes out free of crumbs and looks clean.

FUDGE SWIRL SOUR CREAM POUND CAKE

Makes one 10-inch (25-cm) tube cake

Lessons demonstrated in this recipe:

- How to prepare a cake using the creaming method of mixing.
- Blending the sugar, butter, and cream cheese until fluffy and lighter in color incorporates air into the fat; aids the chemical leavening agents; and results in a lighter cake with great height.

MEASUREMENTS				INGREDIENTS
U.S.		METRIC	BAKER'S %	
4 ounces		115 g	28%	unsalted butter, softened
4 ounces		115 g	28%	cream cheese, softened
14½ ounces	2 cups	410g	102%	granulated sugar
14¼ ounces	3 cups	404 g	100%	all-purpose flour
	1 teaspoon	4 g	1.0%	baking powder
	½ teaspoon	3 g	0.5%	baking soda
	½ teaspoon	2 g	0.7%	salt
4 each		188 g	47%	large eggs
	2 teaspoons	10 mL	2.5%	vanilla extract
8 ounces	1 cup	225 g	56%	sour cream
3½ ounces	7 tablespoons	100 g	25%	warm ganache*
			390.7%	Total Fudge Swirl Sour Cream Pound Cake percentage

*See Ganache, Chapter 8.

1. Preheat oven to 350°F (175°C). Spray a 10-inch (25 cm) false-bottom tube pan with nonstick cooking spray and set aside.

2. In the bowl of an electric mixer, using the paddle attachment, cream the butter, cream cheese, and sugar on medium speed until light in color and fluffy (Figure 7–1). This can take up to 3 to 5 minutes. Stop the mixer occasionally and scrape the bowl with a rubber spatula.

3. In a small bowl, whisk together the flour, baking powder, baking soda, and salt. Set aside.

FIGURE 7–1

4. In a small bowl, whisk together the eggs and the vanilla extract.

5. On low speed, add the egg mixture into the creamed butter in thirds, waiting for the mixture to blend together uniformly before adding more egg (Figure 7–2).

6. On low speed, add one third of the flour mixture into the eggs and butter. Blend until combined and add one half of the sour cream.

7. Add another one third of the flour mixture, blending well, followed by the remaining sour cream (Figure 7–3). Stop the machine and scrape down the sides of the bowl.

8. Add the remaining one third flour mixture and mix until well combined. Remove the bowl from the mixer.

9. Using a rubber spatula, scrape around the bottom and sides of the bowl to make sure the mixture is smooth and well combined (Figure 7–4).

10. Pour half of the batter into the prepared pan and smooth it with a rubber spatula (Figure 7–5).

FIGURE 7–2

FIGURE 7–3

FIGURE 7–4

FIGURE 7–5

11. With a spoon, drizzle half of the ganache on top of the batter in a circle, forming a thick chocolate line (Figure 7–6).

12. With a spatula, drop remaining batter over the ganache. Lightly smooth the top of the batter with a spatula, trying not to mix the ganache into the batter (Figure 7–7). Drizzle the remaining ganache over the batter in a circle.

13. Using a small palette knife, make a cut through the center of the pan near the tube to the outside rim of the pan and pull the knife out (Figure 7–8). Rotate pan and repeat cutting into the batter every two inches all around the pan. Do not mix the ganache into the batter. The batter will look marbled.

14. Bake for 1 hour and 10 minutes, or until a cake tester inserted into the center of the cake comes out clean.

15. Cool thoroughly and remove from the pan.

> **TIP** To prepare a plain sour cream pound cake, omit the ganache and pour all of the batter into the prepared pan and bake as directed.

FIGURE 7–6

FIGURE 7–7

FIGURE 7–8

Fudge Swirl Sour Cream Pound Cake

TWO-STAGE GOLDEN YELLOW CAKE

Makes two 9-inch (22.5-cm) round cake layers

Lessons demonstrated in this recipe:

- How to prepare a cake using the two-stage method of mixing.
- A high ratio cake has more sugar than flour by weight.
- Using an emulsified shortening helps high ratio cake batters form an emulsion by holding large amounts of sugar and liquids.
- The fat is added directly to the dry ingredients to coat the protein strands in the flour and prevent gluten from forming.

MEASUREMENTS				INGREDIENTS
U.S.		**METRIC**	**BAKER'S %**	
9 ounces	2¼ cups	255 g	100%	cake flour, sifted if lumpy
10¾ ounces	1½ cups	305 g	119%	granulated sugar
	1 tablespoon	12 g	4.7%	baking powder
	½ teaspoon	3 g	1.2%	salt
4 ounces	½ cup	115 g	45%	emulsified shortening, room temperature
8 fluid ounces	1 cup	240 mL	94%	milk
	1 teaspoon	5 mL	2.0%	vanilla extract
2 each		94 g	37%	large eggs
			402.9%	Total Two-Stage Golden Yellow Cake percentage

1. Preheat oven to 350°F (175°C). Grease, parchment, and flour two 9-inch (22.5-cm) round cake pans.

2. In the bowl of an electric mixer fitted with the paddle attachment, combine the flour, sugar, baking powder, and salt. Blend on low speed until well combined.

3. Add the vegetable shortening to the dry ingredients and blend on low speed for 2 to 3 minutes, or until the dry ingredients are well coated with the shortening (Figure 7–9).

FIGURE 7-9

4. Place the milk, vanilla, and eggs into a mixing bowl. Whisk to blend.

5. Slowly add one half of the liquid ingredients into the dry ingredients and blend well, scraping down the sides of the bowl.

6. Slowly add the rest of the liquid ingredients, blending and scraping down the sides each time an addition is made (Figure 7–10).

7. Turn the speed up to medium and blend for 30 seconds, or until a smooth batter is made. Divide the batter between the two prepared pans (Figure 7–11).

8. Bake 30 to 35 minutes or until a cake tester inserted into the centers of the cakes comes out clean. Cool until lukewarm and remove layers onto wire rack to cool completely.

FIGURE 7–10

FIGURE 7–11

FUDGY CHOCOLATE CAKE

Makes two 9-inch (22.5-cm) round cake layers

Lessons demonstrated in this recipe:

- How to prepare a cake using the one-stage method, the simplest cake mixing method.
- This method produces a cake with a tender, fine crumb.
- It is important to add liquid ingredients gradually, taking time to scrape the bowl often to keep the batter smooth.
- Buttermilk and cocoa powder are acids that will react with the baking soda, resulting in a neutralization reaction and the creation of carbon dioxide gas which will help leaven the cake.
- Extra baking soda gives a darker, richer color to the cake.

MEASUREMENTS				INGREDIENTS
U.S.		METRIC	BAKER'S %	
7¼ ounces	1 cup	206 g	88%	granulated sugar
7½ ounces	1 cup	212 g	91%	light brown sugar (packed, if measuring by volume)
8¼ ounces	1¾ cups	234 g	100%	all-purpose flour
2½ ounces	¾ cup	70 g	30%	unsweetened cocoa powder
	1½ teaspoons	6 g	2.6%	baking powder
	2 teaspoons	8 g	3.4%	baking soda
	1 teaspoon	6 g	2.6%	salt
4 fluid ounces	½ cup	120 mL	51%	canola oil
2 each		94 g	40%	large eggs
8 fluid ounces	1 cup	240 mL	103%	buttermilk, well shaken
	2 teaspoons	10 mL	4.3%	vanilla extract
	1 tablespoon	15 mL	2.6%	instant espresso powder
1 cup			103%	boiling water
			621.5%	Total Fudgy Chocolate Cake percentage

1. Preheat oven to 350°F (175°C). Grease, parchment, and flour two 9-inch (22.5-cm) round cake pans.

2. Combine the sugar, brown sugar, flour, cocoa powder, baking powder, baking soda, and salt in the bowl of an electric mixer fitted with the paddle attachment. Mix at low speed until blended.

3. In a separate mixing bowl, whisk together the oil, eggs, buttermilk, and vanilla. Slowly add the liquid ingredients into the dry ingredients on low speed until just blended (Figure 7–12). Stop the machine and scrape around the bowl with a rubber spatula. Scrape the bottom and around the sides of the bowl to make sure that all the ingredients are blended thoroughly and there are no lumps.

4. Mix the espresso powder into the boiling water to make coffee and pour coffee into the batter (Figure 7–13). Blend well on low speed for 10 to 20 seconds.

5. The batter will be thin. Divide the batter between the two prepared cake pans.

6. Bake for 30 to 35 minutes or until a cake tester placed into the center of the cakes comes out clean. Cool until lukewarm and remove the two round layers from the pans onto wire racks to cool completely.

FIGURE 7–12

FIGURE 7–13

HAZELNUT GENOISE

Makes two 9-inch (22.5-cm) round cake layers

Lessons demonstrated in this recipe:

- How to prepare a whole egg-foam cake using the sponge method of mixing.
- The air beaten into whole eggs leavens the cake.
- Heating the eggs and sugar produces a more stable foam.
- To reduce gluten and toughness, a low-protein flour is used in combination with cornstarch.
- Genoise cakes often use ground nuts to flavor the cake while the fat in the nuts creates tenderness.
- A small quantity of melted butter is added to create some tenderness.
- The butter is mixed in quickly to a small amount of batter to lighten it, then it is folded into the remaining batter.

MEASUREMENTS				INGREDIENTS
U.S.		METRIC	BAKER'S %	
3 ounces	¾ cup	85 g	155%	confectioners' sugar
3½ ounces	¾ cup	100 g	182%	toasted hazelnuts
8 each		376 g	684%	large eggs, room temperature
4¾ ounces	⅔ cup	130 g	236%	granulated sugar
2 ounces	½ cup	55 g	100%	cake flour
2½ ounces	½ cup	70 g	127%	cornstarch
1 ounce	2 tablespoons	30 g	55%	melted butter, kept warm
			1539%	Total Hazelnut Genoise percentage

1. Preheat oven to 375°F (190°C).

2. Spray two 9-inch (22.5-cm) round cake pans with nonstick cooking spray. Line with parchment circles. Spray again and flour. Tap out excess flour.

3. In the bowl of a food processor, pulverize confectioners' sugar and toasted hazelnuts until nuts are ground to a fine powder (Figure 7–14). Pour into a medium bowl and set aside.

4. Place a large saucepan filled with 1 inch (2.5 cm) of water over medium-high heat and bring it to a simmer.

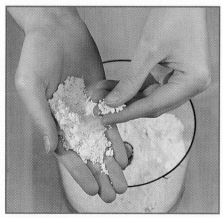

FIGURE 7–14

5. In the bowl of an electric mixer using a handheld whisk, whisk eggs and granulated sugar constantly over the pan of simmering water until the eggs are very warm and foamy, and a thermometer placed in the bowl of eggs (not touching the bottom of the bowl) registers 110°F (43°C) (Figure 7–15).

6. Remove the bowl from the water and place it on the base of the electric mixer. Using the whip attachment on high speed, whip the egg mixture until it has tripled in volume and is cool to the touch (Figure 7–16). This will take several minutes. It should look like marshmallow fluff. As the batter falls from the whip, it should form a ribbon as it drops into the bowl before it dissolves into the rest of the batter (Figure 7–17).

7. Place the melted butter into a small bowl and place it over the simmering water to keep it warm.

8. Place a sieve over a small mixing bowl and sift cake flour and cornstarch into it. Add the flour mixture to the hazelnut and sugar powder.

9. Pour the batter into a very large mixing bowl.

10. Sprinkle one third of the dry ingredients over the egg mixture and gently fold it in using a rubber spatula.

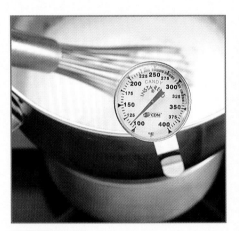

FIGURE 7–15

FIGURE 7–16

FIGURE 7–17

11. Repeat sprinkling another one third of the dry ingredients into the egg mixture, gently folding it in. Sprinkle in the remaining one third of the dry ingredients and fold in gently (Figure 7–18). Do not overmix!

12. Remove the butter from the water bath. Take a dollop of batter and mix it into the butter to lighten it (Figure 7–19).

13. Pour the lightened butter mixture over the batter and fold it in quickly but gently with a rubber spatula.

14. Pour the batter into the prepared cake pans, filling each more than half full.

15. Place the cake pans on a sheet pan so that they are not touching each other, and bake for approximately 18 to 22 minutes, or until the cake is lightly browned and cake tester inserted in the center comes out clean. Remove the cakes from the oven.

16. Using a thin, sharp paring knife, gently cut around the edge of each pan to separate the cake from the sides of the pan while it is still hot.

17. Cool cakes in the pan until the pans feel lukewarm. Remove the cakes from the pans and place on cake racks to cool completely.

FIGURE 7–18

FIGURE 7–19

CHOCOLATE SPONGE CAKE ROLL

Makes 1 cake roll

Lessons demonstrated in this recipe:

- How to prepare a separated egg-foam cake using the sponge method of mixing.
- Egg-foam cakes tend to be less tender than cakes high in fat and are flexible enough to be rolled when hot without breaking.
- The air beaten into the egg whites leavens the cake.
- An acid such as cream of tartar helps stabilize the meringue by partially denaturing the egg proteins.
- Superfine sugar dissolves more quickly, preventing any grittiness.

MEASUREMENTS				INGREDIENTS
U.S.		**METRIC**	**BAKER'S %**	
1⅔ ounces	½ cup	50 g	333%	Dutch processed cocoa powder
	1 teaspoon	2 g	13%	instant coffee powder
	1 teaspoon	5 mL	33%	vanilla extract
3⅔ fluid ounces	⅓ cup + 2 tablespoons	110 mL	733%	boiling water
6 each		114 g	760%	large egg yolks, room temperature
3¾ ounces	½ cup	110 g	733%	superfine sugar
	2 tablespoons	15 g	100%	cornstarch, sifted
	2 tablespoons	15 g	100%	cake flour, sifted
6 each		168 g	1120%	large egg whites, room temperature
	¾ teaspoon	1.5 g	10%	cream of tartar
3¾ ounces	½ cup	110 g	733%	superfine sugar
	1 to 2 tablespoons	8 to 16 g	80%	cocoa powder
			4748%	Total Chocolate Sponge Cake Roll percentage

1. Preheat oven to 350°F (175°C).

2. In a small bowl, whisk together the cocoa, coffee, vanilla, and boiling water until smooth. Allow mixture to cool.

3. Grease a jelly roll pan (17 by 12 inches; 42.5 by 30 cm) and line the bottom with parchment paper. Grease the parchment paper and flour the bottom and sides of the pan, knocking out the excess.

4. In the bowl of an electric mixer using the whip attachment, beat the yolks and $3\frac{3}{4}$ ounces ($\frac{1}{2}$ cup; 110 g) of sugar on high speed until light and fluffy (Figure 7–20). This will take about 5 minutes.

5. On low speed, gradually add the cocoa mixture and blend well (Figure 7–21). Pour the batter into a large mixing bowl and set aside.

6. Slowly add the sifted cornstarch and cake flour. Blend well.

7. In a clean, dry bowl of an electric mixer using the whip attachment, beat the egg whites until foamy (Figure 7–22).

8. Add the cream of tartar and beat until soft peaks form (Figure 7–23).

9. On high speed, gradually add the remaining $3\frac{3}{4}$ ounces ($\frac{1}{2}$ cup; 110 g) superfine sugar and beat until stiff peaks form (Figure 7–24 A and B).

FIGURE 7–20

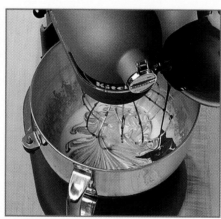

FIGURE 7–21

FIGURE 7–22

FIGURE 7–23

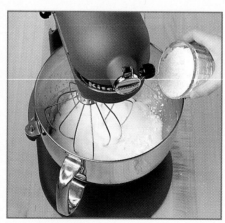

FIGURE 7–24A

FIGURE 7–24B

10. Using a whisk, mix one fourth of the beaten egg whites into the reserved chocolate batter to lighten up the batter (Figure 7–25).

11. Scoop the remaining egg whites on top of the lightened chocolate batter and fold them in gently using a rubber spatula until well combined (Figure 7–26). Do not overmix.

12. Quickly pour the batter into the prepared pan.

13. Bake for approximately 18 minutes or until the cake is puffy and springs back when touched lightly with a finger.

14. While the cake is baking, place a clean kitchen towel on a work surface. Place the cocoa powder in a sieve and dust the kitchen towel evenly with it (Figure 7–27).

15. When the cake is done, immediately flip it onto the prepared kitchen towel. Remove the pan and gently peel off the parchment paper and discard it (Figure 7–28).

16. With the help of the cocoa-dusted kitchen towel, begin to roll the cake from the long end into a tight spiral, then allow it to remain this way until it is completely cooled (Figure 7–29).

17. The cake roll can be unrolled and filled with a variety of fillings.

FIGURE 7–25

FIGURE 7–26

FIGURE 7–27

FIGURE 7–28

FIGURE 7–29

CITRUS CHIFFON CAKE

Makes one 10-inch (25-cm) tube cake

Lessons demonstrated in this recipe:

- How to prepare an egg foam cake using the chiffon method of mixing.
- A liquid fat is added to create tenderness.
- A chemical leavening agent is used to create additional height.

MEASUREMENTS				INGREDIENTS
U.S.		METRIC	BAKER'S %	
9 ounces	2¼ cups	255 g	100%	sifted cake flour
9 ounces	1¼ cups	255 g	100%	granulated sugar
	1 tablespoon	12 g	4.7%	baking powder
	½ teaspoon	3 g	1.2%	salt
4 fluid ounces	½ cup	120 mL	47%	canola oil
5 each		95 g	37%	large egg yolks, room temperature
4 fluid ounces	½ cup	120 mL	47%	cold, freshly squeezed orange juice
1 fluid ounce	2 tablespoons	30 mL	12%	freshly squeezed lemon juice
1 fluid ounce	2 tablespoons	30 mL	12%	freshly squeezed lime juice
	2 teaspoons	12 g	4.7%	orange zest
	2 teaspoons	12 g	4.7%	lemon zest
	2 teaspoons	12 g	4.7%	lime zest
	1 teaspoon	5 mL	2.0%	lemon extract
8 each		224 g	88%	large egg whites, room temperature
	½ teaspoon	1 g	0.4%	cream of tartar
3½ ounces	¼ cup	50 g	20%	granulated sugar
			485.4%	Total Citrus Chiffon Cake percentage

1. Preheat oven to 325°F (160°C).

2. In the bowl of an electric mixer fitted with a paddle attachment, blend the flour, 8¾ ounces (1¼ cups; 250 g) sugar, baking powder, and salt on low speed.

3. Continue at low speed and blend in the oil, egg yolks, orange juice, lemon juice, lime juice, orange zest, lemon zest, lime zest, and lemon extract into the dry ingredients. Blend until smooth (Figure 7–30). Pour batter into a large bowl. Set aside.

4. Place the egg whites into the clean, dry bowl of an electric mixer and, using the whip attachment, beat them only until foamy. Add the cream of tartar and beat until soft peaks form.

5. Slowly add the 3½ ounces (¼ cup; 100 g) sugar and beat until stiff peaks form.

6. Using a whisk, gently mix one fourth of the egg whites into the reserved batter to lighten it up (Figure 7–31).

7. Place the remaining egg whites on top of the lightened batter and fold them in gently with a rubber spatula (Figure 7–32).

8. Pour the batter into an ungreased, false-bottom 10-inch (25-cm) tube pan with metal tabs.

9. Bake for 50 to 60 minutes or until a cake tester inserted into the center comes out clean.

10. When the cake is done, remove it from the oven and invert it so it rests upside down on metal tabs. If the pan has no metal tabs, balance the upside down cake on an overturned ramekin placed under the tube and allow it to cool completely (Figure 7–33).

FIGURE 7–30

FIGURE 7–31

FIGURE 7–32

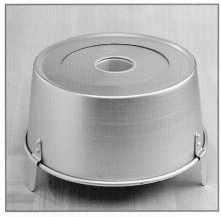

FIGURE 7–33

Citrus Chiffon Cake

COCONUT ANGEL FOOD CAKE

Makes one 10-inch (25-cm) tube cake

Lessons demonstrated in this recipe:

- How to prepare an egg foam cake using the angel food mixing method.
- Only egg whites are used, creating a cake very low in fat.
- Air beaten into the egg whites is the only leavening agent.
- An acid such as cream of tartar provides stability by partially denaturing the egg protein.
- Cooling the cake upside down prevents gravity from causing it to collapse.

MEASUREMENTS				INGREDIENTS
U.S.		METRIC	BAKER'S %	
1 ounce	½ cup	30 g	35%	sweetened shredded coconut (packed, if measuring by volume)
5½ ounces	¾ cup	155 g	182%	superfine sugar
3 ounces	¾ cup	85 g	100%	cake flour
	2 tablespoons	16 g	19%	cornstarch
	¼ teaspoon	1 g	1.2%	salt
12 each		336 g	395%	large egg whites, room temperature
	1½ teaspoons	3 g	3.5%	cream of tartar
3¾ ounces	½ cup	106 g	125%	superfine sugar
	1 tablespoon	15 mL	18%	coconut extract
¼ ounce	2 tablespoons	7.5 g	9%	sweetened, shredded coconut
			887.7%	Total Coconut Angel Food Cake percentage

1. Preheat oven to 325°F (160°C).

2. In the bowl of a food processor, place the 1 ounce (½ cup; 30 g) coconut, the 2¾ ounces (¾ cup; 80 g) sugar, cake flour, cornstarch, and salt. Pulse until the coconut is ground into very fine particles. Transfer the coconut mixture into a medium bowl and set aside.

3. In the bowl of an electric mixer, using the whip attachment, beat the egg whites at high speed until foamy.

4. Add the cream of tartar and continue beating at high speed until soft peaks form.

5. Gradually add the remaining 3¼ ounces (½ cup; 110 g) sugar, tablespoon by tablespoon, followed by the coconut extract, beating until stiff, unwavering peaks form (Figure 7–34).

6. Using a rubber spatula, carefully pour the beaten egg whites into a large mixing bowl, trying to avoid overhandling them.

7. Gradually sprinkle the reserved coconut sugar mixture on top of the beaten whites and fold in gently until all the dry ingredients are blended in (Figure 7–35). Do not overmix.

8. Pour the batter into an ungreased 10-inch (25-cm) tube pan, preferably with metal tabs sticking up from the top of the pan, and smooth with a spatula. Sprinkle the ¼ ounce (2 tablespoons; 7.5 g) coconut evenly over the top (Figure 7–36).

9. Bake for 1 hour or until a cake tester inserted in the center of the cake comes out clean and the coconut is brown on top.

10. Turn the pan upside down and allow it to cool, balancing on the metal tabs. If there are no metal tabs on the pan, balance the upside down cake on an overturned ramekin placed under the tube. Cooling upside down prevents gravity from pulling the cake down and falling.

11. When the cake is cold, using a long, sharp knife, carefully cut all around edges and inner tube portion of the pan, and remove the cake.

FIGURE 7–34

FIGURE 7–35

FIGURE 7–36

Coconut Angel Food Cake

TWO-STAGE GOLDEN YELLOW CAKE FROSTED WITH BOILED COCONUT FROSTING

Makes one 9-inch (22.5-cm) round double layer cake

Additional Ideas That Use the Recipes in This Chapter

STEP A

Make one recipe of Two-Stage Golden Yellow Cake baked in two 9-inch (22.5-cm) round cake pans. Level each layer using a serrated knife if necessary.

STEP B

Make one recipe of Boiled Coconut Frosting (Chapter 8), and toast 5 ounces, (2 cups; 140 g) shredded coconut.

STEP C: ASSEMBLY

1. Place one cake layer on a cardboard cake circle upside down. This creates a flat surface for the most level cake.

2. Spread one third of the filling onto the cake and sprinkle 1¼ ounces (½ cup; 35 g) toasted coconut over the filling and place the second cake layer on top, right side up (Figure 7–37).

3. Spread the top and sides of the cake with the remaining frosting (Figure 7–38). Sprinkle the remaining toasted coconut all over the top and sides (Figure 7–39). Chill the cake in the refrigerator until ready to serve.

FIGURE 7–37

FIGURE 7–38

FIGURE 7–39

Two-Stage Golden Yellow Cake Frosted with Boiled Coconut Frosting

FUDGY CHOCOLATE CAKE FROSTED WITH VANILLA BUTTERCREAM

Makes one 9-inch (22.5-cm) round double layer cake

STEP A

Make one recipe of Fudgy Chocolate Cake baked in two 9-inch (22.5-cm) round cake pans and level each layer using a serrated knife if necessary.

STEP B

Make one recipe of Vanilla Buttercream (Chapter 8) to frost the cake. Make an additional recipe if you wish to decorate.

STEP C: ASSEMBLY

1. Place one cake layer upside down on a cardboard cake circle. Spread one third of the buttercream evenly on the cake (Figure 7–40).

2. Place the second cake layer on top of the buttercream, right side up. Frost the top and sides of the cake with the remaining buttercream (Figure 7–41). Chill the cake until the buttercream is firm.

FIGURE 7–40

FIGURE 7–41

Fudgy Chocolate Cake Frosted with Vanilla Buttercream

CHOCOLATE HAZELNUT MOUSSE TORTE

Makes one 9-inch (22.5 cm) torte

STEP A

Make one recipe of Hazelnut Genoise baked in two 9-inch (22.5-cm) round cake pans.

STEP B

Make one recipe of Hazelnut Praline (recipe follows). Makes 6½ ounces (1½ cups; 170 g).

MEASUREMENTS			INGREDIENTS
U.S.		METRIC	
3½ ounces	½ cup	100 g	granulated sugar
	1 tablespoon	15 mL	light com syrup
2½ ounces	½ cup	70 g	whole, skinned hazelnuts, toasted in a 375°F (190°C) oven for 10 to 15 minutes

1. Combine the sugar and corn syrup in a small heavy saucepan.

2. Bring the mixture to a boil, stirring only until the sugar is dissolved. Using a pastry brush dipped in cold water, wash down any sugar crystals that may have formed on the sides of the pan. If the liquid boils up, remove the pan from the heat temporarily and then return it to a boil.

3. Continue to boil the mixture without stirring until the melted sugar syrup turns a light caramel color (Figure 7–42). This takes approximately 5 to 6 minutes. Watch the caramel carefully or it can burn.

4. Remove the mixture from the heat and add the hazelnuts.

5. Using a spatula, quickly pour out the caramel and hazelnut mixture onto a parchment-lined baking sheet or onto a silicone baking mat that has been placed inside a sheet pan (Figure 7–43). Allow mixture to cool completely. It will harden as it cools.

FIGURE 7–42

FIGURE 7–43

6. Break up the praline and place it in the bowl of a food processor. Process until it is finely ground (Figure 7–44). Store in an airtight container in the freezer until ready to use.

STEP C

Make one recipe of Hazelnut Syrup (recipe follows). Makes 1⅛ cups (33 mL).

MEASUREMENTS			INGREDIENTS
U.S.		METRIC	
4 fluid ounces	½ cup	120 mL	water
3½ ounces	½ cup	100 g	granulated sugar
3 fluid ounces	6 tablespoons	90 mL	hazelnut liqueur

(*Note:* When equal parts of water and sugar are brought to a boil until the sugar dissolves, it is referred to as a *simple syrup.* Simple syrups are often flavored.)

1. Combine the water and sugar in a small saucepan. Bring the mixture to a boil and stir until the sugar dissolves (Figure 7–45). Remove syrup from the heat and add the hazelnut liqueur. Cool and set aside.

STEP D

Make one recipe of Chocolate Hazelnut Mousse Filling (recipe follows). Makes approximately 4 cups (946 mL).

MEASUREMENTS			INGREDIENTS
U.S.		METRIC	
8 fluid ounces	1 cup	240 mL	heavy cream
12 ounces	2 cups	340 g	semisweet chocolate, coarsely chopped
½ ounce	1 tablespoon	15 mL	hazelnut liqueur
7 fluid ounces	¾ cup + 2 tablespoons	210 mL	heavy cream
	5 tablespoons	75 g	superfine sugar
3¾ ounces	¾ cup	105 g	hazelnut praline, finely ground (Step B)

1. Place the heavy cream into a medium saucepan and bring it to a boil.

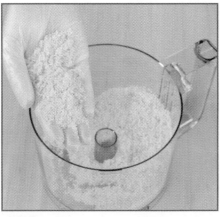

FIGURE 7–44

FIGURE 7–45

2. Remove it from the heat and whisk in the chopped chocolate and liqueur until the mixture is melted and smooth. Set aside to cool for 1 hour at room temperature (Figure 7–46 A and B).

3. In the bowl of an electric mixer, beat the ¾ cup + 2 tablespoons (7 fluid ounces; 210 mL) heavy cream using the whip attachment until soft peaks form.

4. Gradually add the sugar and continue beating until stiff peaks form.

5. On low speed, gradually add the cooled chocolate and cream mixture and blend for 10 seconds only. Remove the bowl from the base of the mixer, and, using a rubber spatula, gently fold in the chocolate until completely incorporated (Figure 7–47A and B). Blend in the praline (Figure 7–48).

STEP E

Make one recipe of Ganache (Chapter 8), cooled to room temperature.

STEP F: ASSEMBLY

1. Split each genoise cake layer in half horizontally for a total of four layers (Figure 7–49). Place one layer in the bottom of a 9-inch round false-bottom pan such as a spring form pan or in a 9-inch (22.5-cm) round 3-inch (7.5-cm) high metal cake ring. If using the metal cake ring, set it on a 9-inch (22.5-cm) cardboard cake circle that has been placed onto a sheet pan.

FIGURE 7–46A

FIGURE 7–46B

FIGURE 7–47A

FIGURE 7–47B

FIGURE 7–48

FIGURE 7–49

2. Using a pastry brush, brush some hazelnut syrup evenly over the cake layer to moisten it (Figure 7–50). Spread one third of the mousse evenly over the moistened layer (Figure 7–51). Place another genoise layer on top. Repeat two times, moistening with syrup and spreading with another one third of the mousse each time.

3. Place the last genoise layer on top and moisten with the remaining syrup (Figure 7–52). Cover with plastic wrap and chill for 3 to 4 hours.

4. Remove the sides of the pan and place the cake on a cake rack placed over a sheet pan (Figure 7–53). Spread room temperature ganache over the top and sides of the torte (Figure 7–54). Garnish the torte with any remaining crushed praline. Chill.

FIGURE 7–50

FIGURE 7–51

FIGURE 7–52

FIGURE 7–53

FIGURE 7–54

Chocolate Hazelnut Mousse Torte

CHOCOLATE SPONGE CAKE ROLL FILLED WITH SWEET RASPBERRY CREAM

Makes approximately 2 cups (473 mL)

STEP A

Make one recipe of the Chocolate Sponge Cake Roll.

STEP B

Make one recipe of the Raspberry Cream Filling (recipe follows).

| MEASUREMENTS | | | INGREDIENTS |
U.S.		METRIC	
	½ teaspoon	1.5 g	unflavored gelatin
	1 tablespoon	15 mL	freshly squeezed cold orange juice
3 ounces	⅓ cup	85 g	seedless raspberry preserves
	2 teaspoons	10 mL	raspberry liqueur like Chambord (or orange juice or water)
8 fluid ounces	1 cup	240 mL	cold heavy cream
1 ounce	¼ cup	30 g	confectioners' sugar, sifted

1. In a small heat proof bowl, sprinkle the gelatin over the cold orange juice (Figure 7–55). Let the mixture sit for 5 minutes to soften.

2. Set a small sauté pan filled with one half inch of water on the stove on medium-high heat and bring it to a simmer. Shut off the heat and place the bowl with the gelatin into the pan to melt it (Figure 7–56). Remove the gelatin from the water and allow it to cool but not to solidify again. The cream will not whip if the melted gelatin is too warm.

3. In another small bowl, mix the preserves and the liqueur until well blended.

4. In the bowl of an electric mixer using the whip attachment, beat the heavy cream on high speed and gradually add the sugar only until it becomes slightly thicker and the whip begins to leave marks in the cream (Figure 7–57).

FIGURE 7–55

FIGURE 7–56

FIGURE 7–57

5. Slowly add the melted and cooled gelatin into the cream and beat on high speed until soft peaks form (Figure 7–58). Gradually add the preserves (Figure 7–59).

STEP C: ASSEMBLY

Use immediately to fill the cooled chocolate cake roll by gently unrolling the cake roll and spreading the cream filling over the entire inside of the roll using an offset spatula (Figure 7–60 A and B). Re-roll the cake and dust it with confectioners' sugar (Figure 7–61).

FIGURE 7–58

FIGURE 7–59

FIGURE 7–60A

FIGURE 7–60B

FIGURE 7–61

Chocolate Sponge Cake Roll Filled with Sweet Raspberry Cream

Professional Profile

BIOGRAPHICAL INFORMATION

Wendy Schonberg
Chef Instructor
Hudson County Community College
Jersey City, NJ

1. Question: *When did you realize that you wanted to pursue a career in baking and pastry?*

Answer: *I was at the University of Wisconsin, not in any culinary related field, and realized that what I wanted to do was go into baking and pastry.*

2. Question: *Was there a person or event that influenced you to go into this line of work?*

Answer: *My mother was from Europe and she was just an amazing baker. We always had fresh baked bread at home. Everything was from scratch; nothing we ate was canned. She really inspired me.*

3. Question: *What did you find most challenging when you first began working in baking and pastry?*

Answer: *I went to school at the Culinary Institute of America in Hyde Park and I took to the program from the start. The only real challenge was learning the discipline that is necessary. It takes much more discipline to bake on a commercial level than baking at home or even for a caterer.*

4. Question: *When and where was your first practical experience in a professional baking setting?*

Answer: *My first position was with The Manor in New Jersey. I worked under a phenomenal chef, Freddy Mayer. As a new graduate I was the "low man on the totem pole" but the chef allowed me to try everything. He taught me about working with chocolate and sugar and to do some sophisticated pieces. He gave me the opportunity to learn so much.*

5. Question: *How did this experience affect your later professional development?*

Answer: *I learned early on that being a woman in this environment meant I had to push a little harder, try a little more.*

6. Question: *Who were your mentors when you were starting out?*

Answer: *Certainly Chef Mayer at The Manor and my mother were both important to shaping my career. But, people like Joe Amendola who really stressed that the basics and the science behind baking were also important. Amendola really knew his stuff. He was great.*

7. Question: *What would you list as your greatest rewards in your professional life?*

Answer: *It has to be teaching. To be with students who are so eager and excited to learn is very fulfilling.*

8. Question: *What traits do you consider essential for anyone entering the field?*

Answer: *Anyone wanting to make this a career must have patience. You just can't give up. You have to keep trying until you get it perfect. You also have to be a stickler for details.*

9. Question: *If there was one message you would impart to all students in this field what would that be?*

Answer: *Don't rush. Take it slow. Start with the basics and then work up to the more difficult skills. Don't try doing sugar work if you can't bake a cake. Once you gain one skill, though, keep going. Try different media and styles. Continue to create.*

CHAPTER 8

Frostings

After reading this chapter, you should be able to:

- Define a frosting.

- State the four reasons to use frostings.

- List the seven categories of frostings.

- Correctly pair cakes with an appropriate frosting.

- Identify the basic tools needed to properly frost a cake.

- Frost a cake.

- Prepare the frosting recipes in this chapter.

Frostings and icings are literally the topping on the cake. Armed with one or two quality cake recipes, a professional baker can create many different looks and flavors for a cake just by changing the frosting. Frostings can be thick and spreadable or thin and pourable. Frostings can take an already wonderful cake to a new level, or transform a plain dessert into an extraordinary work of art.

Frosting Defined

A frosting is actually a sweet topping or covering that is used to fill or coat the top or sides of a cake. The word *frosting* can be used interchangeably with the word *icing*. Frostings can also be mixed and matched with other frostings so that one type of frosting can be used as a filling and another type can be used to frost the top and sides of the cake.

Four Reasons to Use Frostings

> **Four Reasons to Use Frostings**
> - Increases eye appeal
> - Improves overall flavor
> - Enhances texture
> - Slows staling

There are basically four reasons to coat a cake with frosting:

- Increases eye appeal. The addition of frosting raises a plain unfrosted cake to a new level. The cake becomes something special to look at. Eye appeal is very important to the professional baker.
- Improves overall flavor. The frosting adds sweetness and contributes to the overall flavor theme of the cake. For example, if a lemon cake has a mild taste, adding a zesty lemon cream cheese frosting with bits of lemon zest will add to the overall flavor profile, intensifying the flavor so the taste of lemon really stands out.
- Enhances texture. The texture can also be enhanced by using a frosting. A frosting can add moistness to a dry cake or lighten the texture of a heavy, rich cake. Texture refers to how a food feels in the mouth. A frosting can add a smooth, creamy feel to an otherwise plain, dry, crumbly cake.
- Slows staling. Frosting a cake can slow down the staling process by preventing it from drying out. As a cake is exposed to air, moisture evaporates from it, causing it to become dry and hard. Creating a protective barrier by using a layer of frosting helps slow down this process. Because most frostings contain a great deal of fat in them, the richer the frosting, the better the barrier.

The Seven Categories of Frostings

There are seven categories of frostings. They include simple or flat icings, glazes, royal icings, buttercreams, egg-foam or boiled frostings, rich chocolate confectionary frostings, and rolled-out frostings. (See Table 8–1, The Seven Categories of Frostings.)

Simple Icings or Flat Icings

Simple or flat icings are simple to make and are used more as glazes than frostings. Recipes usually start with confectioners' sugar to which water, cream, milk, maple syrup, citrus juices, or corn syrup are added along with flavorings to make a thin, pourable icing that can be

Table 8–1 The Seven Categories of Frostings

CATEGORY	TYPE	COOKED	UNCOOKED
Simple or flat			x
Glazes	Sugar syrup	x	
	Jams, preserves	x	
Royal			x
Buttercream	Simple		x
	French	x	
	Italian	x	
	Pastry cream–based	x	
Egg-foam or boiled	Italian meringue–based	x	
	Swiss meringue–based	x	
Chocolate confectionery	Fudge style	x	
	Ganache style	x	
Rolled-out	Rolled fondant (usually purchased ready-made)	x	
	Modeling chocolate		x
	Marzipan		x

drizzled over coffee cakes, scones, cookies, or sweet yeast breads. Simple icings can be made thicker by adding less liquid. Thicker icing is best for drizzling over warm coffee cakes, Danish pastry, or yeast breads because the warmth of the baked good melts down the icing even more. A thicker icing is also more visible. If the icing is too thin, it tends to blend into the pastry. Ideally, simple icings should be prepared immediately before they are used; but if prepared ahead, they can be warmed over a double boiler until melted. Simple icings dry within several minutes to a dull, hard coating.

A special type of icing that resembles a simple icing can be made from a crystallized sugar syrup known as *poured fondant.* Poured fondant is a sugar syrup that is allowed to crystallize enough to form a thick sugar paste. It is then melted over a hot water bath to approximately 100°F (38°C) and poured over baked goods such as petit fours, napoleons, cakes, cookies, and other small pastries. It is usually purchased commercially in a ready-to-use paste form or in a powder that can be blended with water because it can be difficult to prepare from scratch. Fondant is also available in a form that can be rolled out, called *rolled fondant* (see Rolled-Out Frostings). Poured fondant dries to a shiny, hard coating.

RECIPE

Simple or Flat Icing (This chapter, page 238)

Glazes

A glaze tends to be very thin and has little color. Glazes are very high in sugar content and consist of a sugar syrup, preserves, or jams that have been thinned and melted down. Glazes are usually brushed onto the baked good instead of drizzled or spread. The water in the glaze evaporates quickly, leaving a thin, shiny coating. Glazes tend to be used for doughnuts, Danish pastry, fruit tarts, and some cakes. Glazes for fruit tarts often contain gelatin and are brushed onto the fruit only. After refrigeration, the glaze firms up, coating the fruit and extending its shelf life.

RECIPE
Clear Glaze (Chapter 4, page 137)

Royal Icing

Royal icing is similar to a meringue and consists of confectioners' sugar, an acid (e.g., lemon juice or cream of tartar), and egg whites. Commercially prepared meringue powders require only confectioners' sugar and warm water to create a royal icing. If preparing royal icing from scratch, pasteurized powdered egg whites may be used instead of fresh egg whites if *Salmonella* contamination is a concern. Read the package directions carefully for specific directions on how to rehydrate powdered egg whites. Royal icing resembles a fluffy meringue, but it dries to a very hard state, making it perfect for piped decorations on cakes and cookies, or as a "glue" for a gingerbread house. Royal icing is not used to frost cakes because, once dried, the egg whites make it too hard and unpalatable. Because royal icings dry out when exposed to air, they should be covered with moistened paper towels or plastic wrap when not being used. Decorations of all kinds can be piped onto parchment lined sheet pans, allowed to dry at cool room temperature, and peeled off gently. They then can be placed on cakes or cookies using a dab of icing as glue. Decorations made with royal icing can be kept in airtight containers at cool room temperature for several months.

RECIPE
Royal Icing (This chapter, page 239)

Buttercream Frostings

There are different types of buttercream frostings. Buttercream frostings may consist of butter or vegetable shortening; granulated sugar, corn syrup, or confectioners' sugar; and sometimes whole eggs, egg yolks, or egg whites. Various flavorings can then be added to customize the frosting to the cake. Butter is the best choice of fat because it has a creamy texture and flavor. However, vegetable shortening may be used when a pure white buttercream is desired or the temperature is warm. Buttercreams should be used at room temperature, preferably right after preparation, while still having a spreadable consistency, because the fat hardens when chilled.

RECIPE
Vanilla Buttercream (This chapter, page 244)

SIMPLE BUTTERCREAM

The easiest buttercream to make is called a *simple buttercream*. It is made by combining confectioners' sugar, butter, milk or cream, and flavorings (e.g., chocolate or extracts) until a spreadable consistency is reached. Buttercreams are stable and keep well.

FRENCH BUTTERCREAM

Egg yolks are beaten with an electric mixer using the whip attachment until they are pale yellow. Boiling sugar syrup is slowly beaten into the yolks. The mixture is beaten until cool. Softened butter is slowly added until a light, creamy consistency is reached. French buttercream should be used at room temperature. Once chilled, if it becomes too hard to spread or pipe, it must be brought back to room temperature before using. If it is too soft to spread, it may be chilled for approximately 1 to 2 hours, or placed over an ice water bath, stirring frequently, so the frosting can become more firm and spreadable.

ITALIAN BUTTERCREAM

Italian buttercream, similar to an Italian meringue, begins with beaten egg whites to which a boiling sugar syrup is slowly added. When the mixture is cool, softened butter and flavorings are added.

PASTRY CREAM–BASED BUTTERCREAM

Finally, some buttercreams are made with a stirred custard thickened with a starch, known as a pastry cream, to which softened butter has been incorporated.

Egg-Foam or Boiled Frostings

Egg-foam or boiled frosting consists of nothing more than Italian and Swiss meringues. They are very light and airy. A boiled icing made from an Italian meringue starts out with a boiling sugar syrup that is slowly added to beaten egg whites (an egg foam) in which the high temperature of the syrup partially cooks the egg, producing a thick, fluffy frosting. A boiled frosting made with an Italian meringue is similar to an Italian buttercream but without the butter.

In boiled frosting using a Swiss meringue, egg whites and sugar are whisked and warmed together over a hot water bath. The warm mixture is then beaten until it triples in volume to a thick, shiny frosting.

Boiled frostings should be used immediately and do not keep very well. Sometimes boiled frostings are folded in with other ingredients like pastry cream or whipped cream for added richness. Some recipes include a stabilizer such as gelatin to increase the shelf life.

RECIPE

Boiled Coconut Frosting (This chapter, page 243)

Rich Chocolate Confectionery Frostings

There are two types of chocolate confectionery frosting that are based on two rich chocolate candies: fudge and truffles. Fudge is a rich candy made from a cooked sugar mixture enriched with milk, cream, butter, and flavorings that is allowed to crystallize. Chocolate is one of the most popular flavorings for fudge. Truffles, also a rich confection, are prepared differently; they are based on a mixture of cream and chocolate called ganache. Often the word *fudge* is used indiscriminately when describing any rich chocolate frosting. Many frosting recipes that have the word *fudge* in their title are not true fudge at all; in fact, they are ganache-style frostings. Fudge and ganache frostings taste quite different. Fudge frostings taste sugary and dense, whereas ganache frostings have a creamier mouthfeel because of the higher fat content in the cream and chocolate. Many professional bakers prefer to make a ganache-style frosting over a fudge-style frosting because of its ease of preparation.

FUDGE-STYLE

A fudge-style frosting begins with a rich sugary confection known as *fudge.* To make a fudge-style frosting, an enriched sugar mixture is boiled to the soft ball stage, after which butter and flavorings are added. It is then cooled to lukewarm and beaten until it turns from glossy and shiny to just beginning to dull. The agitation from beating creates small sugar crystals that form a smooth frosting. Beating the frosting while it is too warm encourages larger crystal formation, producing a grainy texture. Fudge-style frostings must be used immediately while still warm or

---

Although recipes vary, generally the chocolate to cream ratios (by volume) are:

1:1 for a sauce

2:1 for a thicker sauce that, when chilled, firms up to a thick frosting

3:1 for candy (truffles)

they will set. Hot water can be used to thin the frosting if it becomes too stiff to spread. Fudge frostings keep well. If the frosting has become too firm to spread, it can be placed in a bowl over a pot of simmering water until it melts. Allow the frosting to cool to a spreadable consistency before using.

RECIPE
Chocolate Fudge Frosting (This chapter, page 240)

GANACHE-STYLE

Ganache-style frostings are the basis for a rich chocolate confection known as a truffle. They are simple to make and avoid any problems that may result from a fudge-style recipe such as over-beating or underbeating, or worrying about correct temperatures. The procedure to make a ganache-style frosting is similar to preparing a candy truffle. A ganache is nothing more than heavy cream that has been brought to a simmer. Some recipes call for butter, corn syrup, or granulated sugar to be heated with the cream. The cream is then taken off the heat, and chopped chocolate or flavorings are added and stirred until the chocolate melts.

Ganache can be used in a number of ways. For instance, warm ganache can be used as a dessert sauce. It can be poured over a cake as a glaze that will harden to a smooth, firm coating if allowed to set in the refrigerator. The liquid ganache can be refrigerated for a few hours where it will firm up to a spreadable or pipeable consistency. Or, it can be beaten to a light, almost mousse-like consistency.

Ganache frostings keep well and can even be frozen for several months in an airtight container. They can be melted down over a double boiler to reuse. The proportion of heavy cream to chocolate is most important in determining how dense a ganache will become. The greater the proportion of cream to chocolate, the thinner the ganache. A lesser proportion of cream to chocolate, the thicker the ganache. A ratio of 1:2 of heavy cream to chocolate works best for a ganache that will set up as a frosting. Truffle candies use less heavy cream and a much higher ratio of chocolate in order to give the truffle a firmer texture. In that case, the ratio of cream to chocolate may be as high as 1:3.

RECIPE
Ganache (This chapter, page 242)

Rolled-Out Frostings

There are three types of rolled-out frostings. They include rolled fondant, modeling chocolate, and marzipan. Frostings that are the consistency of a dough made from various ingredients can be rolled out and fitted onto a cake after it is covered with a thin layer of preserves or buttercream to help it adhere.

The first type of rolled-out frosting is known as *rolled fondant*. Rolled fondant is made using a cooked sugar mixture that is cooled, beaten, and then kneaded like a dough until pliable. It is usually purchased ready-made for the commercial baker. It can then be dyed any color. It is also available flavored with chocolate.

The second type of rolled-out frosting is called *modeling chocolate*. Modeling chocolate is nothing more than melted chocolate and corn syrup. It is kneaded like a dough and allowed to stand for several hours before it is ready to use. It can be rolled out in a similar manner as rolled fondant but tends to be used more for cut-out decorations and flowers.

The third type of rolled-out frosting is called *marzipan*. Marzipan is made from finely ground almonds and sugar, which becomes a thick, sticky, dough-like paste. It, too, can be rolled and fitted over a cake. Sometimes, another frosting is applied to cover the marzipan layer. Marzipan can also be used to create flowers and other decorations to go on top of a cake. Melted jam or preserves brushed over the cake help the rolled out marzipan to adhere. Marzipan is available commercially in a ready-to-use state. Marzipan gives cakes an almond flavor that can complement the cake's taste and texture.

Rolled-out frostings dry out quickly and must be wrapped tightly in plastic and stored in an airtight container.

Pairing the Cake with the Appropriate Frosting

Knowing which frosting to use to cover a cake is most important. The general rule of thumb is that lighter cakes like angel food and chiffon should be paired with lighter frostings like boiled frostings, or drizzled with a simple icing or glaze.

Heavier cakes like a dense chocolate or yellow butter cake are best paired with richer frostings like buttercream, ganache, or fudge-style.

Basic Tools to Properly Frost a Cake

A few simple tools are necessary to aid the baker in frosting a cake:

- Turntable. A turntable is a rotating elevated plate similar to a serving plate on a pedestal. The cake can be turned to facilitate the frosting process.
- Offset palette knife. An offset palette knife also known as an offset spatula, is a long, round-tipped knife with a slight bend in the blade, just before the handle. It is most helpful when frosting a cake and getting into hard-to-reach areas.
- Palette knife. A palette knife is the same as an offset palette knife without the bend near the handle. It, too, is helpful in spreading frostings evenly.
- Pastry brush. A pastry brush looks like a paintbrush. It can be useful in soaking a cake with a flavorful sugar syrup or coating a cake with preserves before frosting. Pastry brushes come in many widths.
- Pastry bag. A pastry bag is a cone-shaped piece of plastic, parchment, or canvas with a large opening at one end with which to fill the cone and a smaller opening at the other end that can be fitted with various pastry tips used to pipe out decorations.
- Pastry tips. Pastry tips are plastic or metal cone shapes, open at both ends, with decorative cuts in the smaller end. Pastry tips are fitted into a pastry bag with or without a coupler (a short plastic tube with a plastic top that screws onto the outside). Couplers are used to allow the tip to fit more securely to the bag. Pastry bags with the tip securely attached should be filled only half full. They are then squeezed from the top so that frosting pushes out through the decorative tip, forcing the frosting into various shapes and designs.
- Icing comb. An icing comb looks like a rectangular or triangular piece of hard plastic or metal in which grooves or edges have been cut out at regular intervals. The icing comb can be gently scraped against the sides of a newly frosted cake where it will leave decorative grooves and ridges.
- Metal cake rings or torte rings. Metal cake rings come in various sizes and shapes without a bottom or top in which layers of cake and soft fillings or mousses can be placed to maintain

their shape. The cake ring is placed onto a sheet pan or cardboard cake circle before being filled. After the filling sets, the ring is then gently lifted off and the cake can then be frosted or glazed on the outside.

- Metal false-bottom tart pan bottom. The metal bottoms from a false-bottom tart pan can be used to separate two cut cake layers. After a layer of cake is cut in half horizontally, the metal bottom is slid between the cake layers to help remove the top layer, supporting it and maintaining its shape. Tart pan bottoms come in several diameters to accommodate different diameter cakes. A cardboard cake circle can also be used but tart pan bottoms are thinner and easier to insert between cake layers.

- Acetate. Acetate (clear sheets of plastic) are used to line the inner sides of a metal cake ring. Once the cake ring is removed, the acetate protects the sides of the cake from frostings or glazes that may ooze over the sides of the cake. Once the frosting or glaze hardens, the acetate sheet can be peeled off. Strips of acetate can also be coated in melted chocolate and placed around the outer perimeter of a cake or torte. Once the chocolate hardens, the acetate is peeled off, leaving a thin chocolate coating.

How to Frost a Layer Cake

1. Remove the cake layers from the cake pans while they are still warm, peeling off and discarding the parchment paper. Place the cakes on metal cake racks and allow them to cool completely before frosting.

2. Use a serrated knife to even off the top of each layer if the cakes have risen unevenly and formed a dome. If additional layers are desired, use a serrated knife to split the layers in half horizontally, rotating the cake with one hand as the cake is cut with the other.

3. Use the bottom of a false-bottom tart pan or a cardboard cake circle to separate the cut layers by slowly sliding the circle through the cut area until it holds the top layer of cake. If the cake has been cut into additional layers, slide a cake circle under each layer to keep it intact before beginning to frost and rebuild each layer.

4. Make the top of the cake the bottom layer by flipping it over onto a same sized cardboard cake circle so that the layer is cut side up. This provides a flat, even surface for the filling or frosting, and provides an even foundation on which to build the layers. Place the cake circle with the cake layer onto a turntable. Some pastry chefs place a dab of frosting in the middle of the turntable under the cake circle so it does not slide.

5. Spread the filling or frosting evenly across the cake layer. A pastry bag fitted with a large plain tip may be used to pipe a uniform thickness of filling or frosting onto the layer. If the filling is not the same as the frosting that will be spread on the top and sides of the cake, pipe a band of frosting around the outer perimeter of the cake layer to act as a barrier between the filling and frosting. This prevents the filling from oozing out the sides of the cake.

6. Place the bottom cake layer over the filling or frosting so the layer is cut side down. The cut side should be down to minimize any cake crumbs that may get into the frosting. Press down gently to help the top layer adhere.

7. Frost the top and sides smooth by rotating the turntable and using a palette knife to spread the frosting. To prevent crumbs from being incorporated into the frosting, use enough frosting so that the palette knife never actually touches the cake itself.

8. Another way to frost a cake is to first spread a thin layer of frosting all over the cake to allow the crumbs to adhere. This is known as a *crumb coating*. Then the top and sides of the cake are frosted again with a thicker layer of frosting.

9. To smooth out irregularities in the frosting, dip the palette knife into very hot water, quickly wipe it clean, and then glide it gently over the top and sides of the cake to barely melt any irregularities or bumps, smoothing out the frosting. This works best with butter creams and ganache-style frostings and may not work for every type of frosting.

10. Additional decorations can be added or piped on at this time.

SIMPLE OR FLAT ICING

Makes approximately 4 fluid ounces ($\frac{1}{2}$ cup; 120 mL)

Lessons demonstrated in this recipe:

- How to prepare a simple or flat icing.
- This type of frosting is the simplest to make.
- Many different flavors of icing can be created depending on what kind of liquid is added to the sugar.

MEASUREMENTS			INGREDIENTS
U.S.		METRIC	
4 ounces	1 cup	110 g	confectioners' sugar, sifted if lumpy
	4 teaspoons	20 mL	a liquid such as milk, cream, water, lemon juice, orange juice, or maple syrup (plus more, if needed)
	$\frac{1}{2}$ teaspoon	2.5 mL	vanilla extract, optional

1. Whisk together the sugar and the desired liquid until the mixture is thick, yet pourable (Figure 8–1). If the frosting is too thick, dilute it with more liquid. Add vanilla extract, if desired.

2. Drizzle or spread on coffee cake, Danish pastry, cookies, or yeast breads. If it is not to be used immediately, cover the frosting with plastic wrap to keep it from drying out. Stir just before using.

FIGURE 8–1

Royal Icing

Makes approximately 8 fluid ounces (1 cup; 240 mL)

Lessons demonstrated in this recipe:

■ How to prepare a royal icing.

■ Pasteurized egg whites are used instead of fresh egg whites to inhibit *Salmonella* contamination because the icing receives no further cooking; however, fresh egg whites may be used.

MEASUREMENTS			INGREDIENTS
U.S.		**METRIC**	
10 ounces	2½ cups	285 g	confectioners' sugar
	⅛ teaspoon	.25 g	cream of tartar
2 each		56 g	large egg whites or the equivalent of 2 pasteurized egg whites made according to manufacturer's directions. (Brands vary. Make sure the manufacturer's label states that the product can be beaten to a foam.)
			Warm water, if needed

1. In the bowl of an electric mixer using the whip attachment, blend the confectioners' sugar and the cream of tartar until well blended.

2. Add the egg whites and beat on medium-high speed until fluffy and the icing forms stiff peaks when the beater is lifted from the bowl (Figure 8–2A and B).

3. If the icing is too thick, thin it down with a small amount of warm water.

4. To prevent the icing from drying out, keep it covered with a moist paper towel or plastic wrap. Use immediately.

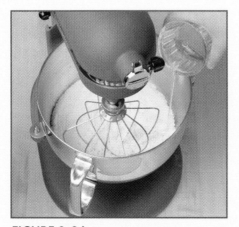

FIGURE 8–2A

FIGURE 8–2B

CHOCOLATE FUDGE FROSTING

Makes approximately
20 fluid ounces (2½ cups;
590 mL)

Lessons demonstrated in this recipe:

- How to prepare a rich chocolate confectionery frosting.
- Corn syrup interferes with crystallization, creating a smoother fudge.
- Fat from the heavy cream gives a smoother texture to the fudge, preventing any curdling of the milk proteins.
- Allowing the fudge to cool to 120°F (54°C) before beating creates many small crystals of sugar, forming a smooth candy frosting.
- Adding hot water helps thin down the frosting if it becomes too thick.
- There is only a small window of time in which to spread the frosting before it firms up and hardens.

MEASUREMENTS		METRIC	INGREDIENTS
4 ounces		115 g	unsweetened chocolate, coarsely chopped
8 ounces	⅔ cup	225 g	light corn syrup
6 fluid ounces	¾ cup	180 mL	heavy cream
2 fluid ounces	¼ cup	60 mL	milk
14½ ounces	2 cups	400 g	granulated sugar
7½ ounces	1 cup	220 g	light brown sugar
	⅛ teaspoon	1 g	salt
2 ounces	4 tablespoons	57 g	unsalted butter, cut into fine pieces
	1½ teaspoons	7.5 mL	vanilla extract
			very hot water, if needed, to thin down the frosting

1. Place the chocolate, corn syrup, cream, milk, both sugars, and the salt in a large, heavy saucepan. Stir lightly with a wooden spoon, scraping the edges of the pan to blend the mixture thoroughly. Wash down the sides of the pan with a pastry brush dipped in cold water.

2. Place the mixture over medium-high heat until the chocolate is melted. Turn the heat to high and boil until a candy thermometer placed into the mixture registers 235°F (113°C) or when a drop of the mixture placed into a glass of cold water can be rolled into a soft ball (Figure 8–3A, B, and C). This is known as the *soft ball stage*.

FIGURE 8–3A

FIGURE 8–3B

FIGURE 8–3C

3. Remove the mixture from the heat and pour it into the bowl of an electric mixer. Place the butter pieces on top of the mixture and allow them to melt on the top without stirring. Keep the mixer off and allow the mixture to cool until a candy thermometer inserted into it registers 120°F (54°C). Any agitation such as stirring when the fudge is so hot can cause large sugar crystals to form, causing the frosting to become grainy.

4. Using the paddle attachment on low speed, blend the mixture while adding the vanilla extract.

5. Increase the speed to medium and beat the mixture for 1 to $1\frac{1}{2}$ minutes until it thickens and just begins to lose its shine, becoming slightly dull (Figure 8–4). Do not overbeat. If the fudge becomes too thick, dribble in a little hot water to thin it down.

6. Quickly pour or spread the fudge frosting onto a cake or a cooled sheet pan of brownies, or place small spoonfuls on top of cookies or cupcakes. There is only a small window of time in which to spread the fudge frosting before it hardens. If it hardens before it can be spread, melt it down over a hot water bath, allowing it to cool to 120°F (54°C) and then beat it again until it cools to a spreadable consistancy and just begins to thicken. Use immediately.

FIGURE 8–4

GANACHE

Makes approximately
28 fluid ounces (3⅓ cups;
830 mL)

Lessons demonstrated in this recipe:

- How to prepare ganache, a rich chocolate confectionery frosting.
- Chopped chocolate is added to simmering cream to create a versatile frosting that can be used as a glaze when warm or as a spreadable frosting when chilled.

MEASUREMENTS			INGREDIENTS
U.S.		METRIC	
12 fluid ounces	1½ cups	355 mL	heavy cream
½ fluid ounce	1 tablespoon	15 mL	light corn syrup
1 ounce	2 tablespoons	30 g	unsalted butter
1 pound	2½ cups	450 g	semisweet chocolate, finely chopped
	1 tablespoon	8 g	Dutch processed cocoa powder

1. Bring the cream, corn syrup, and butter to a simmer, just below the boiling point. Remove the mixture from the heat and add the chocolate and cocoa powder. Some chefs may prefer to place the chocolate and cocoa powder into a bowl and pour the hot cream over it. The first method is chosen here to make sure the chocolate melts completely when placed in the hot pan.

2. Allow the mixture to stand for 5 minutes, then whisk gently until smooth. Place the ganache in a bowl and chill for 2 to 3 hours stirring every 20 minutes until it reaches a spreadable consistency (Figure 8–5). Alternatively, the bowl of ganache can be cooled more quickly by placing it over an ice water bath while stirring frequently until it thickens to a spreadable consistency.

FIGURE 8–5

BOILED COCONUT FROSTING

Makes approximately 128 fluid ounces (8 cups; 3.8 L)

Lessons demonstrated in this recipe:

- ■ How to prepare a boiled or egg-foam frosting using an Italian meringue.
- ■ Boiling sugar syrup is added slowly to beaten egg whites to produce a light, airy frosting.
- ■ Whipped cream is folded in for richness.

MEASUREMENTS		METRIC	INGREDIENTS
U.S.			
4 each		112 g	egg whites, room temperature
	4 tablespoons	60 mL	light corn syrup
6 ounces	⅔ cup + 2 tablespoons	170 g	granulated sugar
2 fluid ounces	¼ cup	59 mL	water
	4 teaspoons	20 mL	coconut extract
12 fluid ounces	1½ cups	350 mL	heavy cream
	4 tablespoons	60 mL	cream of coconut

1. Place the egg whites in the bowl of an electric mixer fitted with the whip attachment.

2. In a small saucepan, combine the corn syrup, sugar, and water and bring the mixture to a boil. While the mixture is boiling, spray a liquid heatproof measuring cup with nonstick cooking spray and set aside. This will ensure that all the sticky sugar syrup slides out easily.

3. When the sugar syrup reaches 230°F (110°C), start to beat the whites at high speed.

4. When the sugar syrup reaches 240°F (115°C) (the soft ball stage), pour the syrup into the greased measuring cup. Slowly pour the syrup down the side of the bowl into the egg whites with the mixer on high speed, trying not to pour the liquid directly onto the whip (Figure 8–6). If the syrup hits the whip, lumps may form.

5. Beat the Italian meringue at high speed until the bowl is cool and stiff peaks have formed (Figure 8–7A). Blend in the coconut extract. Place the meringue in a mixing bowl and set aside.

6. In a clean, dry bowl of an electric mixer, using the whip attachment, beat the heavy cream and the cream of coconut until stiff peaks form. Fold the whipped cream mixture into the meringue using a rubber spatula (Figure 8–7B).

FIGURE 8–6

FIGURE 8–7A

FIGURE 8–7B

VANILLA BUTTERCREAM

Makes approximately 32 fluid ounces (4 cups; 946 mL)

Lessons demonstrated in this recipe:

- How to prepare a French buttercream frosting.
- A hot sugar syrup is poured into beaten egg yolks, followed by softened butter.

MEASUREMENTS			INGREDIENTS
U.S.		**METRIC**	
7¼ ounces	1 cup	200 g	granulated sugar
¼ cup	4 tablespoons	60 mL	light corn syrup
2 fluid ounces	¼ cup	60 mL	water
6 each		114 g	large egg yolks
1 pound		450 g	unsalted butter, very soft (but not melted)
	2 teaspoons	10 mL	vanilla extract

1. Spray a liquid heatproof measuring cup with nonstick cooking spray to ensure that all the sticky sugar syrup will slide out easily.

2. In a saucepan, combine the sugar, corn syrup, and water and bring the mixture to a rolling boil, stirring just until the sugar dissolves.

3. While the sugar syrup is cooking, beat the egg yolks on high speed in the bowl of an electric mixer using the whip attachment until the color lightens to a pale yellow (Figure 8–8).

4. Immediately pour the syrup into the greased measuring cup (Figure 8–9).

5. On high speed, slowly add the hot syrup to the egg yolks, pouring it down the sides of the mixing bowl and not directly onto the whip. Keep beating until the bowl feels cool to the touch.

TIP Pouring the hot syrup from a measuring cup with a spout instead of directly from the saucepan helps prevent the syrup from hitting the whip directly and forming lumps of hardened sugar.

FIGURE 8–8

FIGURE 8–9

TIP Extra buttercream can be stored in an airtight container in the refrigerator for 1 week or frozen for 3 to 4 months. When ready to use, thaw the buttercream and allow it to become softened at room temperature. This may take 3 to 4 hours, depending on the quantity of buttercream. Mix the buttercream to a creamy consistency with the paddle attachment of an electric mixer on low speed. Do not overbeat.

6. Gradually add the softened butter, a few tablespoonfuls at a time, until it is all incorporated, blending in each addition of butter thoroughly (Figure 8–10). The buttercream may appear curdled until all the butter has been incorporated. Add the vanilla extract and blend well. Use at once. If the mixture is too soft, it can be refrigerated for 1 to 2 hours or placed over an ice water bath, stirring frequently with a whisk until it mounds and is of a spreadable consistency.

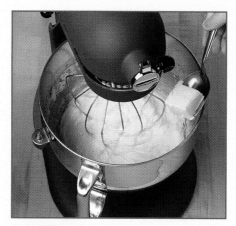

FIGURE 8–10

Another major role that eggs play is as a leavening agent.

A leavening agent is an ingredient within a recipe that helps the final product rise and expand. Eggs, whether whole or in their component parts (whites or yolks), can hold air within their structure. This air produces a light, airy texture in many pastries and desserts. Picture a fluffy soufflé or a light angel food cake and you can already visualize the power of eggs as leaveners.

This section discusses the role eggs play as leavening agents and the role they play in creating different types of meringues. The preparation and the various uses of meringues are also discussed.

Egg Foams Defined

When eggs are beaten, air becomes trapped within the proteins, which then expand when heated. This trapped air pushes on the batter as it heats up. The pressure increases as the heat increases, and the food expands and rises. Eggs with air beaten into them are known as egg foams and can be the basis for cakes, soufflés, meringues, frostings, mousses, and even candies.

In the last chapter, we discussed eggs denaturing or setting and becoming firm as they are heated. Beating eggs will also denature them, or break them into fragments, allowing the fragments to hold trapped air and moisture between them while foaming. As the beating of the whites continues, the denatured proteins surround each air bubble, much like children gathering together to hold hands in a circle. This network of proteins helps to prevent the air from escaping, thus maintaining the structure of each air bubble without collapsing. Picture an air bubble surrounded with loosely bonded protein fragments (Figure 8–11). Whole egg foams and egg white foams can be used to leaven baked goods such as sponge cakes and soufflés.

Structure of Egg Foam
Beating egg whites causes the proteins to break down into fragments, trapping bubbles of air between them, producing a foam.

FIGURE 8–11

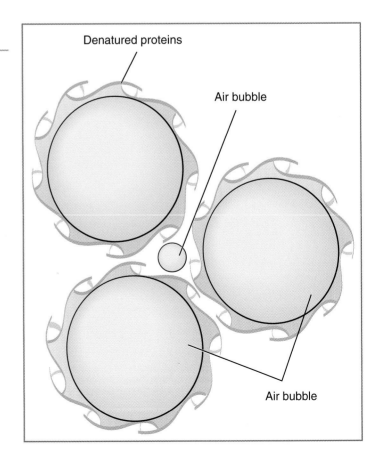

Denatured proteins

Air bubble

Air bubble

The following recipes deal with egg white foams, which are egg whites beaten using the whip attachment of an electric mixer or a whisk. Mixtures of egg white foams and sugar are known as meringues. Egg whites are the natural choice for leavening many types of baked goods because they hold more air than whole eggs or yolks. Meringues are versatile and can be used as a leavener for cakes and soufflés, as a topping on pies and Baked Alaska, or as a frosting for cakes, or they can be used in the preparation of marshmallows or baked at lower temperatures into crisp layers for cakes and tortes.

Three Types of Meringues

The texture of meringues can range anywhere from soft and airy to firm and crisp. The amount of sugar added to an egg foam and the degree to which it is heated determines how hard or firm a meringue is produced. In general, 1 part sugar to 1 part egg whites or a 1:1 ratio is used with 1 ounce (2 tablespoons; 30 g) sugar per 1 egg white for softer meringues and 2 parts sugar to 1 part egg whites or a 2:1 ratio is used with 2 ounces (4 tablespoons; 60 g) sugar per 1 egg white for firmer meringues.

There are three types of meringues. Each one has a different level of stability and stiffness. They are French meringues, Swiss meringues, and Italian meringues. (See Table 8-2, Comparison of the Three Types of Meringues.)

French Meringues

French meringues are also known as common meringues and are the simplest of the three types. French meringues can be used to top desserts like lemon meringue pie or for a Baked Alaska. They can also be used to leaven soufflés and cakes, or they can be baked and used as a crisp meringue cookie or cake layer. A soft French meringue uses a 1:1 ratio of sugar to egg whites; a firmer meringue uses a 2:1 ratio of sugar to egg whites.

Procedure to Prepare a French Meringue

1. In the bowl of an electric mixer using the whip attachment, beat the egg whites with an acid, such as cream of tartar or lemon juice, until thick and foamy.
2. At high speed, slowly add the granulated sugar over a period of 3 to 4 minutes. Continue beating until stiff peaks form.

Swiss Meringues

Start a Swiss meringue by warming the egg whites and sugar over a hot water bath before beating them. The heat helps the sugar crystals dissolve, forming a thick liquid that surrounds and protects each air bubble within the meringue. The heat also denatures the proteins in the egg whites, forming a more stable meringue than a French meringue. Swiss meringues are generally used to prepare cake frostings such as buttercreams and cookies.

Procedure to Prepare a Swiss Meringue

1. Egg whites and granulated sugar are placed in the bowl of an electric mixer over simmering water (as in a double boiler or bain marie) and heated to 120°F (49°C) while being whisked constantly to avoid cooking the whites.
2. Once the correct temperature is reached, the bowl is removed from the heat and the mixture is beaten at high speed until it has cooled and stiff peaks have formed.

Italian Meringues

Italian meringues are central to many frostings, cookies, and confections, such as marshmallows. In an Italian meringue, boiling sugar syrup is heated to 240° to 250°F (115° to 122°C). It is then beaten into the egg whites to help them expand in volume and grow fluffy and light. The boiling sugar syrup cooks the proteins in the egg whites so that they become set, which helps make the Italian meringue the most stable type of the three meringues.

PROCEDURE TO PREPARE AN ITALIAN MERINGUE

1. Have the egg whites in the bowl of an electric mixer with the whip attachment ready to beat.
2. A sugar syrup consisting of granulated sugar, corn syrup, and water is brought to a boil. When the temperature reaches 230°F (110°C), begin to beat the egg whites on high speed.
3. When the sugar syrup reaches 240°F (115°C), lower the mixer speed to medium and slowly pour the syrup down the side of the bowl into the egg whites. Do not pour the syrup directly onto the whip or else small clumps of hardened sugar will form.
4. Turn up the mixer speed to high and continue to beat until the mixture is cool and forms stiff peaks.

Table 8–2 Comparison of the Three Types of Meringues

FRENCH MERINGUE	SWISS MERINGUE	ITALIAN MERINGUE
Cream of tartar or an acid is added to egg whites, which are beaten to stiff peaks while sugar is gradually beaten in. This meringue is uncooked. The final texture depends on the ratio of sugar to egg whites.	Egg whites and sugar are warmed to 120°F (49°C) over a hot water bath. It is then removed from the heat and beaten until cool to soft or stiff peaks. The texture depends on the amount of sugar that is added.	A sugar syrup that has boiled and reached 240° to 250°F (115° to 122°C) is added to partially beaten egg whites in a slow, thin stream. The mixture is beaten until it is cool and forms stiff peaks.
Least stable.	More stable.	Most stable.

Chocolate Cookie Soufflés

Makes 4 (6 ounce, ¾ cup; 180 mL) individual soufflés

Lessons demonstrated in this recipe:

- How an egg foam is used to leaven a soufflé.
- The three basic parts of a soufflé.
- How to correctly fold an egg foam into other ingredients while preventing the egg foam from deflating.

MEASUREMENTS			INGREDIENTS
U.S.		**METRIC**	
			nonstick cooking spray
	4 teaspoons	5 g	granulated sugar
3 fluid ounces	6 tablespoons	90 mL	heavy cream
½ ounce	1 tablespoon	15 g	butter
4½ ounces	¾ cup	130 g	semisweet chocolate chips
3 each		57 g	large egg yolks, room temperature
¾ fluid ounce	1½ teaspoons	7½ mL	pure vanilla extract
3 each		84 g	large egg whites, room temperature
1 ounce	2 tablespoons	30 g	granulated sugar
			2 sour cream fudge cookies (3½ ounces; 105 g) (from the recipe in Chapter 9) coarsely crushed in the food processor
1½ ounces	¼ cup	45 g	semisweet chocolate chips
1½ ounces	¼ cup	45 g	white chocolate chips mixed together in a small bowl
	About 2 tablespoons	18 g	confectioners' sugar for dusting

1. Preheat oven to 400°F (205°C).

2. Spray four 6-ounce (¾-cup; 180-mL) capacity soufflé cups with cooking spray. Add 1 teaspoon (1 g) of sugar to each cup and rotate to coat all sides with sugar (Figure 8–12).

3. Prepare the custard base: In a medium saucepan, heat the cream and the butter to a simmer (small bubbles appear around the edge of the pan and steam forms). Simmering is usually just under a boil. Remove the mixture from the heat.

FIGURE 8–12

4. Finishing and flavoring the custard: Whisk the chocolate chips into the hot heavy cream mixture until melted and smooth (Figure 8–13). Allow the custard to cool until lukewarm. Pour chocolate mixture into a medium mixing bowl. Whisk in 3 egg yolks and the vanilla (Figure 8–14).

5. To prepare beaten egg whites: In the bowl of an electric mixer, beat the egg whites with the whip attachment until soft peaks are reached (Figure 8–15). Gradually add granulated sugar and continue to beat until stiff peaks form (Figure 8–16A and B). Do not overwhip or the whites will look dry, like soap suds. Stop the mixer frequently to check the peaks as they stiffen. Stop the mixer and lift the whip out of the whites to see the peaks form. They should stand unwavering like a mountain peak.

6. Add ⅓ of the egg whites into the custard whisking them in thoroughly (Figure 8–17). The mixture is now lightened, so that the custard will not deflate the rest of the egg whites.

FIGURE 8–13

FIGURE 8–14

FIGURE 8–15

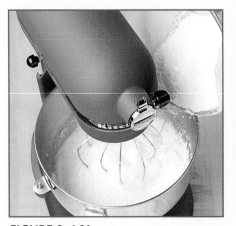

FIGURE 8–16A

FIGURE 8–16B

FIGURE 8–17

FIGURE 8–18

FIGURE 8–19

FIGURE 8–20

> **TIP** To know whether your soufflé cups can hold 6 ounces (¾ cup; 180 mL) of liquid, fill one with that amount of water. If the water reaches the top of the soufflé cup without overflowing, the ramekin has a 6-ounce (¾-cup; 180 mL) capacity.

7. Switch to using a rubber spatula. Gently and swiftly fold the rest of the egg whites into the lightened base (Figure 8–18). Spoon enough batter into each of 4 sprayed sugared soufflé dishes, filling each one halfway (Figure 8–19). Sprinkle the soufflés with half of the cookie crumbs and half of the mixed chips. Spoon the remaining soufflé batter over the cookies and chips, filling to within ¼ inch (6 mm) from the top. Sprinkle the tops of the soufflés with the remaining cookie crumbs, followed by the remaining semisweet and white chocolate chips (Figure 8–20).

8. Place the soufflé cups on a cookie sheet and place in the hot oven. Bake for 16 minutes. They will still shimmy a bit when shaken. Remove them from the oven and use a metal spatula to gently place each soufflé (still in the soufflé cup) on a dessert plate. Dust with confectioners' sugar and serve immediately!

Chocolate Cookie Soufflé

INDIVIDUAL BAKED ALASKAS

Makes approximately 12 3-inch (7½ cm) Individual Baked Alaskas

Lessons demonstrated in this recipe:

- How to prepare a French meringue.
- *Salmonella* is killed using two techniques to prepare a safe meringue: one using pasteurized egg whites and the other using fresh shell egg whites.
- Superfine sugar dissolves more easily, preventing a grittiness to the meringue.
- An acid such as cream of tartar is used to stabilize the meringue.

Baked Alaska, a flaming ice cream cake and meringue dessert, was first invented in the 1920s at a restaurant in New York City named Delmonico's Steak House. Traditionally, Baked Alaska was ice cream packed into a round mold with a layer of sponge cake cut to fit and placed on top. It was frozen solid and then taken out of the mold and flipped over so the sponge cake was on the bottom. A meringue was piped on top and the entire dessert was browned lightly under the broiler or brought flaming to the table after being doused with an alcoholic beverage that had been lit with a match.

STEP A: PLACING THE ICE CREAM ON TOP OF THE CAKE INGREDIENTS

¾ recipe Fudgy Chocolate Cake (Chapter 7) baked in a greased and floured half sheet pan 13 × 18 inches (33 × 46 cm)
12 scoops of Pistachio Ice Cream (Chapter 11) made at least 1 day ahead

1. **Cut 12 3-inch (7½-cm) rounds out of the cake using a round cookie cutter (Figure 8–21). Set the rounds aside on a sheet pan lined with parchment paper. Scraps of cake can be saved for another use.**

2. **Place 1 large scoop of ice cream onto each cake round and freeze until the ice cream is hard (Figure 8–22).**

FIGURE 8–21

FIGURE 8–22

STEP B: PROCEDURE FOR MAKING MERINGUE

Procedure for Making a Meringue Using Pasteurized Egg Whites

MEASUREMENTS			INGREDIENTS
U.S.		METRIC	
6 fluid ounces	¾ cup	180 mL	pasteurized egg whites (the equivalent of 6 large egg whites)
	½ teaspoon	1 g	cream of tartar
10¾ ounces	1½ cups	305 g	superfine sugar
	1 teaspoon	5 mL	vanilla extract

TIP If using pasteurized dried egg whites, be sure to rehydrate them according to the package directions. Be sure that the package states that the egg whites are capable of being whipped to a foam.

1. In the bowl of an electric mixer using the whip attachment, beat the whites and the cream of tartar until the volume has increased and the mixture is thick and foamy.

2. Slowly beat in the sugar ½ to 1 ounce (1 to 2 tablespoons; 15 to 30 g) at a time until stiff peaks form. Beat in the vanilla extract until well incorporated.

3. Remove the ice cream-topped cake rounds with ice cream from the freezer and pipe stars or bands of meringue around the ice cream, using a pastry bag fitted with a large star tip, covering the ice cream completely.

4. Place the meringued cakes in the freezer until they are firm—several hours, or overnight.

Procedure for Making a Safe Meringue Using Fresh Shell Egg Whites

MEASUREMENTS			INGREDIENTS
U.S.		METRIC	
6 each		168 g	large egg whites, left at room temperature for 1 hour
10¾ ounces	1½ cups	305 g	superfine sugar
1 fluid ounce	2 tablespoons	30 mL	water
	½ teaspoon	1 g	cream of tartar
	1 teaspoon	5 mL	vanilla extract

TIP If fresh shell egg whites are used, a procedure is used to destroy any *Salmonella* that may be present in the egg whites. This is done because the egg whites receive no further cooking. Although it looks like a Swiss meringue recipe, it is actually a procedure to cook the egg proteins and prepare a safe meringue.

1. Heat 1 inch (2½ cm) of water in a large saucepan and bring it to a simmer (not quite to a boil). Have a thermometer in a small cup of hot water ready to test the temperature of the egg whites. This helps the thermometer come up to temperature more quickly than if you had placed the thermometer into the egg whites directly. This saves time and prevents the egg whites from overcooking.

2. Place the egg whites, half the sugar, the water, and the cream of tartar into the bowl of an electric mixer. Set over a pot of simmering water. Make sure the bottom of the mixing bowl does not touch the water. Whisk constantly until a thermometer placed into the mixture reaches 160°F (71°C) (Figure 8–23A and B).

3. When the whites reach 160°F (71°C), remove the bowl from the water and place it on the base of the electric mixer. Using the whip attachment, beat until the mixture becomes very thick and foamy.

4. Slowly add the remaining sugar, about ½ to 1 ounce (1 to 2 tablespoons; 15 to 30 g) at a time until stiff peaks form. Add the vanilla extract. Beat until well incorporated.

5. Remove the cake rounds with ice cream from the freezer and pipe stars or bands of meringue around the ice cream, using a pastry bag fitted with a large star tip, covering the ice cream completely.

6. Place the meringued cakes in the freezer until they are firm—several hours, or overnight.

STEP C: ASSEMBLY

Just before serving, preheat the broiler and broil the Baked Alaskas for only a few moments to get the top of the meringue lightly browned. Make sure the top of the broiler is several inches higher than the oven rack so the Baked Alaskas do not burn. A blow torch can be used instead of the broiler (Figure 8–24). Serve at once.

Variation: Baked Meringues

Either one of the above meringue recipes can be baked on their own after being piped onto a parchment-lined sheet pan and served as a cookie. Pipe into walnut-sized mounds onto the prepared sheet pan and bake at 225°F (108°C) for approximately 1 hour or until firm to the touch, but not browned. Remove and cool completely. They can be stored in an airtight container for up to 2 days.

FIGURE 8–23A

FIGURE 8–23B

FIGURE 8–24

Individual Baked Alaska

MARSHMALLOWS

Makes 110 1- to 1½-inch
(2.5- to 3.75-cm) diameter
marshmallows

Lessons demonstrated in this recipe:

- How to prepare an Italian meringue stabilized with gelatin.
- A hot sugar syrup cooks the egg proteins, creating the most stable of the three meringue types.
- Air beaten into the meringue triples its volume.

MEASUREMENTS			INGREDIENTS
U.S.		METRIC	
⅔ ounce	2 tablespoons	20 g	unflavored powdered gelatin
2⅔ fluid ounces	⅓ cup	80 mL	cold water
9 ounces	1¼ cups	270 g	granulated sugar
2 fluid ounces	¼ cup	60 mL	light corn syrup
2⅔ fluid ounces	⅓ cup	80 mL	water
2 each		56 g	large egg whites placed in a small bowl and left at room temperature for 1 hour, or placed in warm water bath to bring to room temperature more quickly
	2 teaspoons	10 mL	pure vanilla extract, or 1 teaspoon (5 mL) vanilla extract and 1 teaspoon (5 mL) coconut extract
			shredded coconut, toasted
			finely chopped semisweet chocolate
			finely chopped pistachio nuts
			confectioners' sugar for dusting

1. Place 2 silicone baking mats over 2 large sheet pans and set aside.

2. Soften the powdered gelatin by sprinkling it over 2⅔ fluid ounces (⅓ cup; 80 mL) cold water in a small bowl, stirring and allowing it to bloom and swell into a jelly-like mass.

3. In the meantime, combine the sugar, light corn syrup, 2⅔ fluid ounces (⅓ cup; 80 mL) water in a heavy saucepan. Bring to a boil, stirring just until the sugar is dissolved. Stop stirring and use a pastry brush dipped in cold water to wash down the sides of the pan if sugar crystals form.

4. Place a candy thermometer into the syrup and boil undisturbed (no stirring) (Figure 8–25). As the water in the syrup comes to its boiling point, the syrup will bubble up quickly. If this happens, turn down the heat to medium and then, as the water boils off, turn the heat back up to high. As the temperature of the sugar syrup climbs to 230°F (110°C), place the egg whites in the bowl of an electric mixer fitted with the whip attachment. Beat the whites on high speed until thick and foamy.

5. Continue beating at high speed and wait until the sugar syrup in the saucepan reaches 240°F (115°C). Turn down the speed of the beater if the whites are getting past the soft peak stage. When the sugar syrup reaches 245°F (118°C), add the softened gelatin into the hot sugar syrup and whisk quickly to melt (Figure 8–26). Remove the mixture from the heat.

6. Slowly pour the hot sugar syrup into the mixing bowl (pouring to the side of but not directly onto the beater) while the beater is turned on medium speed (Figure 8–27). After all the sugar syrup has been added, turn the mixer up to high speed. Do not be concerned if some sugar crystals form as they hit the beater and get splattered onto the sides of the bowl. Add vanilla or coconut extract. Beat until the volume of the mixture triples (Figure 8–28). This should take about 6 to 10 minutes. The mixture will look soft and fluffy.

7. Immediately scoop the mixture into a large pastry bag fitted with a large round or star tip and pipe out large walnut-sized rounds about 1 to 1½ inches (2.5 to 3.75 cm) onto the prepared sheet pans (Figure 8–29).

FIGURE 8–25

FIGURE 8–26

FIGURE 8–27

FIGURE 8–28

FIGURE 8–29

8. Sprinkle each marshmallow with a pinch of toasted coconut, chopped chocolate, or pistachios, if desired. Leave the marshmallows to cool at room temperature, uncovered until firm, at least 3 hours, or overnight.

9. Marshmallows are best eaten up to 1 day after they are made. However, they can be stored in an airtight container in between sheets of parchment paper that have been dusted with confectioners' sugar for up to 2 weeks.

Marshmallows

Professional Profile

BIOGRAPHICAL INFORMATION

Bill Lassiter, CEC, CCE
Chef Instructor
Central Piedmont Community College
Charlotte, NC

1. Question: *When did you realize that you wanted to pursue a career in baking and pastry?*

Answer: *I was 16 or 17 years old and took a part-time job cooking at the college dining hall in Chapel Hill. In those days it was all done by scratch and I got to work in everything from hot food to baking. The fact that the dining hall was where all the basketball players—my heroes—ate, made it the best place in the world for me.*

2. Question: *Was there a person or event that influenced you to go into this line of work?*

Answer: *It was the similarities between cooking and sports that did it for me. In both, you have to be quick on your feet. Being in charge in the kitchen is like calling plays in the huddle. To be great in sports and to be great in any of the culinary areas you have to be able to improvise and be creative.*

3. Question: *What did you find most challenging when you first began working in baking and pastry?*

Answer: *I had experience on the line but I really wanted to learn baking and pastry. Development of the basic skills took some time but I really wanted to master them. I realized that I couldn't hesitate. I just had to get in there and do it.*

4. Question: *When and where was your first practical experience in a professional baking and pastry setting?*

Answer: *I was working the line at a fine dining restaurant in my home town. I saw the manager doing some writing on a cake and it was beautiful. I said to myself, "I want to do that." The manager encouraged me and taught me the basics. When I went to culinary school I had this great training already and it put me ahead of the other students.*

5. Question: *How did this experience affect your later professional development?*

Answer: *It impressed on me the fact that you can always go further. It instilled in me the desire to take desserts to a higher level . . . to keep striving. If you find something you want to do you have to give it a try.*

6. Question: *Who were your mentors when you were starting out?*

Answer: *Matt Henney, the manager at a fine dining restaurant, was an early mentor. He could do just about anything from front-of-the-house to back-of-the-house. Later, I began to really admire the work of Ewald Notter and his team. They are just incredible. I have the utmost respect for them.*

7. Question: *What would you list as the greatest rewards of your professional life?*

Answer: *The reward is in being able to pass on your knowledge. If you don't have that innate desire to share what you know then you can't enjoy being in education.*

8. Question: *What do you consider essential traits for anyone entering the field?*

Answer: *Anyone who wants to be successful in this area has to have a sense of urgency. There is no "later." If you aspire to greatness you must also remember that image is everything. Dress the part. Believe in the part.*

9. Question: *If there was one message you would impart to all students in this field what would it be?*

Answer: *Work smarter, not harder. When you watch the very best in the field you see that it is the way that they are doing things as much as the outcome that is important. The difference between burning out and keeping that creative spark going is how you work. Networking, too, is essential if you want to make your career in this field.*

CHAPTER 9

Cookies

After reading this chapter, you should be able to:

- Define a cookie.

- List the eight categories of cookies.

- Explain the three basic cookie mixing methods.

- Manipulate ingredients to attain desirable cookie characteristics.

- Prepare the recipes at the end of this chapter.

Memories of cookie jars full of homemade cookies always seem to evoke feelings of caring and love. Everyone loves cookies and there are about as many cookie recipes and varieties of cookies as there are personalities in people. Cookies can range anywhere from soft and chewy to crisp and crunchy.

Cookies are defined as a diverse group of small, sweet cakes or pastries that are described and categorized by how the dough is prepared and shaped for baking.

Cookies are similar to cakes in several respects. Just as with cakes, some categories of cookies rely on chemical leaveners to help them rise and some cookie recipes share certain of the same cake mixing methods, such as the creaming, one-bowl, and egg-foam methods.

Although there are clear-cut mixing methods for cakes, the same is not always true for cookies. There are so many varieties of cookies and preparation techniques that recipes can vary greatly even among cookies within the same category. It is this diversity of recipes within each category that makes cookies so much fun to prepare and so popular.

This chapter discusses the different categories of cookies and how to manipulate ingredients to attain desirable cookie characteristics.

Categories of Cookies

There are eight categories of cookies, but for the professional pastry chef, just a few of these categories can be mass produced easily without being too labor intensive. The eight categories of cookies are:

- drop cookies
- refrigerator cookies
- molded cookies
- bar cookies
- sheet cookies
- rolled cookies

Drop Cookies

Drop cookies tend to be made from a soft, moist dough that is dropped from a spoon or ice cream scooper onto a sheet pan. Drop cookies need to be spaced out to allow them to spread during baking (e.g., Sour Cream Fudge Cookies).

Refrigerator Cookies

Refrigerator cookies tend to be made from stiffer doughs. The dough is rolled into logs, or other shapes, then wrapped in plastic or parchment paper and chilled or frozen before being sliced and baked (e.g., Brown Sugar Pecan Refrigerator Cookies).

Molded Cookies

Molded cookies are made from a moderately stiff dough that can be molded and shaped. For example, molded cookie dough may be rolled into balls and coated with sugar or chopped nuts.

The balls are then placed on a sheet pan and a weight of some sort, such as a flat-bottomed drinking glass, is pressed onto the balls to mold them into flattened circles. Other molded cookies may be left to bake as balls that are not flattened (e.g., Truffled Peanut Butter Cookies).

Bar Cookies

Bar cookies consist of a stiff dough formed into long, flattened, rectangular bars. These are baked and then cut into slices. Biscotti is a bar cookie that is sliced and baked again for a short time to make each slice crunchy (e.g., Espresso Almond Biscotti I and Espresso Almond Biscotti II).

Sheet Cookies

Sheet cookies are one of the easiest and least labor intensive of any category. A batter, which can be thin or stiff, is made up and evenly spread into a pan with sides. Some recipes include a cookie or crumb base that is pressed onto the bottom of a pan before a batter is poured over it. After baking, the cookies are cut into squares or even triangles or diamonds. There are endless varieties (e.g., Fudge Brownies).

Rolled Cookies

Rolled cookies tend to be made from stiff dough that, very often, needs chilling to harden the fat within the dough. The dough is then rolled out and cookie cutters are used to cut the dough into shapes before baking (e.g., Golden Coconut Cutouts).

Piped Cookies

Piped cookies are made of soft doughs that can be easily pushed through a pastry bag fitted with a pastry tip to form cookies of various shapes (e.g., Citrus Butter Rings and Chocolate Almond Lady Fingers).

Wafer Cookies

Wafer cookies, also known as *tuiles* (French for "roof tiles"), are much more labor intensive. A thin batter, consisting of egg whites, sugar, flour, butter, and sometimes heavy cream, often referred to as a *tulipe* or stencil batter, is spread over stencils placed on a silicone baking mat or parchment-covered sheet pan or dropped from a spoon and spread into a circle. This stencil is carefully removed and the cookies are baked. The cookies retain their shape without spreading. They bake very quickly and must be monitored carefully so that they do not burn. While hot, they are very malleable and can be molded over cups, cut into shapes, or layered over a rolling pin to create curves and other shapes. As the cookie cools, the melted sugar within the batter recrystallizes, creating a very thin, crisp, delicate cookie. Often, wafer cookies are so thin they are translucent (e.g., Polka-Dotted Pirouettes).

There is another type of wafer cookie batter that spreads during baking. This type of batter contains corn syrup, brown sugar, butter, flour, and sometimes nuts. It is prepared on the stove and the batter is spooned onto sheet pans. Each mound of batter must be spaced out because the batter spreads. While the cookies are still warm, they may be trimmed or a cookie cutter can be used to cut out various shapes before the cookie cools and hardens. These cookies can also be molded around various objects such as bowls and wooden spoons.

The Creaming Method

1. Cream fat and sugar until well combined in an electric mixer using the paddle attachment.

2. Add eggs gradually.

3. Blend in the dry ingredients until just combined.

4. Fold in solid ingredients such as nuts, chocolate chips, and raisins by hand using a rubber spatula or spoon.

Three Basic Cookie Mixing Methods

Cookies are really nothing more than small cakes, so the mixing methods for preparing cookies are very similar to those for mixing cakes (Chapter 7). These methods include the creaming method, the one-bowl method, and the egg-foam method.

Because there are so many variations of cookies, it is sometimes difficult to identify a mixing method within a specific category. For example, Chocolate Almond Lady Fingers are piped onto a sheet pan. They are prepared in a similar manner to a sponge cake in that they rely on beaten eggs, yolks, and/or egg whites for leavening. So how are they to be categorized? They are a piped cookie that uses the egg-foam method of mixing. (*Note:* Not all cookie recipes have a clearly defined mixing method.)

Creaming Method

The creaming method is the same one used for cakes in Chapter 7. A solid fat (e.g., butter or shortening) is blended in the bowl of an electric mixer with some form of sugar and mixed with the paddle attachment until the mixture is well blended. Air bubbles that are beaten into the butter become enlarged when combined with chemical leaveners. It is important to read directions carefully because overcreaming and incorporating too much air will cause cookies to puff, rise too much and then fall. Undercreaming and simply blending the fat and the sugar yields a more compact, denser cookie that will not spread as much. Recipes for cookies using the creaming method vary greatly.

One-Bowl Method

The one-bowl method is the easiest mixing method for cakes and is also used successfully for cookies. All ingredients are placed in the bowl of an electric mixer and blended using the paddle attachment. Avoid overmixing the ingredients as this will cause gluten to develop. On the other hand, in recipes in which chewiness is a desirable characteristic, gluten development may be encouraged.

Egg-Foam Method

The egg-foam method used for cookies is similar to the egg-foam method used for cakes. Whole eggs, yolks, or whites are beaten to a foam using the whip attachment of an electric mixer and then folded gently into other ingredients. Care must be taken to maintain as much air as possible within the batter. Warmer eggs at room temperature make superior egg foams.

Understanding the Characteristics of Cookies and How to Manipulate Them

Some bakers may prefer to tweak or fiddle with a recipe to make it their own unique version. Perhaps their customers crave a crisper chocolate chip cookie versus a softer version or a thicker cookie compared to a thinner one. How can a baker manipulate a recipe to produce cookie characteristics that are customized to his or her needs?

The answer lies in understanding what factors produce certain characteristics within a cookie. There is a certain amount of science in baking, and by understanding the role each ingredient plays, the baker can manipulate the ingredients to produce chosen characteristics in the final product. *Example:* Look at the chocolate chip cookie recipes in Table 9–1. The recipe on the left produces a chewy cookie. The same recipe with a few minor changes in the ingredients on the right will produce a crisper version of the same cookie.

The One-Bowl Method

1. Place dry ingredients into the bowl of an electric mixer using the paddle attachment.

2. Combine liquid ingredients into the dry ingredients.

3. Fold in solid ingredients such as nuts, chocolate chips, or raisins by hand using a rubber spatula or spoon.

The Egg-Foam Method

1. Beat egg yolks and part of the sugar in an electric mixer using the paddle attachment until thick and light in color.

2. Blend dry ingredients into the egg yolk mixture.

3. Beat egg whites with the remaining sugar in the bowl of an electric mixer using the whip attachment until stiff peaks form.

4. Whisk a small portion of the egg whites into the yolk mixture to lighten it.

5. Fold the remaining egg whites into the batter using a rubber spatula.

Table 9–1 Comparison between a Chewy and Crispy Chocolate Chip Cookie Recipe

CHEWY	CRISPY
4 ounces (115 g) unsalted butter, softened	9 ounces (255g) unsalted butter, softened
4 ounces (115 g) solid vegetable shortening	
8 ounces (225 g) light brown sugar	3 ounces (85 g) light brown sugar
4 ounces (115 g) confectioners' sugar	8 ounces (225 g) granulated sugar
13 ounces (370 g) bread flour	13 ounces (370 g) bread flour
1 teaspoon (6 g) baking powder	
¼ teaspoon (1 g) baking soda	1 ½ teaspoons (9 g) baking soda
1 teaspoon (6 g) salt	1 teaspoon (6 g) salt
2 large eggs	1 large egg
	1 large egg white
1½ teaspoons (7.5 mL) vanilla extract	1½ teaspoons (7.5 mL) vanilla extract
12 ounces (340 g) semisweet chocolate chips	12 ounces (340 g) semisweet chocolate chips
Bake at 375°F (190°C) for 10 to 12 minutes.	Bake at 350°F (175°C) for 12 to 14 minutes.
For a chewier cookie:	**For a crisper cookie:**
Some solid vegetable shortening will decrease spread, producing a chewier cookie.	Using a fat with a low melting point such as butter will cause more spread.
	High granulated sugar content enhances crispness.
The protein in the eggs will produce a chewier cookie through the coagulation of proteins.	One less egg decreases chewiness.
Baking powder will help the cookie to rise, producing a thicker cookie, while a small amount of baking soda neutralizes some of the acidity of the brown sugar, allowing some browning but not too much.	
Bread flour is used to absorb more water, reducing spread and increasing chewiness.	Bread flour is used to absorb moisture for a drier, crisper cookie.
Confectioners' sugar prevents spread resulting in a thicker cookie.	Replacing the liquid from the additional egg with an egg white will produce a drier, crisper cookie.
More brown sugar will keep the cookie softer after baking.	A large amount of baking soda will neutralize any acidity from the brown sugar and allow the cookie to brown and crisp.
	Making the cookies thinner and smaller will increase crispness.
Underbaking produces a chewier, softer cookie.	Baking at a lower temperature for a longer time will allow more spread before the cookie sets, producing a crisper cookie.

When describing characteristics of a cookie, desirable adjectives like "crisp," "soft," "chewy," "brown," "pale," "thin," and "thick" come to mind. Generally, cookies do not have only one characteristic such as "soft" or "crisp"; often they share many characteristics within the same type of cookie. The following discussion explores how these adjectives can be translated directly into a recipe using specific ingredients.

Crisp

A crisp cookie is generally produced from a dough with little moisture or liquid in it. It also contains a large amount of granulated sugar and fat. Wafer cookies have more liquid batters but they contain high amounts of sugar and fat that help them to become crisp as they cool and the sugar re-crystallizes. Cookies baked with large amounts of butter versus shortening tend to spread out more, creating thinner and crisper cookies. This is because butter has a lower melting point. Flours containing a higher protein content absorb more liquid from the dough, thereby producing a crisper cookie. Keep the cookie thin. Rolled-out cookies should be no thicker than ¼ inch (6 mm) and refrigerator cookies should be sliced very thin to produce the most crispness. Higher amounts of baking soda in a cookie dough will also produce a crisper cookie. The baking soda weakens gluten strands, causing the cookie to spread out more during baking. More spread means more surface area of the dough (batter) is exposed to the heat; this causes the moisture to evaporate and results in a crisper cookie.

Soft

Soft cookies and cakes have a similar texture. Using hygroscopic sugars such as brown sugar, molasses, corn syrup, and honey that easily absorb moisture from the air, produces a softer cookie much like cake. A lower granulated sugar and fat content will also soften a cookie. A batter that has some acidity in it will not spread out or brown as well and will produce a softer cookie. Adding acidic ingredients such as sour cream or yogurt helps toward this end. A cookie that is slightly underbaked will also be softer. Softer cookies are created when a low-protein flour is used because low-protein flours do not bind with or absorb as much water as a high-protein flour. This unbound moisture or water in the dough forms steam in the oven, causing the cookies to become puffy and soft.

Chewy

Cookies made with high-protein flour are chewy because of the gluten development. Cookie doughs that contain greater amounts of liquid ingredients and granulated sugar with less fat also produce chewiness. Using more eggs causes the proteins to coagulate, yielding a chewier cookie.

Brown

The easiest way to increase the browning of a cookie is to increase the oven temperature or the baking time. Cookies containing only a small amount of corn syrup (containing glucose not fructose) brown at a lower temperature than granulated sugar. Browning occurs also when a higher proportion of baking soda is used. Acidic batters do not brown well and baking soda, a base, neutralizes the acidity of the dough, allowing the cookie to brown. Flours that are high in protein also produce cookies that brown better than flours lower in protein.

Pale

Underbaking is one way to prevent a cookie from browning. To prevent a cookie from coloring, acidity in the dough must be maintained. To maintain this acidity, the amount of baking soda is kept to a minimum while the amount of baking powder is increased. Baking powder, which already contains an acid, does not neutralize any acidity in the batter and browning is inhibited. The lower the protein content of a flour used in a cookie dough, the less browning occurs. Flours that are bleached and acidic, such as cake flour, also produce paler cookies.

Spread

Varying ingredients of a cookie dough can also affect how much the dough will spread in the oven.

Increased Spread

A thinner cookie dough with a great deal of liquid in it produces a cookie with more spread. A high amount of granulated sugar also increases spread. Using a higher amount of baking soda weakens gluten strands, neutralizes acidity, and causes the cookie to spread. Flours with low protein produce more spread. For cookies using the creaming method of mixing, overcreaming creates more spread. Using a fat with a lower melting point, such as butter, produces a cookie with more spread. Baking at a lower oven temperature gives the dough more time to spread. Placing the dough on a greased cookie sheet also increases spread.

Decreased Spread

Using flours with high-protein contents decreases spread by binding with more liquid within the dough, creating more structure. When using the creaming method, cream the fat and sugar only until combined and until there are no lumps. The less air incorporated into the mixture, the less the cookie dough will puff up and spread out. Using confectioners' sugar instead of granulated sugar also decreases spread because confectioners' sugar contains cornstarch, which absorbs moisture and produces a drier, stiffer dough. Using baking powder, which leavens without decreasing acidity, reduces spread because acidic doughs set more quickly. Baking at higher temperatures reduces spread because the cookie dough sets faster. Placing the cookie dough on an ungreased sheet pan reduces spread. Thicker doughs using a fat with a higher melting point such as vegetable shortening produce a cookie with less spread. Chilling the dough before baking also decreases spread.

Table 9–2 illustrates how different ingredients can achieve desired characteristics.

Tips for Making Successful Cookies

◼ While they are still slightly warm, remove cookies from sheet pans that have been greased or parchment lined using an offset spatula.

◼ Do not overbake cookies. Often, cookies appear soft and underbaked after the allotted time in the oven, but they firm up as they cool.

◼ For cookies that are not meant to rise, it is important for the dough to remain the same shape after baking as it was when it went into the oven. Because little or no chemical leaveners are used in these recipes, it is important not to overmix the dough, which would cause air to be incorporated and the cookie to rise.

Table 9–2 Desired Characteristics of Cookies Using Various Ingredients

USE	CRISP	SOFT	CHEWY	BROWN	PALE	INCREASED SPREAD	DECREASED SPREAD
High-protein flour	✓		✓	✓			✓
Low-protein flour		✓			✓	✓	
Eggs			✓				
Egg whites	✓						
Less liquid	✓						✓
More liquid		✓	✓			✓	
Baking powder (does not neutralize acidity)		✓			✓		✓
Baking soda (neutralizes acidity)	✓			✓		✓	
Less granulated sugar and fat		✓					
More granulated sugar, less fat			✓				
More granulated sugar and fat	✓					✓	
Hygroscopic sugars		✓					
Confectioners' sugar							✓
Fats with high melting points							✓
Fats with low melting points						✓	

Note: The same ingredients may affect a cookie's characteristics in different ways depending on the other ingredients added to them.

SOUR CREAM FUDGE COOKIES

Makes approximately 3 dozen cookies

(*Note:* To make giant cookies, use a 2-ounce [¼ cup; 60-mL] ice cream scoop and bake the cookies for 2 minutes longer.)

Lessons demonstrated in this recipe:

- ▪ How to prepare a drop cookie.
- ▪ The creaming method of mixing provides air bubbles that, in conjunction with chemical leaveners, help the cookies to rise.
- ▪ Baking soda neutralizes acidic ingredients like brown sugar and sour cream to create carbon dioxide gas for leavening.
- ▪ Decreasing the acidity of the batter allows better browning.
- ▪ A high-protein flour increases chewiness.
- ▪ Sour cream adds richness and tenderness.
- ▪ A higher quantity of sugar with less fat creates chewiness.

MEASUREMENTS				INGREDIENTS
U.S.		METRIC	BAKER'S %	
6¼ ounces	1¼ cups	175 g	100%	bread flour
	1 teaspoon	4 g	2.3%	baking powder
	1 teaspoon	4 g	2.3%	baking soda
	½ teaspoon	3 g	1.7%	salt
2½ ounces	¾ cup	70 g	40%	Dutch processed cocoa powder
9 ounces	1½ cups	255 g	146%	semisweet chocolate, chopped
3 ounces		85 g	49%	unsweetened chocolate, chopped
3 ounces	6 tablespoons	85 g	49%	unsalted butter, softened
5½ ounces	¾ cup	155 g	89%	granulated sugar
5½ ounces	¾ cup	155 g	89%	light brown sugar (packed, if measured by volume)
2 each		94 g	54%	large eggs
4 ounces	7 tablespoons	115 g	66%	sour cream
	1 teaspoon	5 mL	2.9%	pure vanilla extract
9 ounces	1½ cups	255 g	146%	semisweet chocolate chips
				nonstick cooking spray
			837.2%	Total Sour Cream Fudge Cookies percentage

1. Preheat the oven to 325°F (163°C). Sift together the flour, baking powder, baking soda, salt, and cocoa powder into a bowl. Whisk the mixture and set it aside.

2. Melt the semisweet and unsweetened chocolate in a double boiler or in a bowl placed over a pot of simmering water. Cool to room temperature and set aside.

3. In the bowl of an electric mixer using the paddle attachment, cream the butter and sugars together on medium speed for approximately 1 to 2 minutes, or until the mixture becomes somewhat light (Figure 9–1).

FIGURE 9–1

FIGURE 9–2

FIGURE 9–3

4. On low speed, add the eggs, one at a time, incorporating each one well before adding the next. Stop the machine occasionally and scrape down the sides of the bowl.

5. With the machine still on low speed, slowly add the cooled melted chocolate and blend well (Figure 9–2). Stop the mixer and scrape down the sides of the bowl with a rubber spatula. Add the sour cream and vanilla extract, blending well (Figure 9–3).

6. On low speed, add the flour and cocoa mixture and blend only until combined (Figure 9–4). Remove the bowl from the electric mixer and, using a rubber spatula or metal spoon, blend in the chocolate chips by hand.

7. Using a 1-ounce (⅛ cup; 30-mL) ice cream scoop, place scoopfuls of batter onto a sheet pan covered with parchment paper; space each scoopful 3 inches (7.5 cm) apart (Figure 9–5). With a moistened hand, flatten each cookie slightly (Figure 9–6). Bake for about 18 minutes. The fudge cookies will be soft but will firm up as they cool. Allow the cookies to cool on the sheet pan for 5 minutes before removing and placing them on cooling racks.

FIGURE 9–4

FIGURE 9–5

FIGURE 9–6

BROWN SUGAR PECAN REFRIGERATOR COOKIES

Makes approximately 100 cookies

Lessons demonstrated in this recipe:

- How to prepare a refrigerator cookie using the creaming method.
- A small amount of brown sugar adds flavor without softening the cookie.
- Confectioners' sugar prevents too much spread.
- Baking soda neutralizes any acidity from the brown sugar, producing a browner and crisper cookie.
- Using some bread flour reduces spread while producing more crispness.
- Chilling the dough reduces spread.
- Ground pecans add flavor.

MEASUREMENTS				INGREDIENTS
U.S.		**METRIC**	**BAKER'S %**	
8 ounces		226 g	34%	unsalted butter
3¾ ounces	½ cup	106 g	16%	light brown sugar (packed, if measuring by volume)
3 ounces	¾ cup	85 g	13%	confectioners' sugar
12¾ ounces	2 ½ cups + 2 teaspoons	360 g	54%	bread flour
10¾ ounces	2 ½ cups + 2 ½ tablespoons	305 g	46%	cake flour
	1 teaspoon	6 g	0.9%	salt
	1 teaspoon	4 g	0.6%	baking soda
5½ ounces	1½ cups	160 g	24%	pecans, finely ground
2 each		94 g	14%	large eggs
2 teaspoons		10 mL	1.5%	vanilla extract
				1 large egg white to be used as glue
				pecan halves for garnish
			204%	Total Brown Sugar Pecan Refrigerator Cookies percentage

FIGURE 9–7

FIGURE 9–8

FIGURE 9–9

1. Preheat the oven to 375°F (190°C). In the bowl of an electric mixer using the paddle attachment, cream the butter and sugars until the mixture just begins to lighten in color.

2. In another bowl, whisk together the flours, salt, baking soda, and pecans. Set aside.

3. Slowly add the eggs and vanilla to the butter and sugar mixture and mix on low speed until well blended.

4. On low speed, add the flour mixture, mixing until just combined.

5. Divide the dough in half.

(Note: After step 6, the logs of dough can be wrapped airtight and stored in the freezer for 2 to 3 months. Thaw the dough in the refrigerator before slicing.)

6. Form each half into a log 13 inches (33 cm) long (Figure 9–7). Wrap each log in plastic wrap and chill for at least 3 hours or until firm.

7. Slice each log crosswise into ¼-inch (6-mm) thick slices and place them on a sheet pan covered with parchment paper (Figure 9–8). Using a pastry brush, brush some egg white in the middle of each cookie. Place 1 pecan half on top of each cookie, pressing gently to adhere (Figure 9–9).

8. Bake the cookies for about 9 to 12 minutes or until the bottoms and edges are light brown. Allow the cookies to cool on the sheet pan for 5 minutes before removing and placing them on cooling racks.

TRUFFLED PEANUT BUTTER COOKIES

Makes approximately
40 3-inch (7.5-cm) cookies

Lessons demonstrated in this recipe:

- How to prepare a molded cookie using the creaming method.
- The high proportions of fat in the form of butter and peanut butter add to the tenderness of the cookie.
- Equal parts of granulated and brown sugars balance out the crispness, so the cookie is firm, yet tender.
- A high-protein flour binds with liquid in the dough, forming some chewiness and less spread.
- The whole-wheat flour adds a nutty texture, playing up the peanut butter theme while adding to the nutritional value.
- The addition of baking soda neutralizes the acidity in the brown sugar, resulting in better browning.
- Adding a larger quantity of baking powder helps the cookies to puff up and rise.
- Chilling the dough before baking prevents spread, forming a thicker cookie.

STEP A: TRUFFLE FILLING

MEASUREMENTS			INGREDIENTS
U.S.		METRIC	
8 ounces	1 cup	240 mL	heavy cream
24 ounces	4 cups	680 g	semisweet chocolate, finely chopped
	1 teaspoon	5 mL	vanilla extract

1. In a heavy saucepan, bring the cream to a boil. Remove it from heat.

2. Add the chocolate to the cream and stir the mixture gently with a whisk until the chocolate melts. Whisk in the vanilla and blend well.

3. Pour the truffle mixture into a 15-inch by 10-inch (38-cm by 25-cm) rectangular pan with sides lined with foil (Figure 9–10). Chill the truffle mixture for approximately 2 hours or until it is very firm.

4. Remove the pan from the refrigerator and pull the foil with the truffle mixture out of the pan and place on a work surface. Using a 1¾-inch (4.5-cm) fluted, round cookie cutter, cut out as many rounds as possible (Figure 9–11). Reserve the rounds on a parchment-lined sheet pan. Chill the rounds until needed.

FIGURE 9–10

FIGURE 9–11

FIGURE 9–12

FIGURE 9–13

FIGURE 9–14

STEP B: PEANUT BUTTER COOKIES

MEASUREMENTS				INGREDIENTS
U.S.		**METRIC**	**BAKER'S %**	
8 ounces		225 g	58%	unsalted butter at room temperature
7 ounces	¾ cup + 2½ tablespoons	200 g	51%	granulated sugar
7½ ounces	1 cup	212 g	54%	light brown sugar (packed, if measuring by volume)
9 ounces	1 cup + 2 tablespoons	255 g	65%	smooth peanut butter
10¼ ounces	2 cups + 1 tablespoon	290 g	74%	bread flour
3½ ounces	¾ cup	100 g	26%	whole-wheat flour
	1 tablespoon	12 g	3.1%	baking powder
	1 teaspoon	4 g	1.0%	baking soda
	½ teaspoon	3 g	0.8%	salt
2 each		94 g	24%	large eggs
	2 teaspoons	10 mL	2.6%	vanilla extract
6 ounces	1¼ cups	175 g	45%	salted, roasted peanuts
				granulated sugar for rolling
			404.5%	Total Truffled Peanut Butter Cookies (Step B) percentage

1. Preheat the oven to 375°F (190°C). In the bowl of an electric mixer using the paddle attachment, cream the butter, sugars, and peanut butter for 1 to 2 minutes or until the mixture lightens in color.

2. In another bowl, thoroughly whisk together the flours, baking powder, baking soda, and salt; set the mixture aside.

3. On low speed, add the eggs, one at a time, to the peanut butter mixture, blending to incorporate one egg before adding the next. Add the vanilla extract.

4. On low speed, add the flour mixture to the peanut butter and blend until just combined. Add the peanuts. Scrape the dough into a large bowl. Cover and chill for at least 2 hours or overnight.

5. Using a 1-ounce (30-mL) ice cream scoop, scoop the dough into rough ball shapes. Smooth each ball by rolling it between the palms of your hands. Roll each ball into a bowl of granulated sugar and place them spaced 2 inches (5 cm) apart on a parchment lined sheet pan (Figure 9–12).

6. Using the flat bottom of a drinking glass or a solid measuring cup, flatten each ball into a circle (Figure 9–13).

7. Bake for 10 to 12 minutes or until the cookies are very light brown. Do not overbake.

8. Remove the cookies from the oven and immediately place 1 of the reserved chocolate truffle disks in the center of each cookie, pressing down slightly to help them adhere (Figure 9–14). Allow the cookies to cool for 5 minutes before removing them from the sheet pan and cooling on racks.

ESPRESSO ALMOND BISCOTTI I

Makes approximately 36 to 40 biscotti

Biscotti in Italian means "twice baked." First, a log of dough is baked and sliced. The slices are then baked again for a shorter time just to lightly toast them. Traditionally, biscotti cookies are so hard that the only way to eat them without breaking teeth is to dip them into wine or coffee. The first biscotti recipe results in a more traditional, hard biscotti whereas the second biscotti recipe that follows results in a crisp, yet more tender and crumbly cookie that will not break teeth.

Lessons demonstrated in this recipe:

- How to prepare a bar cookie that is twice baked using the one-bowl method.
- Very little liquid in the dough produces crispness.
- A high-protein flour absorbs any liquid, creating a crisper cookie.
- A simple one-bowl method of mixing is used.
- Coarsely ground coffee beans and lemon zest produce a cookie that mimics the beverage for which this cookie is named. Orange zest is added for extra citrus flavor.
- A small amount of baking powder is used for leavening and a larger amount of baking soda is used to neutralize any acidity, which aids browning and crisping.
- Little fat in the dough produces a hard cookie.
- The crisp texture is beneficial in that whole almonds can be used to produce attractive slices that will not crumble when sliced.
- Cutting the logs into thinner slices will produce a crisper biscotti when baked again.

MEASUREMENTS				INGREDIENTS
U.S.		METRIC	BAKER'S %	
10 ounces	2 cups	285 g	80%	bread flour
2½ ounces	½ cup	70 g	20%	whole wheat flour
3¾ ounces	½ cup	85 g	24%	light brown sugar (packed, if measuring by volume)
3½ ounces	½ cup	100 g	28%	granulated sugar
	1 teaspoon	6 g	1.7%	grated lemon zest
	1 teaspoon	6 g	1.7%	grated orange zest
	2 tablespoons	10 g	2.8%	instant espresso powder
¼ ounce	2 tablespoons	10 g	2.8%	coffee beans, coarsely ground
	1 teaspoon	4 g	1.1%	baking soda
	½ teaspoon	2 g	0.6%	baking powder
	½ teaspoon	3 g	0.8%	salt
3 each		140 g	39%	large eggs
½ fluid ounces	1 tablespoon	15 mL	4.2%	vegetable oil
	1 teaspoon	5 mL	1.4%	vanilla extract
	½ teaspoon	2.5 mL	0.7%	almond extract
5 ounces	1 cup	140 g	39%	whole natural almonds (with the skin on), lightly toasted
	1 teaspoon	5 mL	1.4%	egg wash, as needed
				granulated sugar for sprinkling
			249.2%	Total Espresso Almond Biscotti I percentage

1. Preheat the oven to 325°F (163°C). In the bowl of an electric mixer using the paddle attachment, blend the flours, sugars, the zest, espresso powder, ground coffee beans, baking soda, baking powder, and salt on low speed until combined.

2. On low speed, add the eggs, one at a time, then the oil followed by the vanilla and almond extracts. Mix just enough to form a dough.

3. Using a rubber spatula or a spoon, blend in the almonds by hand.

4. Form the dough into two logs approximately 15 inches (37.5 cm) long by 3½ inches (8.7 cm) wide, and place them on a parchment-covered sheet pan, leaving 4 inches (10 cm) between them.

5. Brush the logs with egg wash and sprinkle each one generously with granulated sugar. Bake for 30 minutes or until light brown.

6. Remove the logs from the oven and allow them to cool in the pan for 5 to 10 minutes. Using a serrated knife, cut each log crosswise, on the diagonal, into slices ¼ inch (6 mm) thick and place the slices cut-side down back on the baking sheet.

> **TIP** This particular recipe for biscotti keeps for several weeks in an airtight container.

7. Return the biscotti to the oven for 6 to 8 minutes. Turn the biscotti over and return them to the oven for another 6 to 8 minutes. Cool.

ESPRESSO ALMOND BISCOTTI II

Makes approximately
40 biscotti

Lessons demonstrated in this recipe:

- How to prepare a bar cookie that is twice baked using the one-bowl method.
- Changing from bread flour to all-purpose flour decreases gluten formation, making a more crumbly cookie.
- Adding some brown sugar along with granulated sugar attracts moisture to produce a softer cookie.
- Beating the eggs and sugar until light creates air in the dough, producing a lighter, less dense cookie, which aids leavening.
- An increased amount of baking powder provides leavening while decreasing spread and browning, producing a thicker cookie.
- A small amount of baking soda is used to neutralize some acidity and provide some leavening.
- Because the biscotti are so tender, using slivered almonds instead of whole prevents crumbling the biscotti as they are sliced.
- A higher oven temperature causes the dough to set faster, thereby producing a thicker cookie.
- Adding a large quantity of oil coats gluten strands and produces a tender biscotti.
- Cutting the logs into thicker slices also produces a more tender biscotti.

MEASUREMENTS				INGREDIENTS
U.S.		**METRIC**	**BAKER'S %**	
16½ ounces	3½ cups	470 g	78%	all-purpose flour
4¾ ounces	1 cup	135 g	22%	whole wheat flour
	2 tablespoons	10 g	1.7%	instant espresso powder
¼ ounce	2 tablespoons	10 g	1.7%	coffee beans, coarsely ground
¼ ounce	2 teaspoons	8 g	1.3%	baking powder
	½ teaspoon	2 g	0.3%	baking soda
	½ teaspoon	3 g	0.5%	salt
11 ounces	2½ cups	310 g	51%	slivered almonds
5½ ounces	¾ cup	155 g	26%	light brown sugar (packed, if measuring by volume)
3½ ounces	½ cup	100 g	17%	granulated sugar
	1 teaspoon	6 g	1%	grated lemon zest
	1 teaspoon	6 g	1%	grated orange zest
4 each		188 g	31%	large eggs
8 fluid ounces	1 cup	240 mL	40%	vegetable oil
	½ teaspoon	2.5 mL	0.4%	vanilla extract
	1 teaspoon	5 mL	0.8%	almond extract
	1 teaspoon	5 mL	0.8%	egg wash, as needed
			274.5%	Total Espresso Almond Biscotti II percentage

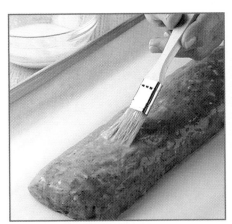

FIGURE 9–15

FIGURE 9–16

FIGURE 9–17

> **TIP** If time is short, the slices can be placed upright (not on their sides) and baked once for 6 to 8 minutes.

1. Preheat the oven to 350°F (177°C). In a mixing bowl, whisk together the flours, espresso powder, coffee beans, baking powder, baking soda, salt, and almonds. Set the mixture aside.

2. In the bowl of an electric mixer using the paddle attachment, blend the two sugars, the zests, and the eggs on medium speed until the mixture is slightly thickened.

3. On low speed, gradually add the oil, vanilla, and almond extracts.

4. On low speed, add the dry ingredients and blend until just combined.

5. Divide the dough in half and form each half into a log approximately 15 inches (37.5 cm) long by 3½ inches (8.7 cm) wide. Place each log onto a sheet pan covered with parchment paper. The logs will rise and spread as they bake.

6. Brush each log with egg wash using a pastry brush and generously sprinkle each one with granulated sugar (Figure 9–15).

7. Bake the logs for approximately 30 to 32 minutes or until they are firm but have little color. The top of the logs should look crackled.

8. Allow the logs to cool for 5 to 10 minutes and then cut each log crosswise on the diagonal into ¾-inch (2-cm) slices using a serrated knife (Figure 9–16). Place the slices cut side down on the same sheet pan.

9. Return the slices to the oven and bake for 6 to 8 minutes on each side (Figure 9–17). Cool completely.

FUDGE BROWNIES

Makes one 11-inch by 15-inch (27.5-cm by 37.5-cm) pan *or* two 8-inch (20-cm) square pans

Lessons demonstrated in this recipe:

- How to prepare a sheet cookie.
- Beating the eggs and sugar incorporates air into the batter, which works in tandem with the baking powder for leavening power.
- Brown sugar increases moistness because of its hygroscopic qualities, producing a fudgy brownie.
- Bread flour, a high-protein flour, increases chewiness.
- Slightly underbaking brownies increases chewiness.

MEASUREMENTS				INGREDIENTS
U.S.		METRIC	BAKER'S %	
12 ounces		340 g	166%	unsalted butter
5 ounces		145 g	71%	unsweetened chocolate, coarsely chopped
5 ounces		145 g	71%	semisweet chocolate, coarsely chopped
	2 tablespoons	10 g	4.9%	instant coffee powder
3½ ounces	¾ cup	100 g	49%	all-purpose flour
3¾ ounces	¾ cup	105 g	51%	bread flour
	½ teaspoon	2 g	1%	baking powder
	½ teaspoon	3 g	1.5%	salt
15 ounces	2 cups	425 g	207%	light brown sugar (packed, if measuring by volume)
14½ ounces	2 cups	410 g	200%	granulated sugar
8 each		376 g	183%	large eggs
	1½ teaspoons	7½ mL	3.7%	vanilla extract
7 ounces	2 cups	200 g	98%	walnuts, chopped
6 ounces	1 cup	170 g	83%	semisweet chocolate, coarsely chopped
			1190.1%	Total Fudge Brownies percentage

Continued

1. Preheat the oven to 350°F (175°C). Grease an 11-inch by 15-inch (27.5-cm by 37.5-cm) rectangular pan or two 8-inch (20-cm) square pans. Set aside.

2. In a bowl placed over a pot of simmering water, melt the butter, the two chocolates, and the coffee. Set the mixture aside to cool.

3. In a mixing bowl, whisk the two flours, the baking powder, and the salt. Set aside.

4. In the bowl of an electric mixer using the paddle attachment, beat the sugars on low speed and gradually add the eggs and vanilla. Beat at medium speed until well blended and slightly thickened.

5. On low speed, slowly add the melted chocolate and butter mixture to the sugar and eggs. Blend the mixture until well combined (Figure 9–18).

6. Blend in the flour mixture on low speed and mix only until blended. Remove the bowl from the mixer. Blend in the chocolate and the walnuts by hand using a rubber spatula or spoon until just combined (Figure 9–19).

7. Pour the batter into the prepared pan and spread it evenly (Figure 9–20).

8. Bake for 32 to 35 minutes or until the brownies are set and firm on top but still slightly soft in the center. (Note: A cake tester inserted into a brownie will not come out clean.)

9. Cool completely and chill the brownies for 1 hour in a refrigerator to make cutting them into bars easier.

FIGURE 9–18

FIGURE 9–19

FIGURE 9–20

GOLDEN COCONUT CUTOUTS

Makes approximately 3½ dozen cookies

Lessons demonstrated in this recipe:

■ How to prepare a rolled cookie using the creaming method.
■ Using confectioners' sugar creates less spread.
■ Using some low-protein flour creates tenderness.
■ Rolling the dough thin produces a crisp cookie.
■ This dough resembles a pâte sucrée with coconut added.
■ Less liquid in the dough contributes to crispness as does a high sugar and fat content.

MEASUREMENTS				INGREDIENTS
U.S.		METRIC	BAKER'S %	
7 ounces	1½ cups	200 g	63%	all-purpose flour
4 ounces	1 cup	115 g	37%	cake flour
	½ teaspoon	1 g	0.3%	ground cinnamon
	¼ teaspoon	1 g	0.3%	salt
3 ounces	1 cup	85 g	27%	shredded, sweetened coconut, lightly toasted and then finely ground in a food processor
8 ounces		225 g	71%	unsalted butter, softened
4 ounces	1 cup	115 g	37%	confectioners' sugar, sifted
1 each		19 g	6%	large egg yolk
	1 teaspoon	5 mL	1.6%	coconut extract
				extra confectioners' sugar, for dusting
				extra all-purpose flour, for dusting
1 each		28 g	9%	large egg white
1½ ounces	½ cup	40 g	13%	shredded, sweetened coconut, coarsely chopped in a food processor, not toasted
			265.2%	Total Golden Coconut Cutouts percentage

1. Preheat the oven to 325°F (163°C). In a bowl, sift together the flours, cinnamon, and salt. Add the ground, toasted coconut and mix until well blended. Set aside.

2. In the bowl of an electric mixer using the paddle attachment, cream the butter and sugar on medium speed until light. Do not overcream. Add the egg yolk and the extract to the butter mixture (Figure 9–21).

3. On low speed, add the flour and coconut mixture, mixing until just incorporated.

4. Gather the dough into a ball. Wrap it in plastic wrap and chill for 2 to 3 hours or until it is firm enough to roll out (Figure 9–22). The dough can be placed in the freezer for a short period of time to speed up the chilling process.

FIGURE 9–21

FIGURE 9–22

5. Cut the dough in half, placing one half back into the refrigerator. On a surface that has been lightly dusted with flour and confectioners' sugar, roll out the dough to a ¼-inch (6-mm) thickness. Using a scalloped or fluted, round, 3-inch (7.5-cm) cookie cutter, cut out shapes (Figure 9–23). Cookie cutters in different shapes also work. Place the cookies on parchment-lined sheet pans. Repeat rolling and cutting the other half of the dough.

6. Using a pastry brush, glaze each cookie with the reserved egg white and sprinkle with some of the coarsely chopped untoasted coconut (Figure 9–24). Bake for 10 to 14 minutes or until lightly browned.

FIGURE 9–23

FIGURE 9–24

CITRUS BUTTER RINGS

Makes 18 ring cookies, approximately 3 inches (7.5 cm) in diameter

Lessons demonstrated in this recipe:

- How to prepare a piped cookie using the creaming method.
- The creaming method of mixing produces a cookie that is light in texture while not over-creaming to ensure less spread.
- A low-protein flour is used for tenderness and to prevent overbrowning.
- An egg white dries out the dough, adding crispness and little color.
- Confectioners' sugar and cornstarch both prevent spread.
- No chemical leavener is used, so these cookies should look the same size going in the oven as they do coming out.

MEASUREMENTS				INGREDIENTS
U.S.		METRIC	BAKER'S %	
7 ounces		200 g	118%	unsalted butter, softened
2 ounces	½ cup	55 g	32%	confectioners' sugar, sifted
6 ounces	1½ cups	170 g	100%	cake flour
1¼ ounces	¼ cup	35 g	21%	cornstarch
1 each		28 g	16%	large egg white
½ fluid ounce	1 tablespoon	15 mL	9%	orange juice, no pulp
	1 teaspoon	5 mL	2.9%	lemon extract
	½ teaspoon	2.5 mL	1.5%	vanilla extract
	½ teaspoon	3 g	1.8%	lime zest
	½ teaspoon	3 g	1.8%	lemon zest
				confectioners' sugar for dusting
			304%	Total Citrus Butter Rings percentage

Continued

FIGURE 9–25

FIGURE 9–26

1. Preheat the oven to 350°F (175°C). In the bowl of an electric mixer using the paddle attachment, cream the butter and sugar only until they form a smooth paste.

2. In another bowl, sift together the flour and cornstarch. Set aside.

3. On low speed, add the egg, orange juice, extracts, and zests to the butter and sugar mixture. Mix until well combined.

4. Slowly add the flour mixture. Mix only until a dough forms.

5. Place the dough into a large pastry bag fitted with a medium star tip.

6. Pipe the dough into 3-inch (7.5-cm) diameter ring shapes onto a parchment-lined sheet pan (Figure 9–25).

7. Bake the rings for 15 minutes or until the cookies are set and the bottoms are light brown. The cookies should be pale and have little color.

8. Cool and dust each cookie with confectioners' sugar (Figure 9–26).

CHOCOLATE ALMOND LADY FINGERS

Makes approximately 2½ dozen lady fingers

Lessons demonstrated in this recipe:

■ How to prepare a piped cookie using the egg-foam method.

■ Beating warm eggs gives better volume.

■ The egg-foam mixing method is used to produce a light, sponge cake-like cookie.

■ Lady fingers follow the mixing method similar to an egg foam cake.

■ Air beaten into the eggs leavens the cookie.

■ An acid like cream of tartar provides stability to the meringue.

MEASUREMENTS				INGREDIENTS
U.S.		METRIC	BAKER'S %	
3 each		57 g	143%	large egg yolks
1¾ ounces	¼ cup	50 g	125%	granulated sugar
1¼ ounces	⅓ cup	40 g	100%	cake flour
		1 g	1%	pinch salt
	2 tablespoons	16 g	40%	Dutch processed cocoa powder
	1 teaspoon	5 mL	13%	vanilla extract
	½ teaspoon	2.5 mL	6%	almond extract
3 each		84 g	210%	large egg whites, room temperature
	¼ teaspoon	0.5 g	1.3%	cream of tartar
2 ounces	¼ cup	55 g	138%	granulated sugar
				sliced almonds for sprinkling
				confectioners' sugar for dusting
			777.3%	Total Chocolate Lady Fingers percentage

FIGURE 9–27

FIGURE 9–28

FIGURE 9–29

1. Preheat the oven to 350°F (175°C). On a sheet pan covered with parchment paper, using a pencil and ruler, draw 2 lines down the length of the sheet pan, spaced 3 inches (7.5 cm) apart (Figure 9–27). Directly below those, make two more lines 3 inches (7.5 cm) apart 1 inch away from the first lines. Turn the side with the pencil marks facing down so the marks are now on the underside but can still be seen through the paper (Figure 9–28). Set aside.

2. In the bowl of an electric mixer using the paddle attachment, mix the egg yolks and sugar on medium-high speed until the eggs turn pale yellow and appear thickened (Figure 9–29). This takes about 5 minutes.

3. In another bowl, sift the cake flour, salt, and cocoa powder together. Whisk to blend.

4. On low speed, blend in the flour and cocoa into the egg yolks and then add the extracts until they are well combined. Transfer the batter to a large mixing bowl (Figure 9–30). The batter will be thick.

5. In a clean, dry bowl of an electric mixer using the whip attachment, beat the whites on high until foamy. Add the cream of tartar and continue beating until soft peaks form.

6. Slowly add the 2 ounces (¼ cup; 55 g) granulated sugar and beat until stiff peaks form (Figure 9–31). Do not overbeat.

FIGURE 9–30

FIGURE 9–31

FIGURE 9–32

7. Using a rubber spatula, scoop ⅓ of the whites on top of the chocolate batter and whisk in thoroughly to lighten (Figure 9–32).

8. Scoop the remaining whites onto the chocolate batter and, using a rubber spatula, gently fold the whites in until they are well combined (Figure 9–33). Do not overfold.

9. Take the prepared sheet pan and place it so that the long side is facing you. Working quickly, scoop half of the batter into a large pastry bag with ½-inch (1.25-cm) opening and squeeze out 3-inch (7.5-cm) long finger shapes from one line to the next, like the rungs of a ladder, spacing them approximately 1 inch apart using the pencil marks as a guide (Figure 9–34). Refill the bag and repeat forming lines of batter between the second set of lines.

10. Sprinkle each line of batter with sliced almonds and dust with confectioners' sugar (Figure 9–35).

11. Bake the lady fingers for 12 minutes or until they are puffed and cooked through. Remove them from the oven and cool on racks. Peel each lady finger off gently using an offset spatula.

> **TIP** If the lady fingers will not be used immediately they can be cooled and kept right on the parchment paper and stored in an airtight plastic container for 2 to 3 days at room temperature. Keeping them on the parchment paper in an airtight container prevents them from drying out and maintains their spongy texture.

FIGURE 9–33

FIGURE 9–34

FIGURE 9–35

POLKA-DOTTED PIROUETTES

Makes about 3 dozen
pirouette cookies

Lessons demonstrated in this recipe:

■ How to prepare a wafer cookie.

■ High sugar and high fat create a crisp cookie.

■ The cookies are quite malleable when hot and set as they cool, because sugar melts in the oven and recrystallizes upon cooling, forming a crisper cookie.

■ The use of egg whites, which act as drying agents, cause the cookies to be crisp.

■ Wafer cookies should be baked in small numbers to give the baker enough time to roll each cookie into a cylinder before they become too hard.

MEASUREMENTS				INGREDIENTS
U.S.		**METRIC**	**BAKER'S %**	
4 each		112 g	149%	large egg whites
6¾ ounces	1 cup	190 g	253%	superfine sugar
2¾ ounces	⅔ cup	75 g	100%	cake flour, sifted
2 ounces	4 tablespoons	57 g	76%	unsalted butter, melted
	1 teaspoon	5 mL	7%	vanilla extract
	2 teaspoons	5 g	7%	unsweetened cocoa powder, sifted
			592%	Total Polka-Dotted Pirouettes percentage

1. Preheat the oven to 350°F (175°C). In the bowl of an electric mixer using the paddle attachment, combine the egg whites and the sugar. Mix on medium speed until foamy.

2. On low speed, add the flour and mix until well combined.

3. Slowly add the melted butter and vanilla extract (Figure 9–36).

4. Remove 2 ounces (¼ cup; 60 mL) of the batter to a small bowl and whisk in the cocoa powder until smooth and well combined (Figure 9–37). This cocoa batter will be used to make the polka dots. Set aside.

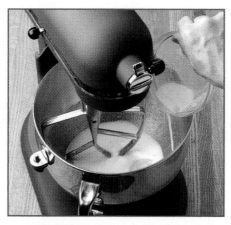

FIGURE 9–36

FIGURE 9–37

FIGURE 9–38

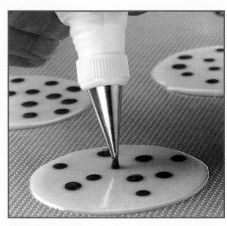

FIGURE 9–39

5. On a parchment-lined or silicone mat-lined half sheet pan, drop 1 teaspoon (5 mL) of batter onto the prepared pan, keeping cookies at least 3 inches (7.5 cm) apart. Do not bake more than 4 to 6 cookies at a time or they will harden before they can be rolled.

6. Using a spoon or an offset spatula, spread the batter into a 3-inch (7.5-cm) circle (Figure 9–38). Repeat this procedure for each cookie.

7. Place the cocoa batter into a pastry bag with a small round tip. Pipe dots of cocoa batter randomly on each circle of batter (Figure 9–39).

8. Bake for 6 to 9 minutes or until the cookies are light brown around the edges.

9. Immediately, flip one cookie upside down using an offset spatula and roll it around the handle of a small wooden spoon, a small dowel rod, or a chopstick until it forms a tight cylinder (Figure 9–40). Gently slide the pirouette off the spoon handle and place it on a rack to cool completely (Figure 9–41).

10. If the cookies become cool and harden before they can be rolled, return the pan to the oven for a few seconds to soften them.

TIP Alternatively, the batter can be thinly spread over a stencil that has been cut into various shapes and then baked as described.

FIGURE 9–40

FIGURE 9–41

Professional Profile

BIOGRAPHICAL INFORMATION

Jenn Solloway-Malvitz
Chef Instructor
Fox Valley Technical College
Appleton, WI

Question: *When did you realize that you wanted to pursue a career in baking and pastry?*

Answer: *My grandmother was in the restaurant business in San Francisco so I've been around a professional kitchen since I was five. It wasn't until I went to City College of San Francisco that I was encouraged to go into baking and pastry.*

Question: *Was there a person or event that influenced you to go into this line of work?*

Answer: *My pastry chef instructor at City College of San Francisco, Chef Henri Cochennec, saw in me abilities and talents that I didn't realize I had. He supported me and helped me in so many ways.*

Question: *What did you find most challenging when you first began working in baking and pastry?*

Answer: *From the very beginning I loved working in baking and pastry. Baking to me is eatable art. My biggest challenge has always been a lack of time. There are not enough hours in the day to do what I want to do.*

Question: *When and where was your first practical experience in a professional baking setting?*

Answer: *My first job after school was pastry chef at a Sheraton Hotel in Phoenix. The property had five dining venues. To take on this type of responsibility and work—just out of school—was a huge step.*

Question: *How did this experience affect your later professional development?*

Answer: *I learned some valuable lessons there, many through hard knocks. I made a coconut macaroon cookie, drizzled with chocolate, that was used for the turndown service. It was beautiful but a lot of work. I had to make about 1,000 of them each night—every night. I learned that just because you can do something doesn't mean you should do it!*

Question: *Who were your mentors when you were starting out?*

Answer: *I had the opportunity to work with Bradley Ogden and learned a lot from him while doing my internship. I also had the privilege of working with Annie Somerville at Greens at Fort Mason in San Francisco. From her I learned so much about the benefit of using whole, natural foods while working in a peaceful kitchen environment. I have really been blessed to work with these people.*

Question: *What would you list as your greatest rewards in your professional life?*

Answer: *I waited a long time to become a full-time chef instructor. For eight years I was an adjunct, running my company and doing teaching on the side. The opportunity to give back and share my knowledge with the students is the greatest reward. The responsibility that has been placed on me to teach Culinary and Pastry Arts is truly a blessing.*

Question: *What traits do you consider essential for anyone entering the field?*

Answer: *Stamina, creativity, passion, and compassion are the four most important traits anyone entering this field needs. Any student who plans to make a career in this area needs to learn compassion. It is only through compassion that a manager can attract and keep good staff. And it is only with a top notch staff that any business can be competitive. Without creativity and passion your work goes flat.*

Question: *If there were one message you would impart to all students in this field what would that be?*

Answer: *You can never think that you are done learning. You can take a break but you can't stop. If you stop then you become stagnant and obsolete. Education is the power that yields creativity.*

CHAPTER 10

Building Blocks with Sugar

After reading this chapter, you should be able to:

- Demonstrate the three basic steps to prepare sugar syrups for desserts and confections.

- Recognize when a sugar syrup is done.

- Describe the process of crystallization.

- List three ways to control crystallization.

- Prepare caramelized sugar using the dry method and the wet method.

- Prepare the recipes in this chapter.

Sugar is one of the most widely used ingredients in baking. It contributes many wonderful characteristics to baked goods and among its most widely used characteristics is its ability to be melted down, cooked, and then cooled to be used in a variety of confections and desserts. Looking at a grain of ordinary granulated sugar, it is nothing more than a crystalline substance. By understanding the process of crystallization, a baker has the power to change the form and chemical structure of this ordinary crystal and use it as a tool to create a wide array of diverse products.

The crystals of granulated sugar cannot be changed without first being dissolved in water and brought to a boil to form a sugar syrup. If the boiling is allowed to continue, the syrup becomes concentrated. If boiling continues even further, the syrup turns to caramel. It is sugar syrups that are truly the foundation for many desserts and confections like Italian meringues, marshmallows, fudge, hard candies, caramels, caramel sauces, and pulled sugar decorations.

This chapter explores the foundations of sugar syrups, the properties of crystals, and how to control the crystallization process.

Three Basic Steps in Preparing Sugar Syrups for Desserts and Confections

Three basic steps in preparing sugar syrups for desserts and confections:
- Dissolve sugar in water.
- Concentrate the sugar syrup.
- Cool the sugar syrup.

There are three basic steps in preparing sugar syrups for desserts and confections:
1. Dissolve sugar in water.
2. Concentrate the sugar syrup.
3. Cool the sugar syrup.

Dissolving Sugar in Water

Creating a sugar syrup is the first step in producing a variety of desserts and confections. A sugar syrup is simply one or more sugars dissolved in water. Sugar syrups begin by dissolving a specific amount of sugar into a specific amount of water. Sugar syrups can also be referred to as *solutions*. There are two parts to a solution. The water is known as the solvent (the liquid that a substance is placed into to help it to dissolve), and the sugar is known as the solute (the substance to be dissolved, usually a solid). Together the solvent and solute are known as a solution and, in this case, the solution is called a *sugar syrup*. Heating the sugar syrup allows the sugar to dissolve and the water to evaporate, which concentrates the solution.

The warmer the water, the more sugar can dissolve. Picture two beverages: a glass of iced tea and a steaming hot cup of coffee. It is common knowledge that if you stir one teaspoon of granulated sugar into either liquid, the sugar will dissolve and sweeten the beverage. This is the same principle behind sugar syrups. You may have noticed, however, that not all the sugar stirred into the iced tea dissolves; instead, some undissolved sugar settles at the bottom of the glass. At the same time, an even greater amount of sugar, say two tablespoons, mixed into an identical amount of hot coffee seems to dissolve completely. This occurs because the warmer the water, the more sugar can be dissolved into it. Heat keeps the sugar molecules moving sufficiently quickly that they never come close enough to each other for long enough to join to form crystals. The same principle applies to sugar syrups in that the hotter the water gets, the greater the amount of sugar it will dissolve.

SIMPLE SYRUPS

The easiest sugar syrup to make is known as a simple syrup. Simple syrups are cooked the least amount of time of any sugar syrup. Sugar syrups are made by combining equal parts of water and granulated sugar by weight and brought to a boil. The mixture is boiled just until the sugar is dissolved. Other flavoring ingredients may be added to the simple syrup such as fruit juices, extracts, coffee, or alcohol.

The concentration of the simple syrup depends on the amount of water added and can vary depending on the baker's preference. For example, some bakers prefer a lighter syrup and use two parts water to one part sugar. Simple syrups are then cooled and can be used in a variety of desserts such as brushing onto cake layers to provide flavor and moistness in the base of frozen desserts such as sorbets, and to glaze the tops of pastries.

Simple syrups can be placed in a covered container and stored in the refrigerator for several weeks.

RECIPES
Hazelnut Syrup (Chapter 7, page 219)
Simple Syrup (This chapter, page 304)

Concentrating the Sugar Syrup

The second step is to intensify or concentrate the sugar syrup. Depending on what is being made will determine how concentrated the sugar syrup will need to be. This is accomplished by continuing to cook or boil the sugar syrup. Boiling accomplishes two things. First, it completely dissolves all the sugar in the water; second, once the temperature reaches 212°F (100°C), the water begins to evaporate.

As more water evaporates, the sugar syrup becomes more concentrated. At the same time, the boiling point of the sugar syrup rises. Eventually, the temperature of the sugar syrup will surpass the boiling point of water. As the temperature climbs, the sugar syrup becomes more and more concentrated. Bakers and confectioners use the temperature of the syrup as a measure of its concentration or density. The concentration depends on how much water is left in the syrup after the boiling process has been completed. The greater the amount of water remaining, the softer the syrup will become when it is cooled. The smaller the amount of water remaining, the harder the syrup will become when it is cooled.

Different desserts and confections require different concentrations of sugar syrup. As a sugar syrup climbs in temperature, chemical changes occur within the sugar causing it to break down into smaller, different sugars. As these chemical changes occur, the sugar syrup caramelizes and changes from a clear to a dark brown color similar to brewed tea. The syrup is now referred to as caramelized sugar. The more dramatic the color change the sugar undergoes, the greater the chemical changes. For example, a pale, amber colored sugar syrup hardens and resembles a sheet of tinted glass. There is still enough of the original sugar left to recrystallize upon cooling. Very dark caramelized sugar remains soft upon cooling because most of the original sugar is gone due to chemical changes with little of the original sugar left to recrystallize and firm up. When preparing caramel, do not allow it to darken too much or it will burn and then ignite.

Cooling the Sugar Syrup

Once the correct temperature of the sugar syrup is reached, the syrup must be used immediately and combined with other ingredients, or it must be cooled and the cooking process stopped. Care must be taken while cooling sugar syrups because any stirring can cause the syrup to crystallize suddenly. Cooling slows down the molecules of sugar so that they get closer together and eventually re-form crystals. When sugar syrups are rapidly cooled and beaten, many large crystals form. When sugar syrups are cooled more slowly without beating or stirring, smaller and fewer crystals form.

Depending on what is being made, there are many paths a sugar syrup can take. If a caramel sauce is being made, heavy cream is usually added slowly to the boiling syrup to stop the cooking process and lower the temperature. If an Italian meringue is being made, for marshmallows or a frosting, the boiling syrup is immediately beaten into egg whites until the meringue triples in volume. If a fudge candy is being made, the sugar syrup needs to cool down, without stirring, before it is beaten to prevent large crystals from forming. For confections, it is preferable to form smaller, finer crystals which give a smoother texture and a nicer mouthfeel. Large crystals may form if a sugar syrup is stirred. This is because stirring causes small crystals in the syrup to bump into each other, repeatedly gathering more and more crystals, much like a snowflake clumps together with other snowflakes to form a snowball. This creates very large crystals that have poor mouthfeel. Larger crystals turn a sugar syrup cloudy and recrystallize into a gritty tasting clump. The size of the crystals that form upon cooling will determine the color, clarity, and texture of the final product.

Two Ways to Determine When a Sugar Syrup Is Done

Getting the sugar syrup to the proper temperature is crucial to the success of the final product. Every confection made from a sugar syrup requires a specific concentration of sugar. For example, sugar syrups that are used to prepare Italian meringues are typically boiled to 235° to 240°F (113° to 115°C) so that the heat of the syrup instantly cooks the egg whites, producing a stable meringue. If the syrup is boiled to a lower temperature, the meringue would be less stable because the egg whites would not be cooked enough. If the syrup is boiled to a higher temperature, it would form small lumps of sugar as it was incorporated into the egg whites or get too hard to become incorporated at all.

There are two ways to determine the correct temperature of the sugar syrup. The first way is to use a candy thermometer, which has a range of 100° to 400°F (38° to 204°C). There is a clip on the side of the thermometer so it can be attached to the side of a pot. A digital thermometer may also be used.

The other way to determine the correct temperature is by using the cold-water test. Once the caramelized sugar reaches what appears to be the correct temperature, a drop of the syrup is placed into a glass of cold water. The cold water cools the syrup instantly, allowing the baker to feel whether the syrup is the correct consistency and texture.

Each temperature range is associated with specific characteristics of the small amount of syrup and its ease in being gathered into a ball. (See Table 10–1, A Description of the Characteristics of the Cold-Water Test versus Sugar Syrup Temperatures.)

There are two ways to determine when a sugar syrup is done:

■ Temperature

■ The cold-water test

Table 10–1 A Description of the Characteristics of the Cold-Water Test versus Sugar Syrup Temperatures

COLD WATER TEST	DESCRIPTION	TEMPERATURES
Thread	The syrup forms thread-like projections when dropped.	230°–235°F (110°–113°C)
Soft ball	The syrup forms a soft ball that flattens easily between the fingers.	235°–240°F (113°–115°C)
Firm ball	The syrup forms a firm, yet malleable ball.	245°–250°F (118°–122°C)
Hard ball	The syrup forms a hard, inflexible ball.	250°–265°F (122°–130°C)
Soft crack	The syrup forms straw-like threads that bend.	270°–290°F (132°–143°C)
Hard crack	The syrup forms threads that are hard and brittle.	300°–310°F (149°–155°C)
Caramel	By observation only, color of syrup is golden brown and remains a liquid.	320°–350°F (160°–177°C)

Understanding the Process of Crystallization

As a sugar syrup is cooled during the last stage of its preparation, it is important to control the size of the crystals that form. For the most part, larger crystals leave a gritty taste on the tongue. Smaller crystals are not detected on the tongue as easily and feel smoother and creamier in texture. Before discussing how to control crystallization, the process of crystallization should be understood.

Granulated sugar is known chemically as sucrose. It is a combination of two smaller, simpler sugars: glucose and fructose. Crystallization is the process in which crystals form—in this case, sugar crystals. In order for a substance to form crystals, it must be pure. In other words, sucrose will only form crystals when sucrose and only sucrose is present. If a foreign substance is introduced, even in the form of a different sugar or another ingredient, few crystals are formed.

Picture a small child playing with square building blocks that are all the same shape. As the child places one block against the other, a wall forms with each square block fitting in exactly next to another. This wall of blocks is what crystallized sugar (sucrose) would look like under a microscope—a repeated shape packed closely together.

If the child then tries to build a wall with two different shaped blocks, alternating a circular set of blocks with the square ones, the structure is too unstable to form. This is similar to what would happen if a foreign substance was introduced into the sucrose. Almost no crystallization would occur because the sucrose is no longer pure and by itself.

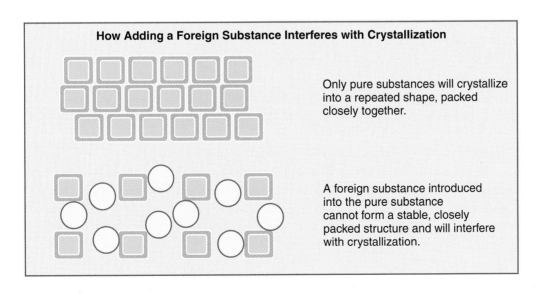

How Adding a Foreign Substance Interferes with Crystallization

Only pure substances will crystallize into a repeated shape, packed closely together.

A foreign substance introduced into the pure substance cannot form a stable, closely packed structure and will interfere with crystallization.

If the process of crystallization is understood, discovering ingredients that can interfere with this process can be useful to the baker or confectioner in controlling the size and number of crystals that form.

Ways to Control Crystallization

Once a sugar syrup becomes highly concentrated, it tends to be unstable. This means that any agitation by stirring or rapid cooling can start a chain reaction wherein a few sugar crystals can quickly join together with others, creating an unusable, solid mass. This is not a problem if you are making a crystalline confection such as rock candy. However, for creamier confections and desserts, crystallization needs to be controlled or interfered with. There are three ways to interfere with the crystallization process. They include adding an interfering agent, keeping stirring to a minimum, and preventing crystals from forming on the side of the pot.

> **Three Ways to Interfere with Crystallization**
> - Adding an interfering agent
> - Keeping stirring to a minimum
> - Preventing crystals from forming on the sides of the pot

Adding an Interfering Agent

Adding an interfering agent can slow down crystallization. Interfering agents, as the name implies, "interfere" with how crystals form by getting in between sugar molecules and preventing them from joining together to create crystals.

Some interfering agents are added to the sugar syrup at the beginning of preparation as the ingredients are combined, whereas others are added after cooking is completed.

INTERFERING AGENTS ADDED DURING THE PREPARATION OF SUGAR SYRUPS

Two different ingredients that can be used to interfere with crystal formation during the preparation of sugar syrups include adding a different type of sugar and adding an acid.

Adding a different type of sugar can interfere with crystallization. When another type of sugar is added to the sucrose, an impure state is reached, preventing crystallization from occurring. Corn syrup is a frequently used sugar that is added to sucrose for this purpose. Corn syrup is actually glucose and impedes crystallization by helping maintain a confection's smooth qualities.

Adding an acid to the ingredients during the preparation of a sugar syrup also prevents crystals from forming. This is due to a process called inversion. As sucrose is heated with an acid the sugar breaks down into its two component sugars—glucose and fructose. Because the sucrose has changed chemically into two different sugars, an impure state is reached and crystallization is hindered. This creates a smoother texture in confections and desserts. Acids commonly used include lemon juice, vinegar, cream of tartar, or brown sugar.

> **TIP**
> Too much of an interfering agent such as corn syrup or an acid can prevent any crystallization from occurring. This could be problematic in certain confections in which a specific amount of crystallization is necessary for the candy to set up and harden.

INTERFERING AGENTS ADDED AFTER COOKING SUGAR SYRUPS

Adding interfering agents after the sugar syrup has completed cooking can also inhibit crystal formation. Interfering agents that are commonly used are high-fat ingredients such as heavy cream and other milk products, chocolate, eggs, and butter. The fat coats sugar molecules, preventing them from joining together to form large, gritty crystals.

For example, recipes for fudge, caramel, and caramel sauces generally call for heavy cream to be added to the hot sugar syrup. Low-fat milk products should not be used because the proteins can denature and curdle due to the high temperature of the syrup. Cream has a much higher fat content and the fat globules protect against the protein's curdling, allowing the mixture to remain smooth.

Other liquids that can be added to a hot sugar syrup include water, fruit juices, flavored teas, and brewed coffee. Alcohol may also be added, once the temperature of the syrup has been reduced.

Keeping Stirring to a Minimum

Another way to control crystals from forming is to keep stirring to a minimum. For example, when preparing the sugar syrup, place the sugar in the pot first and then pour the water around the outer edge of the sugar closest to the sides of the pot. The water immediately dissolves any crystals of sugar that may have attached to the sides. Stirring is not necessary once the boiling process has started. Some chefs stir the sugar syrup just until the sugar has dissolved and the boiling process has begun.

Preventing Crystals from Forming on the Sides of the Pot

The last way to prevent crystals from ruining a sugar syrup is to dissolve them as they are forming. As the sugar comes to a boil, a pastry brush dipped in water is used to brush around the outer edges of the pot near the level of the syrup to wash down any sugar crystals that may not have dissolved initially. The brush should never dip into the syrup, because this could start crystals forming.

Some chefs prefer to cover the pot of sugar syrup as it comes to a boil so, as the steam rises, it hits the cover and condenses back down, washing any crystals from the sides of the pot. The pot is uncovered once the syrup starts to boil to allow water to evaporate, concentrating the syrup. Some chefs even oil the sides of the pot to prevent crystals from sticking.

(*Note:* All it takes is one small undissolved crystal of sugar to start a chain reaction that can ruin the entire batch of sugar syrup.)

Two Methods to Prepare Caramel

Sugar cooks and turns to caramel within the temperature interval of 320° to 350°F (160° to 177°C). There are two methods of preparing caramelized sugar:

- The wet method
- The dry method

> Two Methods of Preparing Caramelized Sugar:
> - The wet method
> - The dry method

The Wet Method

The wet method consists of simply preparing a sugar syrup from sugar and water and boiling it to the caramel stage. Because burning sugar during caramelization is always a concern, the wet method may help reduce this risk in that the sugar is first dissolved in water. When preparing caramel using the wet method, use a much larger pan than is necessary to prevent any boiling over of the sugar syrup. When the sugar and water are combined with other ingredients (e.g., lemon juice, vinegar, cream of tartar, or corn syrup) may be added to prevent crystallization from occurring. The mixture is brought to a rolling boil and, because the liquid is so viscous, it gives the appearance of molten hot lava with many bubbles in the pan popping and reforming quickly in succession. The temperature of the mixture is monitored carefully until the correct color and temperature is reached.

The wet method takes longer than the dry method because it takes time to boil off all the water. Precautions must be taken to prevent crystallization because even a crystal of sugar on the side of the pan could start a chain reaction of crystal formation.

The Dry Method

The dry method consists of granulated sugar being placed in a sauté pan (not nonstick) or heavy saucepan and cooked without adding any water until the caramel stage is reached. A few drops of

TIP When color is the determining factor for when the caramel is done, it is important to remove it from the heat when it turns the shade of very light brewed tea. Although the color is still too light, the cooking process continues for a short time after the pan of caramel is removed from the heat (carryover cooking) and the caramel will continue to darken.

an acid such as lemon juice may be added before cooking begins. The sugar is heated until the crystals melt down and go through chemical changes, producing a piping hot molten syrup. The sugar must be stirred and rotated constantly so the crystals on the bottom do not burn. Color—not temperature—determines the correct degree of doneness. This method is the simpler of the two, yet the chef has little room for error. The sugar can go from caramelized to burnt very quickly. The melted sugar does not boil or bubble in the dry method.

Where Caramel Gets Its Flavor

Part of the deep, rich flavor associated with caramel comes from the many chemical reactions that occur in the upper temperature ranges of caramelization. A reaction that also contributes to flavor in caramel is caused by the Maillard reaction. As discussed in Chapter 1, when sugars and proteins are heated to high temperatures, complex flavors develop. The Maillard reaction occurs when dairy products are added to caramelized sugar syrups, which not only prevents further crystallization but also intensifies the caramel flavor.

Using Caution When Working with Sugar

Because of the extremely high temperatures that are reached during sugar syrup preparation, precautions should be taken: never leave the work area when working with sugar.

- When adding cream or other liquids to a boiling sugar syrup, do so with care because the syrup can boil up quickly. Remove the pan from the heat when adding any liquid, whisking gently.
- Never touch a boiling sugar syrup or caramelized sugar with your bare hands. Sugar is sticky and clings to skin, causing severe burns.
- Keep a bowl of ice water nearby in case of splatters to immediately cool down the sugar syrup that sticks to your skin.

Tips for Preparing Sugar Syrups

- Use a larger pan than necessary because the boiling point of water is reached quickly and the liquid may suddenly boil up. Using a larger pan prevents boiling over of the contents.
- Have all of your mise en place ready because sugar syrups will continue to cook, even when off the heat, and can burn quickly.
- Sugar syrups darken faster in the presence of minerals, so be aware that the syrup may cook more quickly when hard water is used.
- To begin a sugar syrup, heat all of the ingredients gradually until they are dissolved. Then turn the heat up.

Storage of Confections Made from Caramel

Due to the hygroscopic nature of sugar, moisture can cause decorations made with sugar to become soft and sticky. Proper storage is important in maintaining the confection's quality. Confections made from caramel decorations can be stored for at least 2 weeks in an airtight container at room temperature with a desiccant (a drying agent) to prevent moisture from getting in. Confections made with caramel can also be kept in an airtight container in the freezer for up to 2 weeks. The caramel can then be used immediately from the freezer.

Any caramel product containing dairy products such as caramel sauce should be stored in the refrigerator and rewarmed before using.

RECIPES:

Boiled Coconut Frosting (Chapter 8, page 243)
Caramel Maple Sauce (This chapter, page 308)
Chocolate Caramel Nut Brittle (This chapter, page 305)
Chocolate Fudge Frosting (Chapter 8, page 240)
Croquembouche (This chapter, page 312)
Hazelnut Praline (Chapter 7, page 218)
Marshmallows (Chapter 8, page 256)
Spiked Caramel-Dipped Nuts (This chapter, page 316)
Vanilla Butter Cream (Chapter 8, page 244)

SIMPLE SYRUP

Makes approximately
14 fluid ounces (1¾ cups;
414 mL)

Lesson demonstrated in this recipe:

■ How to prepare the simplest sugar syrup with the least amount of cooking.

MEASUREMENTS			INGREDIENTS
U.S.		**METRIC**	
8 fluid ounces	1 cup	240 mL	water
7¼ ounces	1 cup	200 g	granulated sugar

1. Combine the water and the sugar in a heavy medium saucepan.

2. Bring the mixture to a boil and cook for 1 minute or until the sugar is completely dissolved. Cool the syrup completely.

3. Simple syrup can be placed in a covered container and stored in the refrigerator for several weeks.

CHOCOLATE CARAMEL NUT BRITTLE

Makes approximately
½ pound (225 g)

Lessons demonstrated in this recipe:

- How to prepare caramelized sugar using the dry method.
- An acid helps hinder crystallization.
- The sugar resembles wet sand as it melts.
- Stirring frequently helps prevent burning in any one spot.

MEASUREMENTS			INGREDIENTS
U.S.		METRIC	
7¼ ounces	1 cup	200 g	granulated sugar
	¼ teaspoon	1.25 mL	lemon juice
2 ounces	⅓ cup	60 g	semisweet chocolate, chopped
1¼ ounces	⅓ cup	35 g	roasted salted peanuts
1¼ ounces	⅓ cup	35 g	roasted whole almonds
1¼ ounces	⅓ cup	35 g	hazelnuts, whole

1. Preheat oven to 350°F (175°C). Place the sugar and the lemon juice in a heavy medium saucepan or sauté pan (Figure 10–1).

2. Place over medium-high heat while stirring and scraping the bottom constantly with a spoon. Wash down any crystals of sugar from the sides of the pan with a pastry brush dipped in water.

3. In several minutes, the sugar will gradually melt and become a clear liquid (Figure 10–2).

4. Continue to cook the sugar. When it turns a pale brown, remove it from the heat (Figure 10–3).

FIGURE 10–1

FIGURE 10–2

FIGURE 10–3

5. Stir in the chocolate until it is melted. Quickly stir in the nuts (Figure 10–4A and B).

6. Immediately pour the mixture onto a sheet pan lined with a silicone baking mat (Figure 10–5).

7. Place the sheet pan in preheated oven for 5 to 10 minutes or until the caramel has thinned out and the nuts are toasted. Be careful not to let the caramel burn. Cool the brittle completely and break it into pieces when hardened (Figure 10–6). This nut brittle makes an attractive garnish as is or it can be ground in a food processor and used to coat the sides of a cake. It can be stored in an airtight container (with a desiccant) at room temperature for at least 2 weeks or it can be frozen for up to 2 weeks. The brittle can be served immediately from the freezer. Do not thaw the brittle because it will become soft and gooey because of the sugar's hygroscopic nature.

FIGURE 10–4A

FIGURE 10–4B

FIGURE 10–5

FIGURE 10–6

Chocolate Caramel Nut Brittle

CARAMEL MAPLE SAUCE

Makes approximately 14 fluid ounces (1¾ cup; 414 mL)

Lessons demonstrated in this recipe:

- How to prepare caramelized sugar using the wet method.
- Not stirring the sugar syrup during preparation discourages crystals from forming.
- Heavy cream acts as an interfering agent, coating sugar crystals and producing a smooth sauce.

MEASUREMENTS			INGREDIENTS
U.S.		**METRIC**	
7¼ ounces	1 cup	200 g	granulated sugar
4 fluid ounces	½ cup	120 mL	water
6 fluid ounces	¾ cup	180 mL	heavy cream at room temperature
¾ fluid ounce	1½ tablespoons	22.5 mL	maple syrup

1. Place the sugar in a medium-sized heavy saucepan. Pour the water along the edges of the pan so it washes down any sugar crystals that may have stuck to the sides of the pan. Do not stir the mixture.

(Note: The endpoint of this step is determined by the color, not necessarily the temperature.)

2. Heat the mixture to medium heat until the sugar dissolves. Brush the sides of the pan with a pastry brush dipped in water if any sugar crystals appear (Figure 10–7). Turn the heat to high and continue to cook the sugar syrup for approximately 10 to 12 minutes, or until it turns the color of light brewed tea. The temperature should read approximately 320°F (160°C).

3. Remove the pan from the heat and slowly whisk in the heavy cream (Figure 10–8). Use caution because the mixture will boil up. Place the caramel back on low heat and whisk gently until the hard bits of caramel have melted into a smooth sauce.

TIP The caramel maple sauce can be placed in an airtight container and stored in the refrigerator for up to 2 to 3 days. Rewarm it over low heat until it becomes pourable.

4. Remove the sauce from the heat and whisk in the maple syrup (Figure 10–9). Cool down until just warm. Serve immediately.

FIGURE 10–7

FIGURE 10–8

FIGURE 10–9

Ice Cream with Caramel Maple Sauce

VANILLA PASTRY CREAM

Makes 2⅓ cups (18⅔ fluid ounces; 550 mL)

(*Note:* A vanilla bean can be used instead of vanilla extract. Slit one half of a vanilla bean lengthwise and scrape out the seeds with a knife. Place the seeds and the bean into the milk and cream in step 1. Remove the bean before tempering the eggs. Omit the vanilla extract.)

Lessons demonstrated in this recipe:

- How to prepare a custard filling (also known as a pastry cream). A pastry cream is a stirred custard that contains a starch.
- Starches prevent egg proteins from curdling.
- Egg yolks and sugar should not be allowed to sit together without being stirred.
- Bringing the custard to a boil destroys alpha-amylase, an enzyme that breaks down starch.
- Covering the hot custard with plastic wrap prevents casein from forming.

MEASUREMENTS			INGREDIENTS
U.S.		**METRIC**	
8 fluid ounces	1 cup	240 mL	whole milk, removing 1 fluid ounce (2 tablespoons; 30 mL) and setting it aside
4 fluid ounces	½ cup	120 mL	heavy cream
3 each		57 g	large egg yolks
3½ ounces	½ cup	100 g	granulated sugar
¾ ounce	2 tablespoons + 1½ teaspoons	20 g	cornstarch
	1 teaspoon	5 mL	vanilla extract

1. In a heavy medium nonaluminum saucepan, bring the milk (minus 1 fluid ounce; 2 tablespoons; 30 mL) and cream to a boil. Avoid using aluminum pans, which react with milk and egg products, turning them a greenish-gray color.

2. In a heat-proof stainless steel or tempered glass mixing bowl, whisk egg yolks, sugar, cornstarch, and 1 fluid ounce (2 tablespoons; 30 mL) of reserved milk until there are no lumps (Figure 10–10).

3. Slowly drizzle the hot milk mixture into the egg yolk mixture, whisking constantly until all the milk has been added (Figure 10–11). Pour the mixture back into the saucepan and place it over medium-high heat. Bring to a boil, whisking constantly. The mixture will become thick. Boil and whisk for 1 minute. Do not allow the custard to burn.

FIGURE 10–10

FIGURE 10–11

FIGURE 10–12

FIGURE 10–13

4. Remove the mixture from the heat and whisk in the vanilla extract (Figure 10–12). Pour the custard into a bowl and place a piece of plastic wrap directly onto the surface to prevent a skin from forming (Figure 10–13). Place it in the refrigerator until cold, about 3 to 4 hours. The custard filling can be made 1 day in advance and should be kept chilled in the refrigerator.

CROQUEMBOUCHE

Makes 1 croquembouche

A Croquembouche is a tower of cream puffs filled with a cream filling and covered with a crunchy coating of caramel. *Croquembouche* means "crunch in the mouth" in French. It can be made in stages. The day before you intend to serve it, prepare the cream puffs and then the custard filling. The day you intend to serve it, prepare the caramel before beginning the assembly.

STEP A

Prepare one recipe Cream Puffs, baked and cooled, in advance (Chapter 5).

STEP B

Prepare one recipe Vanilla Pastry Cream.

STEP C: THE DAY YOU WANT TO ASSEMBLE THE CROQUEMBOUCHE

Place the vanilla pastry cream into a pastry bag fitted with a small round tip. Insert the tip into the hole of each cream puff, and fill each puff with cream. Set aside. Repeat with the remaining puffs.

STEP D

Prepare one to two recipes of Caramel (recipe follows).

CARAMEL FOR CROQUEMBOUCHE

Lessons demonstrated in this recipe:

- How to prepare caramelized sugar using the wet method.
- Corn syrup acts as an interfering agent to control crystallization.

| MEASUREMENTS | | | INGREDIENTS |
U.S.		METRIC	
7¼ ounces	1 cup	200 g	granulated sugar
	2 tablespoons	30 mL	light corn syrup
2 fluid ounces	¼ cup	60 mL	water
6 ounces	1 cup	170 g	semisweet chocolate, finally chopped, or mini chips, or nonpareils (small discs of chocolate coated in tiny sugar pellets) for sprinkling
1¼ ounces	¼ cup	34 g	finely chopped pistachios for sprinkling

1. Place the sugar and corn syrup in a small heavy saucepan. Pour the water around the edges of the pan. Do not stir. Place the pot over medium-high heat and bring the mixture to a rolling boil. If any crystals form on the inside of the pan, wipe them down with a pastry brush dipped in water. Turn the heat to high if the mixture is not boiling.

2. Boil the mixture undisturbed, watching closely, until the caramel thickens and turns the color of very light brewed tea. Remove from heat. Do not allow the caramel to get too dark. A candy thermometer is not needed for this recipe because color is the determining factor as to when cooking is completed. Allow the caramel to cool for 3 to 5 minutes or until it just begins to thicken.

   The caramel will be used as glue to stick the cream puffs together. Caramel thickens as it cools, so the cream puffs will be dipped into the warmer, thinner caramel to build up the shape of the croquembouche. Then when it has thickened more, the caramel syrup will be drizzled off of a fork over the croquembouche, which will harden and form a hard caramel coating (Figure 10–14).

FIGURE 10–14

(*Note:* The coated doily [which is coated with a waxy substance] or parchment paper makes it easier to peel off the cream puff later.)

TIP To clean the pot, fill the pot of hard caramel with water halfway up and bring it to a boil on the stove. Remove the pan from the heat and let it sit until the caramel melts.

TIP The caramel can also be used to make decorations on a sheet pan covered with a silicone baking mat. Once the caramel reaches the consistency of honey, drizzle it in lines or a pattern of swirls onto the baking mat and allow it to harden. The caramel decorations can be used to garnish various desserts.

STEP E: ASSEMBLY

1. On an 8- to 9-inch (20- to 23-cm) cardboard cake circle or a large serving plate covered with a coated doily or a round of parchment paper, dip each filled puff into the caramel and place it onto the doily or parchment paper, with the glazed side facing the same direction as where the next puff will go to form a small circle of approximately 12 puffs that are stuck together (Figure 10–15).

2. Add another layer of puffs on top of the bottom layer of 12, dipping them into the caramel to help them stick together (Figure 10–16). Add puffs, layer by layer, using fewer as you go up higher to form a hollow cone, finally placing one puff right on the top. All the puffs should be used.

   Alternatively, a croquembouche mold can be used to shape the cream puffs into a cone. A croquembouche mold is a cone-shaped piece of stainless steel with a flat top around which filled cream puffs can be stacked. Once the caramel hardens, the croquembouche is lifted off the mold.

3. As the caramel cools it will thicken and reach the consistency of honey. Drizzle thin strings of caramel using a fork all over the cream puff cone to form a caramel coating (Figure 10–17). Sprinkle chopped chocolate or mini chocolate chips, nonpareils, or nuts all over, if desired (Figure 10–18). (If the caramel has already hardened, make a second batch for drizzling.)

   This makes a spectacular edible centerpiece. The croquembouche should be eaten as soon as possible, but it can be stored in the refrigerator up to 4 to 6 hours. The caramel will absorb moisture from the refrigerator and become soft if left any longer.

FIGURE 10–15

FIGURE 10–16

FIGURE 10–17

FIGURE 10–18

Croquembouche

SPIKED CARAMEL-DIPPED NUTS

Makes 20 Spiked Caramel-Dipped Nuts

Lessons demonstrated in this recipe:

- How to prepare caramelized nuts using the wet method.
- The nuts can be used as a garnish on various desserts to create an impressive visual presentation.
- The caramel is almost the consistency of molasses so that it adheres to the nuts and forms a spike.

STEP A:

Make one recipe of Caramel for Croquembouche.

STEP B: SPIKED CARAMEL-DIPPED NUTS

MEASUREMENTS		INGREDIENTS
U.S.	**METRIC**	
Approximately 20	Approximately 20	whole hazelnuts or almonds

1. Place a toothpick into the ends of each nut just far enough so the nut is secured on the toothpick. Set them aside. (If using almonds, place the toothpick in the flat side of the almond, roughly two thirds to three fourths of the way toward the rounded end of the nut. The pointed end of the nut will be the end from which the caramel drips out. If using hazelnuts, place the toothpick toward the middle of the nut, pointed end down, to allow the caramel to drip from the point.)

2. Place a heavy sheet pan along a work table with the long side lined up with the edge of the table's surface. Secure one of the skewered nuts under the sheet pan with the nut facing out to get an idea as to how far out the skewer should be. It should stick out over the edge of the table yet should be secured enough to stay in place. Lay parchment paper on the floor to catch the dripping caramel. Remove the skewered nut from under the sheet pan and place it with the remaining skewered nuts next to the work area.

3. Follow the Caramel for Croquembouche recipe through step 2.

4. Quickly hold one skewered nut at a time and completely dip the nut into the caramel and lift it out (Figure 10–19). Place the toothpick under the edge of the sheet pan so the nut (point facing down) is facing out over the edge of the table. The caramel will continue to drip (Figure 10–20). After a minute or so, the caramel will cool and harden, leaving a long spike of caramel hanging from the nut.

FIGURE 10–19

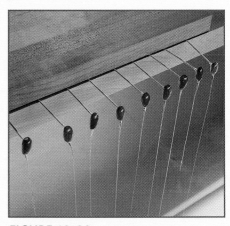

FIGURE 10–20

5. Continue the above process with the remaining nuts.

6. Once cooled and hardened, using kitchen shears, trim the spikes between 1 and 2 inches (2.5 and 5 cm) away from the nuts (Figure 10–21). Detach each nut from its toothpick and place them in an airtight container with a desiccant for at least 2 weeks or in an airtight container in layers separated by parchment paper in the freezer for up to 2 weeks.

Variation: To make caramel dipped nuts without spikes, follow the recipe, but do not skewer the nuts with a toothpick. Dip each nut into the caramel using a fork, allowing the excess to drip off. Place the caramelized nuts on a sheet pan covered with a silicone baking mat to cool and harden. Store as noted above.

FIGURE 10–21

Dessert Garnished with Spiked Caramel-Dipped Nuts

Professional Profile

BIOGRAPHICAL INFORMATION

Jeffrey Ward, CEC, CCE
Dean of Patisserie and Baking
Pennsylvania Culinary Institute
Pittsburgh, PA

1. Question: *When did you realize that you wanted to pursue a career in baking and pastry?*

Answer: *When I was 15, I was helping my mom, aunts, and grandmother make fresh Polish sausage and I enjoyed it so much. I wanted to get into some part of the industry. After I had culinary training, I wanted to learn everything about baking and pastry.*

2. Question: *Was there a person or event that influenced you to go into this line of work?*

Answer: *It was in school when I began to take the baking courses, that I realized how much I enjoyed them. I had done a lot of culinary but the baking and pastry was new. I realized I had to have baking and pastry skills to be well rounded. I also thought that with the baking and pastry I would be much more marketable.*

3. Question: *What did you find most challenging when you first began working in baking and pastry?*

Answer: *The challenging aspect of going from culinary to baking is to change from a free-wheeling approach to a more exact one. There is no "a pinch of this and a pinch of that" in baking.*

4. Question: *Where and when was your first practical experience in a professional baking setting?*

Answer: *In my first year out of school I got a job at a high-volume banquet facility in western New York. One day the pastry chef quit and my boss said that since I had just left school and had taken some baking and pastry courses that I could fill in. I headed the pastry shop there for 7 months.*

5. Question: *How did this experience affect your later professional development?*

Answer: *It taught me how important it is to pay attention to the basic methods and to execute them well. It also taught me the value of organization and to remember all the details.*

6. Question: Who were your mentors when you were starting out?

Answer: *Maurice Clark helped me so much when I was in "vo-tech" school. He really gave me direction. Later, my boss at that banquet facility, Norman Hart, had a tremendous impact on my career.*

7. Question: *What would you list as the greatest rewards in your professional life?*

Answer: *The real reward is when a student calls and says that my being strict with him in class and all the hours in class paid off and that he landed the job he really wanted. I also learn from students every day. That's also one of the rewards of being in education.*

8. Question: *What traits do you consider essential for anyone entering the field?*

Answer: *Patience and organization skills are the number one and two traits that are necessary in this area. And, the most overlooked quality that you have to possess is common sense.*

9. Question: *If there was one message you would impart to your student in this field what would that be?*

Answer: *You must never stop learning. You can learn something from every place you work and from every person you work with. Don't burn any bridges.*

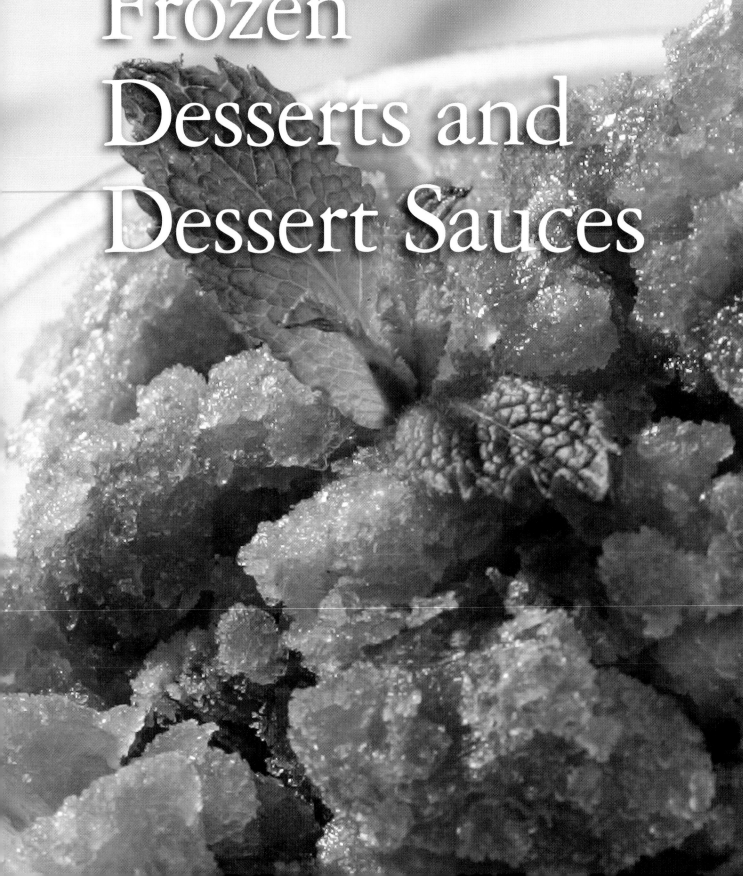

CHAPTER 11

Frozen Desserts and Dessert Sauces

After reading this chapter, you should be able to:

- Explain the difference between churn-frozen and still-frozen desserts.

- Explain the factors that contribute to texture in frozen desserts.

- Describe the roles that sugar and salt play in preparing frozen desserts.

- Demonstrate how to prepare the churn-frozen and still-frozen desserts in this chapter.

Ice creams and other frozen desserts always evoke such wonderful memories either from one's childhood or adulthood. Frozen desserts can be refreshing and simple such as a sorbet or granité, or they can consist of a rich, creamy frozen custard used as a grande finale to a wonderful meal.

There exists a wide variety of frozen desserts and they include not only ice creams and frozen yogurts, but also sorbets, sherbets, granités, and semifreddos. This chapter explores different categories of frozen desserts, how they are prepared, and different factors that affect their texture and taste.

Store-bought frozen desserts are good in a pinch, but pastry chefs with a talent for creating sweet frozen concoctions are in a class all of their own.

The Importance of Good Sanitation When Preparing Frozen Desserts

Many frozen desserts use custard bases in their preparation and that is why good sanitation is most important. Custard bases containing either cooked or uncooked eggs are especially susceptible to bacterial growth. Before and after handling any egg or dairy products, hands should be thoroughly washed. All equipment, work surfaces, and utensils should also be thoroughly cleaned and sanitized before and after handling. The equipment and utensils used in the preparation of frozen desserts should be made of stainless steel or other noncorrosive material. After processing, frozen desserts can be safely stored in the freezer at 6° to 10°F (−14° to −12°C). Storing frozen desserts at lower temperatures of −10° to 0°F (−24° to −18°C) prevents large ice crystals from forming and helps to maintain the proper texture.

Some frozen desserts such as ice cream should be served at a slightly higher temperature to allow for a creamier texture on the tongue.

Two Categories of Frozen Desserts

Frozen desserts can be broken down into two basic categories:

■ Churn-frozen
■ Still-frozen

See Tables 11–1, Various Types of Frozen Desserts, and 11–2, The Make-Up of Churn-Frozen and Still-Frozen Desserts, for a breakdown of the different types of frozen desserts and the composition of each.

Table 11–1 Various Types of Frozen Desserts

CHURN-FROZEN STIRRED WHILE BEING FROZEN	"IN-BETWEEN" STIRRED ONLY A FEW TIMES, THEN LEFT TO FREEZE	STILL-FROZEN LAYS STILL WHILE FREEZING WITH NO STIRRING
Ice cream	Granité (granita)	Mousses
Gelato (Italian ice cream)		Semifreddo
Frozen yogurt		Bombes
Sorbet		Parfaits
Sherbet		

Table 11–2 The Make-Up of Churn-Frozen and Still-Frozen Desserts

CHURN-FROZEN DESSERTS	STILL-FROZEN DESSERTS
Ice cream and gelato — A custard base consisting of cream, milk, egg yolks, sugar, and flavorings	Custard base — A custard base like Crème Anglaise used for churn-frozen desserts and flavorings. Whipped cream is folded in.
Frozen yogurt — Yogurt, sugar, and flavorings	Italian meringue base — An Italian meringue base with a sugar syrup poured into beaten egg whites, added flavorings, and folded into whipped cream
Sorbet — Sugar, water, juices, fruit, fruit purée, and other flavorings	Fruit base — Fruit purée, sugar syrup, and folded into whipped cream
Sherbet — Milk, egg whites, flavoring, and/or gelatin	

Churn-Frozen Desserts

Churn-frozen desserts are desserts that are stirred as they freeze constantly being churned or rotated. Churn-frozen desserts are processed or frozen in an ice cream machine and include ice cream, frozen yogurt, gelato, sorbet, and sherbet. Flavorings are added before freezing but additional solid ingredients like nuts, fruits, or chocolate are usually added after processing.

Ice Cream

Churn-frozen desserts like ice cream usually start with a base. Most ice creams start with a custard base, either cooked or uncooked, using heavy cream and whole milk, eggs, or egg yolks, sugar, and flavorings. (*Note:* When preparing an uncooked custard base, pasteurized eggs should be used to prevent *Salmonella* contamination.) The base is then frozen in an ice cream machine where a paddle or dasher continually stirs it. There are limitless possibilities when it comes to flavoring churn-frozen desserts. By law, ice cream must contain at least 10% milkfat and a maximum of 100% overrun. Overrun is the increase in volume that occurs when air is incorporated into the base through the churning process. For a more detailed discussion on overrun, see "Incorporation of Air" under Factors That Contribute to Texture.

A custard sauce such as Crème Anglaise works well as an ice cream base, and if frozen in an ice cream machine, becomes a rich vanilla ice cream. A number of flavorings can then be added to the custard base either before or after processing.

RECIPE

Pistachio Ice Cream (This chapter, page 331)

Gelato

Gelato is a rich Italian ice cream that is denser than its American counterpart.

Frozen Yogurt

Frozen yogurt is just as the name implies, yogurt that has been frozen. Milk is cultured with bacteria to form a thickened liquid that is sweetened, flavored, and frozen in an ice cream maker. Because frozen yogurt tends to be made with low-fat milk, it contains a lower percentage of milkfat than ice creams, making it relatively low in calories and cholesterol.

Sorbet

Unlike ice cream and frozen yogurt, sorbet contains no dairy products at all and tends to have a harder consistency than fat-laden ice creams. Sorbets are usually made from a base consisting of

 TIP Crème can be chilled and frozen in an ice cream machine to produce a rich, creamy ice cream.

fruit or fruit purée sweetened with a sugar syrup made from granulated sugar and water, usually in equal parts, that is brought to a boil just to dissolve the sugar. The sugar in the base prevents the sorbet from freezing into a block of ice. Because there are very few solids in a sorbet, when it is produced commercially stabilizers may be used. Stabilizers keep the liquid ingredients blended together and act as an emulsifying agent to prevent any separation. Sorbets usually contain very little fat. Often they are served between courses of a meal as a palate cleanser, or *intermezzo*, to whet the appetite. Because sorbets are often made from fruits, they can be a colorful and refreshing dessert in warmer months.

RECIPES

Mango Lime Sorbet (This chapter, page 335)
Spiced Pomegranate Chai Sorbet (This chapter, page 336)
Tangerine Orange Sorbet (This chapter, page 337)

Sherbet

A sherbet is slightly richer than a sorbet because it contains a small amount of milk or cream and sometimes egg whites and gelatin to provide a creamier texture. By law, sherbet must contain less than 2% milkfat and usually has more sugar to compensate for the lack of fat. Sherbets are generally fruit flavored and are not as rich as ice creams.

RECIPE

Limon Sherbet (This chapter, page 339)

Why Churning Is So Important

Churning becomes a crucial step in preparing churn-frozen desserts, ensuring that they have a smooth and creamy texture. Churning contributes a great deal to texture by preventing large ice crystals from forming and incorporating air into the base as it freezes. (The effect of air being incorporated into the base during churning is discussed later in this chapter.) When water is frozen, ice is the result. Because frozen dessert bases very often contain some water-based ingredients (even milk and cream contain a certain percentage of water), how they are frozen will affect their final texture.

Smaller ice crystals feel better on the tongue than larger, grittier ones. This is where churning becomes very important. Rotating and stirring the base ingredients as they are being frozen results in smaller, finer ice crystals that are barely noticed on the tongue. When tasting an ice cream, for example, a sensation of creaminess should be felt, not iciness from large chunks of ice. The active churning never allows any ice crystals that form to grow beyond a certain size, thus keeping the dessert smooth. (See Factors That Contribute to Texture.)

Between Churn-Frozen and Still-Frozen

Before discussing still-frozen desserts, there is an "in-between" category of frozen desserts that falls somewhere between churned and still-frozen known as granité (granita). *Granité* is the French for "granite" and *granita* is the Italian name for this coarse, icy dessert. They consist of a base that is similar to a sorbet base such as sugar and water flavored with fruit, fruit juices, coffee, tea, herbs, or alcohol. They tend to be less sweet so that they will freeze to a coarse, icy consistency. They are stirred only a few times while freezing before being allowed to harden. To serve them, the flavored ice is scraped with a fork and the slushy shards of ice are scooped into dessert glasses and served as a refreshing light dessert or as an intermezzo between courses of a meal. Granité or granita typically do not use an ice cream machine. However, if a large quantity is being prepared, they can be processed in an ice cream machine and poured into rectangular pans such as a hotel pan to be frozen until firm. The base can be broken up into chunks and placed in a food processor to chop up the ice to a coarse consistency instead of scraping it with a fork.

RECIPE
Pomegranate Chai Granité (This chapter, page 341)

Still-Frozen Desserts

Still-frozen desserts often start much the same as churn-frozen desserts, except still-frozen desserts lay still as they freeze with no agitation or churning. They include mousses, semifreddos, bombes, and parfaits. Because still frozen desserts are not churned and stirred while freezing, other ingredients such as beaten egg whites or whipped cream are folded in to lighten the base. The air that is incorporated into the base mixture prevents large ice crystals from forming during freezing. The light, fluffy mixture is then poured into molds, soufflé cups, ramekins, or layered with other ingredients like cake, cookies, lady fingers, praline, nuts, or sauces. The variations for still-frozen dessert bases and flavorings are endless. Many custard bases used for churn-frozen desserts can also be used as still-frozen bases. It is what is done with those bases in the latter stages of preparation, freezing, and molding that distinguishes them into their own categories. For instance, a custard base lightened with meringue or whipped cream and still frozen is re-ferred to as a frozen mousse. A frozen mousse can also be referred to as a semifreddo or frozen soufflé. That same mixture layered with other flavored mousses and poured into a mold is re-ferred to as a bombe.

Yet another still-frozen dessert called a European-style parfait is made from a boiling sugar syrup beaten into egg yolks or whole eggs with whipped cream and flavoring ingredients folded in. The mixture is poured into glasses or molds and still frozen. American parfaits are composed of ice creams layered with sauces, nuts, and whipped cream.

There are three types of bases that can be made into a still-frozen dessert:

- Custard base. A custard base is simply a cooked custard like Crème Anglaise plus flavorings. This is the same type of base used to prepare some churn-frozen desserts such as ice cream. However, in a still-frozen dessert, often the custard base is then folded into whipped cream.
- Italian meringue base. An Italian meringue base starts with a boiling sugar syrup poured into beaten egg whites and beaten until the volume increases and stiff peaks form. Flavoring ingredients are added and folded into whipped cream.
- Fruit base. A fruit base is made using fruit purées with a sugar syrup which is then folded into whipped cream.

> Three Types of Bases That Can Be Used to Make a Still-Frozen Dessert
> - Custard base
> - Italian meringue base
> - Fruit base

RECIPE
Black and White Chocolate Semifreddo (This chapter, page 343)

Factors That Contribute to Texture

There are many reasons why people love ice cream and other frozen desserts. They have characteristic qualities that are highly desirable. The following are five factors that contribute to texture:

Size of Ice Crystals

High-quality frozen desserts such as ice cream should be smooth and creamy in texture and never grainy or icy. Graininess in a frozen dessert like ice cream is directly related to the formation and size of ice crystals during freezing. The smaller the ice crystals formed, the smoother the finished product. Larger ice crystals are coarse and feel grainy on the tongue. To keep the ice crystals small for a churn-frozen dessert, the actual freezing process must be accomplished quickly with constant churning, keeping the temperature as low as possible. Chilling the base before churning helps toward this goal. In still-frozen desserts, the addition of air-filled meringues or whipped cream acts as a barrier and prevents large ice crystals from forming.

Amount of Fat

Another major factor that contributes to texture is the amount of fat used. Typically, the highest quality ice creams are the ones with the most fat. They taste the smoothest and creamiest. As the water in the ice cream base begins to freeze, fats flow between the molecules of water, coating them and keeping them from joining together and enlarging to form large, grainy ice crystals. The more fat used in both churn-frozen and still-frozen desserts, the creamier they taste.

Incorporation of Air

The amount of air incorporated into a frozen dessert base also affects texture. The constant stirring or churning of churn-frozen desserts incorporates air into the base, which makes it light and airy in texture. This increase in volume due to the incorporation of air is referred to as overrun. A certain percentage of overrun helps make an ice cream's texture light and airy. Too much air incorporated into an ice cream produces a foamy texture that is undesirable while too little air produces a hard, dense, coarse-textured ice cream.

Overrun is expressed as a percentage. For example, if after churning the base expands to double its original volume, it has an overrun of 100%. Just because the law allows ice cream to have a maximum of 100% overrun does not mean the ice cream will be of the highest quality. Ice creams of lesser quality tend to contain the minimum amount of fat allowed by law (10%) and the highest percentage of overrun. The highest quality ice creams can contain percentages of fat higher than the minimum set by law, up to twice as much (20%), and the lowest amount of overrun, sometimes as low as 20%.

The amount of overrun produced is dependent upon several factors, including the ice cream machine used, how full the freezer section of the machine is filled, how much fat is contained in the base, and how long the base is churned in the machine.

Emulsifiers and Stabilizers

In commercially produced ice creams, certain ingredients like emulsifiers and stabilizers are added to create a more desirable texture and prevent any separation of ingredients. Egg yolks, commonly part of an ice cream custard base, become a perfect emulsifying agent able to keep the water- and fat-based ingredients of the base in a suspension without separating. Stabilizers may include such ingredients as vegetable gums, gelatin, and agar. These ingredients bind themselves to water drops that may melt out from within the ice cream during variations of temperature during storage. The stabilizer binds itself to the water and prevents it from attaching itself to other ice crystals, which would cause them to form larger, grainier ice crystals.

Aging

Aging describes the act of placing the finished base into a refrigerator to chill for several hours or overnight before freezing it. As the temperature of the base cools down, flavors begin to mingle and changes take place within it. First, the fat begins to become more solid. Second, the proteins within the milk and cream swell and bind with water, giving the mixture more body. This creates thickness, which translates after freezing to a smoother texture in the finished product. Bases can be aged for as little as 4 hours. Ideally, bases should be aged for approximately 12 to 24 hours.

Summing Up Texture with Mouthfeel

Mouthfeel is the term used to describe how a food tastes and feels in the mouth and on the tongue. Mouthfeel is most important when describing the texture and different qualities of foods. Chefs know how important mouthfeel is in frozen desserts. A high-quality ice cream melts into a smooth, creamy liquid in the mouth. There should be no graininess from undissolved sugar or ice crystals detected on the tongue. As previously discussed, all frozen desserts contain ice crystals, but the ones with small crystals taste the least icy and grainy. Often, the higher the fat content, the better the mouthfeel. Lower fat or fat-free ice creams tend to be icy and grainy with poor mouthfeel.

Building Flavors in Frozen Desserts

It is more difficult to flavor a frozen dessert than a dessert that is to be served warm or at room temperature because the molecules in frozen foods move more slowly than the molecules in warmer foods. This slowing down of food molecules tends to dull flavors because flavor components reach nerve endings in the nose and taste buds more slowly. This is important information to have when preparing frozen desserts because flavors need to be bolder and more pronounced. Flavoring ingredients for frozen desserts are only limited by one's imagination and can be complemented with the addition of solid ingredients folded into the base after it has been processed in an ice cream machine. Some of these ingredients include chopped chocolate, nuts, fruits, preserves, cookies, and caramel.

After chilling the base, taste it and adjust the flavor as necessary before processing it in an ice cream machine. For example, if a vanilla bean Crème Anglaise base lacks flavor after chilling, try adding some pure vanilla extract to boost the flavor to where it will be more pronounced, even after freezing.

After processing, pack ice creams and churn frozen desserts into an airtight container, cover, and freeze for several hours to harden and develop their flavor and texture even more.

The Power of Sugar and Salt in Preparing Frozen Desserts

Sugar and salt, both crystalline solids, are very important in the preparation of frozen desserts. Sugar as an ingredient within a frozen dessert and salt as an aid in the actual freezing process, both depress the freezing point of water. This means that a water mixture containing sugar or salt will not freeze to a completely solid state, even below freezing temperatures. This becomes very useful to the pastry chef when preparing frozen desserts.

Sugar

Sugar is one of the key ingredients in a frozen dessert. The amount of sugar added to a base determines how hard or soft the final product is. If a container of water is placed in the freezer, it will freeze rock hard in a matter of hours. This does not make a very appealing dessert. But if sugar is dissolved in water and placed in the freezer, after several hours the mixture will be slushy and can be scooped, never hardening completely. The sugar prevents the water from freezing the base to a rock hard state by keeping it in an icy semisolid state, even at freezing temperatures.

The process of freezing occurs gradually. The sugar within the base actually blocks water molecules from getting close enough to join together to form ice crystals. This causes the freezing point that would normally occur at 32°F (0°C) to drop to 27°F (−3°C). As the temperature continues to decrease, the molecules of water slow down and begin to crystallize by freezing, despite the sugar-containing base. Because some of the water has effectively been removed from the base (through crystallization) the base mixture that remains becomes even more concentrated with sugar. This in turn lowers the freezing point even further.

The ratio of liquid to sugar determines how hard the dessert becomes. The greater the amount of sugar added, the softer the dessert. For example, a recipe for sorbet would generally contain twice the sugar that a recipe for a granité would have. Sorbets tend to be soft and easily scooped, whereas a granité should freeze more solidly into a coarser texture. Recipes vary, so the chef should experiment and vary the amount of sugar to attain the desired consistency. (*Note:* Fruits and fruit juices contain natural sugar that could affect the consistency of the dessert. Adjust the sugar accordingly.) Sugar is the ingredient of choice to add to frozen dessert bases; even though salt also depresses the freezing point of water, the dessert would be inedible.

Salt

Years ago, salt was a crucial aid in the freezing of churn-frozen desserts. An icy brine that consisted of salt, water, and ice cubes surrounded a container holding the base in a machine known as a hand-cranked ice cream freezer. Modern technology has eliminated the need for salt and hand cranking, but some chefs still prepare churn-frozen desserts in these old-fashioned machines. They believe that they have more direct control over the freezing process and the texture of the final product. Typically, rock salt, consisting of large, grayish crystals, also called *ice cream salt*, was used to create the icy brine. Although it is a finer consistency, regular table salt can also be used successfully.

In order to discuss the role of salt and how it relates to frozen desserts, a brief discussion of ice cream machines is necessary. The faster the base freezes, the smaller the ice crystals and the smoother the dessert. Salt is used in certain ice cream machines as a way to control the temperature of the water surrounding the base. The more salt added, the lower the temperature of the water. The lower the temperature of the water, the faster the base will freeze.

There are several types of ice cream machines on the market both for commercial and home use. Both keep liquid ingredients in a frozen dessert base from freezing rock hard. Commercial ice cream machines have a built-in compressor that freezes the base in a tub or can be fitted with a motorized dasher that keeps the base moving continously. These machines are usually made of stainless steel or other noncorrosive material. They are quite large and can freeze several gallons at once without the need for salt and ice.

Some chefs in smaller operations may prefer to prepare frozen desserts in small batches in either a manual or electric ice cream machine. The manual machines use a hand crank to rotate the base, which is surrounded by a mixture of ice, salt, and water. This process can take a great deal of time depending on the size of the machine. Smaller manual machines make small batches in about 30 minutes. Electric ice cream machines are also available to produce very high-quality frozen desserts. Some electric ice cream machines require the empty tub be frozen for several hours before the machine is plugged in; then when the tub starts rotating, the base is poured into the tub for processing. Other electric ice cream makers require that an icy, saltwater brine surround the tub during processing; with these machines it is necessary to start out with the base placed inside a metal can in which a dasher is inserted. The can is placed inside a large, insulated bucket-like opening in the machine. Between the bucket and metal can there is a space large enough to fill with an ice, salt, and water mixture. Before filling the space to begin the freezing process, the machine needs to be turned on so the base is rotating with the dasher in it before any freezing actually takes place. It is important to begin turning the base before the freezing process starts to ensure uniform freezing. This avoids any large ice crystals from forming. To prepare the brine, a small amount of icy water is first placed in the space between the bucket and metal can followed by layers of ice, cubed or chopped, and salt. Placing water and ice down first prevents any salt from gathering at the bottom of the machine. The salt behaves like sugar and depresses the freezing point of the water, allowing the ice to stay liquid at a lower temperature than freezing (32°F; 0°C) and keeping it in a super cold, liquid state. This super, icy cold water flows completely around the can, covering more surface area and freezing the base quickly and more efficiently than ice cubes alone. Processing time for these machines ranges from 20 minutes to 1 hour.

Regardless of which type of ice cream machine is used, it is important to follow the manufacturers' directions.

Tips for Preparing Successful Frozen Desserts

- Always clean and sanitize all equipment, utensils, and work surfaces before and after preparing frozen desserts.
- Be sure that the base is well chilled before processing it in an ice cream machine. A warm base takes much longer to freeze and may develop large ice crystals, becoming icy and grainy.
- Never fill the can of the ice cream machine more than two thirds full. The space allows for expansion while air is incorporated into the base.
- After processing, pack the soft frozen mixture into an airtight container, covering it with plastic wrap and sealing it well to prevent unwanted food odors from seeping in. The mixture should be frozen for several hours or overnight to develop flavors before being served.
- If the base is combined with solid ingredients, fold them in gently after processing but before packing into the airtight container.
- No matter what type of machine is being used for processing, the base should always be rotating before the freezing process begins to ensure even freezing.
- Frozen desserts are best stored at −10° to 0°F (−24° to −18°C) or below. The quality of a frozen dessert can be maintained for up to 2 months, if stored properly.

■ Although alcohol and liqueurs can be used to flavor frozen desserts, they can lower the freezing point of the base to the degree that the dessert is too soft and will never completely freeze.

■ Most sorbet bases can be made into a granité or granita by partially being still-frozen, stirring often until icy chunks form, and then freezing for 3 to 4 hours. If the base is put through an ice cream machine, it becomes a sorbet. (*Note:* In order to achieve an icier consistency for a granité, it may be necessary to decrease the sugar to a ratio of 4 parts water-based liquid to 1 part sugar.)

PISTACHIO ICE CREAM

Makes 64 fluid ounces
(2 quarts)

Lessons demonstrated in this recipe:

- How to prepare ice cream, a churn-frozen dessert.
- The base of an ice cream is a stirred custard without a starch that contains cream, milk, sugar, eggs, and flavorings.
- Eggs coagulate when heated, thickening the custard.
- Tempering the eggs allows them to be gradually brought up to temperature without curdling.
- Heating the custard to a maximum of 185°F (85°C) allows thickening without curdling.
- The thickened custard should coat the back of a spoon.
- Chilling the base thoroughly allows flavors to develop and ensures it will freeze quickly, resulting in a creamier texture.

ICE CREAM BASE

U.S.		METRIC	INGREDIENTS
2 each		94 g	large eggs
5½ ounces	¾ cup	155 g	granulated sugar
8 fluid ounces	1 cup	240 mL	whole milk
16 fluid ounces	2 cups	480 mL	heavy cream
½ fluid ounce	1 tablespoon	15 mL	almond extract
2 to 3 drops		2 to 3 drops	green food coloring, optional
4 ounces	¾ cup	120 g	pistachio nuts, shelled and coarsely chopped (use natural, not red)

1. Set up an ice water bath to cool the hot custard quickly. Fill a large bowl halfway with ice and water. Place a smaller bowl into the larger one and set a strainer over the smaller bowl. Set the ice water bath aside.

2. In a medium mixing bowl (placed on a folded wet kitchen towel to keep it stable), whisk 2 large eggs until they turn thick and a lemon yellow color, approximately 2 to 3 minutes.

3. Gradually whisk in the sugar, a little at a time, until well combined.

4. In a medium saucepan over medium-high heat, scald the milk by heating the milk to just below the boiling point until small bubbles form around the outer edges of the pan.

(Note: The cooked ice cream base, also known as a Crème Anglaise, can be used as a dessert sauce if left unfrozen. This chapter, page 348.)

5. While whisking constantly, temper the eggs by gradually dribbling a little hot milk into the egg mixture (Figure 11–1). Keep adding milk until all of it is in the egg mixture, whisking all the time.

6. Pour the egg-milk mixture back into the saucepan.

7. Now that the eggs have been tempered, place the pan over medium-low heat until the custard sauce thickens a bit. Stir constantly using a wooden spoon and watching carefully. The custard should never come to a boil or the eggs could curdle. For best results, use a kitchen thermometer. When the temperature reaches 185°F (85°C), remove the pan from the heat (Figure 11–2).

   The custard should coat the back of the spoon and leave a path as you pull your finger across it. Do not place the spoon back into the custard.

8. Pour the custard through the strainer of the prepared ice water bath (Figure 11–3). The strainer catches lumps of curdled egg that form, ensuring a smooth custard, while the ice water bath rapidly cools the custard. Allow the mixture to cool for a few minutes.

9. Whisk in the heavy cream, almond extract, and food coloring, if using.

10. Allow the custard to cool, then cover the bowl with plastic wrap and refrigerate several hours or overnight until the mixture is well chilled, or to chill the mixture more quickly, place the bowl over an ice water bath.

FIGURE 11–1

FIGURE 11–2

FIGURE 11–3

FIGURE 11–4

FIGURE 11–5

> **TIP**
> After processing, scoop the soft ice cream into a plastic container and fill it almost to the top. Cover the surface of the ice cream with plastic wrap and cover tightly with a lid; this prevents air from getting into the ice cream and prevents freezer burn and ice crystals from forming. Freeze for several hours until firm.

11. Process the cold base in an ice cream machine according to the manufacturer's directions (Figure 11–4). When the ice cream is finished, stir in pistachio nuts until well blended (Figure 11–5).

Variations: If pistachio is not a favorite, there are many possibilities for different flavorings to be added to the base custard recipe. Add any of the following items to the base before processing, omitting the almond extract, green food coloring, and pistachio nuts:

■ Add 1 to 2 tablespoons (15 to 30 mL) vanilla extract to make vanilla ice cream.
■ Experiment with other flavoring extracts such as peppermint, lemon, or coconut.

You can also add any of the following items to the base after processing:

■ Fold in 6 ounces (1 cup; 180 g) chopped semisweet chocolate
■ 6 ounces (1 cup; 180 g) chopped peanut butter cups
■ Lightly swirl in 8 fluid ounces (1 cup; 240 mL) of room temperature Rich Chocolate Sauce (this chapter, page 351) for a marble, swirl effect.

Pistachio Ice Cream

MANGO LIME SORBET

Makes approximately 32 fluid ounces (1 quart; 1 L)

Lessons demonstrated in this recipe:

- How to prepare sorbet, a churn-frozen dessert.
- Because cold temperatures dull flavors, ripe mangoes add intense fruit flavor and natural sweetness.

MEASUREMENTS			INGREDIENTS
U.S.		METRIC	
1 fluid ounce	2 tablespoons	30 mL	corn syrup
7¼ ounces	1 cup	200 g	granulated sugar
4 fluid ounces	½ cup	120 mL	water
14½ ounces		415 g	very ripe mangoes, slightly less than 1 pound, total
			juice from 1½ limes
			zest from 1 lime
2 fluid ounces	¼ cup	60 mL	fresh orange juice

1. In a heavy saucepan, bring corn syrup, sugar, and water to a boil until the sugar is completely dissolved. Set aside to cool.

2. Peel each mango and cut off as much flesh as possible, avoiding the fibrous areas too close to the pit. Cut the flesh into small chunks.

3. Purée the mangos, lime juice, and zest until completely smooth.

4. Through the top of the food processor, slowly pour the cooled sugar syrup and orange juice over the purée and blend well (Figure 11–6). Taste for seasoning to see if more sugar is needed. Pour the purée into a bowl and chill in an ice water bath or in the refrigerator until it is ice cold.

5. Process the mixture in an ice cream machine. Pour sorbet into an airtight container and freeze for several hours.

FIGURE 11–6

SPICED POMEGRANATE CHAI SORBET

Makes approximately 32 fluid ounces (1 quart; 1 L)

Lessons demonstrated in this recipe:

- How to prepare sorbet, a churn-frozen dessert.
- The spices in the chai tea offset the tartness of the pomegranate.
- Fruit sorbet makes a refreshing intermezzo or a light dessert.

MEASUREMENTS			INGREDIENTS
U.S.		METRIC	
20 fluid ounces	2½ cups	600 mL	pomegranate juice, fresh or bottled
3½ ounces	½ cup	100 g	granulated sugar
2 each		2 each	chai spice black tea bags
			zest from 1 lime
½ fluid ounce	1 tablespoon	15 mL	freshly squeezed lime juice

(*Note:* Bottled pomegranate juice contains some sugar and if fresh pomegranate juice is used, more sugar may be needed, so taste the base before processing.)

1. In a heavy medium saucepan, stir together the pomegranate juice and sugar. Bring the mixture to a boil and stir until the sugar dissolves. Remove the mixture from the heat.

2. Add the 2 tea bags, lime zest, and lime juice and whisk well (Figure 11–7). Pour into a bowl and allow the mixture to cool down to room temperature. Leave the tea bags in the mixture to steep.

3. Remove the tea bags, squeezing out as much liquid as possible, and pour the mixture into a bowl and chill until cold.

4. Process the sorbet base in an ice cream machine.

5. Scrape the sorbet into an airtight container and freeze for several hours.

FIGURE 11–7

TANGERINE ORANGE SORBET

Makes approximately
24 fluid ounces (3 cups;
720 mL)

MEASUREMENTS			INGREDIENTS
U.S.		METRIC	
6 fluid ounces	¾ cup	180 mL	water
4¾ ounces	⅔ cup	130 g	granulated sugar
10 fluid ounces	1¼ cups	295 mL	tangerine juice (from 4 to 5 large tangerines)
10 fluid ounces	1¼ cups	295 mL	freshly squeezed orange juice
½ fluid ounce	1 tablespoon	15 mL	fresh lime juice

1. In a heavy medium saucepan, combine water and sugar. Bring the mixture to a boil, stirring until the sugar dissolves completely. Remove from heat. Cool to room temperature.

2. Pour the cooled sugar syrup into a bowl and blend in the tangerine, orange, and lime juices (Figure 11–8). Chill until ice cold.

3. Process the base in an ice cream machine.

4. Pour the sorbet into an airtight container and freeze for several hours.

FIGURE 11–8

A Sorbet Presentation

LIMON SHERBET

Makes approximately 32 fluid ounces (1 quart; 1 L)

Lessons demonstrated in this recipe:

- How to prepare sherbet, a churn-frozen dessert.
- A combination of lemon and lime provides a pleasant tartness.
- Gelatin is used as a stabilizer to improve texture.
- Beaten egg whites are added to lighten the base before processing.
- A small amount of yogurt and milk are added for creaminess.
- Pasteurized egg whites are used because the egg whites are not cooked.
- The base is not chilled to prevent the gelatin from gelling before other ingredients are added and it is processed.

MEASUREMENTS			INGREDIENTS
U.S.		**METRIC**	
$\frac{1}{4}$ ounce	$2\frac{1}{2}$ teaspoons	7 g	unflavored gelatin
8 fluid ounces	1 cup	240 mL	water
$3\frac{1}{2}$ ounces	$\frac{1}{2}$ cup	100 g	granulated sugar
4 fluid ounces	$\frac{1}{2}$ cup	120 mL	corn syrup
			zest from 2 lemons and 2 limes
2 fluid ounces	$\frac{1}{4}$ cup	60 mL	lime juice
2 fluid ounces	$\frac{1}{4}$ cup	60 mL	lemon juice
8 fluid ounces	1 cup	240 mL	whole milk
4 ounces	$\frac{1}{2}$ cup	120 g	lemon yogurt
2 each		56 g	pasteurized egg whites
$1\frac{3}{4}$ ounces	$\frac{1}{4}$ cup	50 g	granulated sugar

(Note: If using liquid pasteurized egg whites, be certain the egg whites are able to be beaten to a foam. If using dried egg whites, reconstitute according to the manufacturer's directions.)

1. **Place gelatin in a heavy medium saucepan and add water, granulated sugar, corn syrup, and zest. Bring the mixture to a boil until all the sugar is dissolved.**

2. **Remove from the heat and stir in lime and lemon juices. Strain into another bowl and cool to room temperature.**

3. **Whisk in milk and yogurt (Figure 11–9). Set aside.**

FIGURE 11–9

4. In the bowl of an electric mixer, beat the egg whites using the whip attachment until foamy and gradually add the remaining granulated sugar. Beat until soft peaks form.

5. Whisk the beaten egg whites into the reserved mixture (Figure 11–10). Process in an ice cream machine. Pour into an airtight container and freeze for several hours.

FIGURE 11–10

Limon Sherbert

POMEGRANATE CHAI GRANITÉ

Makes approximately 16 fluid ounces (1 pint; 0.5 L)

Lessons demonstrated in this recipe:

- How to prepare a granité, an "in-between" churn- and still-frozen dessert.
- Less sugar in the base allows a coarser, icier texture.
- The same flavors are used as in the pomegranate sorbet, so comparisons in texture can be made between the two recipes.

MEASUREMENTS			INGREDIENTS
U.S.		METRIC	
8 fluid ounces	1 cup	240 mL	water
2½ ounces	⅓ cup	70 g	sugar
8 fluid ounces	1 cup	240 mL	pomegranate juice
	1 teaspoon	5 mL	freshly squeezed lime juice
			1 chai spice black tea bag

1. Bring water and sugar to a boil making sure all of the sugar is completely dissolved. Remove from heat.

2. Whisk in pomegranate and lime juices. Add chai tea bag and allow the mixture to steep until cool.

3. Pour the mixture into a rectangular pan such as a hotel pan. The more spread out the base is, the quicker it will freeze. Freeze for 45 minutes and stir the ice that forms in the corners of the pan and bring them into the center (Figure 11–11). Freeze for 30 to 45 minutes more, stirring and chopping up any large chunks of ice (Figure 11–12). Repeat chopping every 30 minutes until the mixture is comprised of small shards of ice. Alternatively, the frozen mixture can be placed into a food processor and pulsed until the ice chunks are coarsely chopped.

4. Scrape the granité with a fork and scoop into glasses to serve at once (Figure 11–13).

(Note: This recipe can be doubled or tripled. The base should be divided between two or three pans so that the base is never greater than 1 inch [2.5 cm] to facilitate the freezing process.)

FIGURE 11–11

FIGURE 11–12

FIGURE 11–13

Pomegranate Chai Granité

BLACK AND WHITE CHOCOLATE SEMIFREDDO

Makes 1 loaf pan: 9¼ by 5¼ by 2¾ inches (23 by 13 by 7 cm)

Lessons demonstrated in this recipe:

- How to prepare a still-frozen dessert called a *semifreddo*.
- A custard base adds richness.
- Whipped cream incorporates air, creating a light, airy texture.
- Angling the pan during freezing adds visual interest to the finished dessert.

STEP A: GANACHE CENTER

MEASUREMENTS			INGREDIENTS
U.S.		**METRIC**	
6 fluid ounces	¾ cup	180 mL	heavy cream
½ ounce	1 tablespoon	15 g	unsalted butter
½ ounce	1 tablespoon	15 g	granulated sugar
9 ounces	1½ cups	255 g	semisweet chocolate, chopped

1. Place the heavy cream, butter, and sugar into a heavy medium saucepan. Bring the mixture to a boil and remove from the heat.

2. Whisk the chocolate into the mixture, until it is well blended. Set aside and allow to cool to room temperature.

STEP B:

Make one recipe of Crème Anglaise (this chapter, page 348) through step 4. Make sure it is hot.

STEP C: ASSEMBLY

MEASUREMENTS			BAKER'S %	INGREDIENTS
U.S.		**METRIC**		
4½ ounces	¾ cup	130 g		high-quality white chocolate, finely chopped
4½ ounces	¾ cup	130 g		high-quality semisweet chocolate, chopped
16 fluid ounces	2 cups	480 mL		heavy cream, divided

1. Line a metal 9¼- by 5¼- by 2¾- inch (23- by 13- by 7-cm) loaf pan with two pieces of plastic wrap, allowing the excess plastic wrap to hang over the edges of the pan. Set aside.

2. Place approximately 12 fluid ounces (1½ cups; 360 mL) hot Crème Anglaise into each of two bowls, dividing it in half.

3. Using a different whisk in each bowl, whisk the chopped white chocolate into one bowl and the chopped semisweet chocolate into the other bowl until both chocolates are melted (Figure 11–14). Have ready two bowls sitting in two ice water baths with a sieve in each. Pour each mixture through a sieve and whisk each until it is cold, stirring frequently. Once cooled, the dark chocolate Crème Anglaise should be stored until needed in the refrigerator with a piece of plastic wrap placed directly on top of it.

4. In a bowl of an electric mixer, place 8 fluid ounces (1 cup; 240 mL) of the heavy cream. Using the whip attachment, beat the cream until it reaches soft peaks. Using a rubber spatula, fold the whipped cream into the cooled white chocolate Crème Anglaise (Figure 11–15). Place the prepared loaf pan at a 45-degree angle. Pour the mixture into the pan (Figure 11–16). Place it in the freezer, maintaining the 45-degree angle by propping it. Freeze until firm.

5. Spread the reserved ganache evenly over the frozen white chocolate mixture (Figure 11–17). Return to the freezer for 30 minutes, still maintaining the angle.

6. Whip the remaining heavy cream to soft peaks. Stir the chilled dark chocolate Crème Anglaise with a whisk in case any chocolate has settled to the bottom of the bowl. Add the whipped cream to the mixture and gently fold the two together. Remove the loaf pan from the freezer and set it flat on a smooth surface. Pour the dark chocolate mixture into the pan and spread it evenly (Figure 11–18). The mixture should now come to the top of the pan. Cover the pan with plastic wrap and freeze for several hours.

FIGURE 11–14

FIGURE 11–15

FIGURE 11–16

FIGURE 11–17

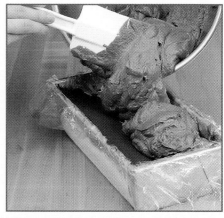
FIGURE 11–18

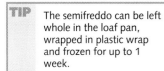

TIP The semifreddo can be left whole in the loaf pan, wrapped in plastic wrap and frozen for up to 1 week.

7. When ready to serve, unmold the semifreddo onto a cutting board. Allow it to sit at room temperature for 10 minutes to warm up slightly. Using a sharp knife, trim both ends off and cut the remainder into slices. Serve immediately.

Black and White Chocolate Semifreddo

CHERRY PORT SAUCE

Makes approximately
12 fluid ounces (1½ cups;
355 mL)

Lessons demonstrated in this recipe:

- How to prepare a starch-thickened sauce.
- Using the canned juice and fruit intensifies the cherry flavor.
- Port complements the cherry flavor and adds depth to the sauce.
- The sauce is boiled for several seconds to thicken it and to cook out any starchy taste.
- Straining the sauce makes it smooth and pipeable for painting a plate.

MEASUREMENTS			INGREDIENTS
U.S.		METRIC	
16 ounces	1 pound	450 g	dark, sweet cherries, canned
2 to 4 fluid ounces	¼ to ½ cup	60 to 120 mL	port
3½ ounces	½ cup	100 g	granulated sugar
	1 teaspoon	6 g	lemon zest
	1 tablespoon	8 g	cornstarch
1 fluid ounce	2 tablespoons	30 mL	port

1. Drain the cherries, reserving the liquid. Pour the liquid into a liquid measuring cup and add enough port to reach the 8 fluid ounces (1 cup; 240 mL) mark (Figure 11–19). Set the mixture aside. In a food processor, purée the cherries. There should be approximately 6 fluid ounces (¾ cup; 180 mL).

2. Pour the puréed cherries into a heavy medium saucepan. Add the reserved juice and port, granulated sugar, and the lemon zest.

3. Bring the mixture to a boil until the sugar is completely dissolved. Pour the mixture through a strainer into a clean saucepan, discarding any solids left in the strainer.

4. In a small bowl, using a whisk, dissolve the cornstarch into the 1 fluid ounce (2 tablespoons; 30 mL) of port until it is completely dissolved and there are no lumps (Figure 11–20). Whisk the cornstarch mixture into the cherry mixture (Figure 11–21). Place the saucepan over medium-high heat and bring it to a boil, whisking constantly. The mixture will thicken.

FIGURE 11–19

FIGURE 11–20

FIGURE 11–21

TIP If the sauce becomes too thick, thin it with some water.

5. Boil the mixture for several seconds to cook the starch while whisking constantly.

6. Remove the sauce from the heat and allow it to cool. The sauce can be served warm or cold. To store it, place the cooled sauce into an airtight container and refrigerate for up to 3 days.

Cherry Port Sauce

CRÈME ANGLAISE

Makes approximately 16 fluid ounces (2 cups; 473 mL)

Lessons demonstrated in this recipe:

- How to prepare a custard sauce where eggs and cream or milk are heated until the egg proteins coagulate and thicken.
- A custard prepared on top of the store is known as a *stirred custard*.
- The custard is cooked until it reaches 180°F (82°C) to prevent curdling of the egg proteins.
- The custard should be thick enough to coat the back of a spoon.
- The cooking process is stopped quickly by pouring the sauce into a sieve set over a bowl in an ice water bath.
- Custard sauces can be flavored in many different ways.

MEASUREMENTS			INGREDIENTS
U.S.		METRIC	
1 each		1 each	vanilla bean, split in half lengthwise
8 fluid ounces	1 cup	240 mL	heavy cream
12 fluid ounces	1½ cups	360 mL	whole milk
5 each		95 g	large egg yolks
3½ ounces	½ cup	100 g	granulated sugar

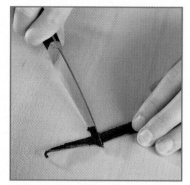

FIGURE 11–22

FIGURE 11–23

1. Using a small, sharp knife, scrape down the length of the inside of the vanilla bean and put the black pulp along with the bean into a heavy medium saucepan (Figure 11–22). Add the heavy cream and the milk. Bring the mixture to a boil. Remove the pan from the heat and cover it to allow the milk and cream to become infused with vanilla flavor.

2. Set up an ice water bath using 2 bowls: 1 large and 1 medium sized. Fill the larger bowl one third of the way with ice. Add cold water to the ice until the ice cubes are just floating. Place the smaller bowl into the ice water and the strainer into the smaller bowl. Set aside.

3. In a heatproof mixing bowl, whisk the egg yolks and the sugar until well combined. Slowly dribble in some of the hot milk mixture to temper the eggs and then gradually add the remaining milk mixture, discarding the vanilla bean.

4. Pour the custard back into the saucepan and place over medium-low heat. Stir constantly with a wooden spoon until the thermometer reaches 180°F (82°C). You may have to move the pot on and off the heat if the custard gets too hot. It should never come to a boil. When the custard is done, it should coat the back of the spoon. To ensure the custard is done, once it coats the back of the spoon, take a finger and drag it across the back of the spoon. If a path remains, the custard has thickened to the correct consistency (Figure 11–23). Do not place the spoon back into the custard sauce.

TIP If a vanilla bean is not available, in step 1, heat the milk and heavy cream to a simmer and set aside. Continue with the recipe as it is written. Add 1½ (7.5 mL) vanilla extract to the custard sauce after it has been strained and cooled in step 5.

5. Remove the custard from the heat and immediately pour the sauce through the strainer into the reserved bowl set over the ice water (Figure 11–24). The strainer will catch any curdled egg protein to ensure a smooth sauce. Stir the sauce until it cools. Pour it into a smaller container and place a piece of plastic wrap directly on the surface of the custard. Refrigerate.

Variation: For white chocolate or semisweet chocolate Crème Anglaise, follow the same instructions through step 4. Before straining the sauce into the ice water bath, gently whisk in 7¼ ounces (1½ cups; 205 g) finely chopped, high-quality white or semisweet chocolate (Figure 11–25). If the sauce becomes too thick as it cools, thin it out with some heavy cream.

Variation: For coconut Crème Anglaise, follow the same instructions as the above recipe through step 4. Before straining the sauce into the ice water bath, gently whisk in 5 ounces (½ cup; 140 g) strained cream of coconut and 1½ teaspoons (7.5 mL) coconut extract.

FIGURE 11–24

FIGURE 11–25

Dessert with Crème Anglaise

RICH CHOCOLATE SAUCE

Makes approximately
14 fluid ounces (1¾ cup; 414 mL)

(*Note:* This sauce can be stored for several days in the refrigerator or frozen in an air tight container for several weeks. It can be melted down over a hot water bath until warm and fluid.)

> **TIP** If the sauce becomes too thick, it can be thinned with heavy cream.

Lessons demonstrated in this recipe:

- How to prepare a simple chocolate sauce, also known as a ganache.
- The sauce is versatile and can be served warm or at room temperature.
- It can be frozen for several months and melted down over a double boiler.
- When chilled, ganache can be piped onto a plate.

MEASUREMENTS			INGREDIENTS
U.S.		METRIC	
8 fluid ounces	1 cup	240 mL	heavy cream
1 ounce	2 tablespoons	30 g	unsalted butter
	2 tablespoons	30 mL	corn syrup
12 ounces	2 cups	340 g	semisweet chocolate, finely chopped
	1 tablespoon	15 mL	cognac or rum

1. Bring the cream, butter, and corn syrup to a boil. Remove the mixture from the heat and add the chocolate (Figure 11–26). Whisk the sauce until the chocolate has melted and the sauce is smooth.

2. Add the cognac or rum (Figure 11–27). The chocolate sauce can be served warm or at room temperature. To store the sauce, allow the sauce to cool to room temperature and place it in an airtight container in the refrigerator.

FIGURE 11–26

FIGURE 11–27

Dessert with Rich Chocolate Sauce

APPLE CIDER BOURBON REDUCTION SAUCE

Makes approximately 8 fluid ounces (1 cup; 240 mL)

Lessons demonstrated in this recipe:

- How to prepare a reduction sauce.
- Boiling reduces flavorful liquids which concentrates flavors and produces an intensely flavorful sauce with a syrupy consistency.

MEASUREMENTS			INGREDIENTS
U.S.		METRIC	
16 fluid ounces	2 cups	480 mL	apple cider
8 fluid ounces	1 cup	240 mL	bourbon
3¾ ounces	½ cup	105 g	light brown sugar (packed, if measuring by volume)
	¼ teaspoon	0.5 g	apple pie spice

1. Combine the apple cider, bourbon, brown sugar, and apple pie spice in a heavy medium saucepan (Figure 11–28). Boil the ingredients until they are reduced to approximately 8 fluid ounces (1 cup; 240 mL) (Figure 11–29). Remove the sauce from the heat.

2. The sauce will thicken as it cools and can be served warm or at room temperature, where it can be drizzled over a variety of desserts.

FIGURE 11–28

FIGURE 11–29

Professional Profile

BIOGRAPHICAL INFORMATION

Charleen Huebner
Chef
Exceptional Events Catering Company
Alexandria, VA

1. Question: *When did you realize that you wanted to pursue a career in baking and pastry?*

Answer: *I think I first considered a career in culinary when I was about 12 or 13 years of age and I started watching Julia Child on PBS.*

2. Question: *Was there a person or event that influenced you to go into this line of work?*

Answer: *My dad was the person who influenced me the most. I remember always cooking with him on weekends and holidays. Being a chef was a career he wished he had pursued.*

3. Question: *What did you find most challenging when you first began working in baking and pastry?*

Answer: *The most challenging aspect was the importance of measuring everything precisely, unlike cooking. And, remembering to say something is in the oven "baking" and not "cooking"!*

4. Question: *Where and when was your first practical experience in a professional baking setting?*

Answer: *My first practical experience was at The Culinary Institute of America when I decided after getting my AOS in the culinary arts to get my degree in baking in pastry. What drew me to this field was the opportunity to work with a good chef, someone who could give me direction, inspiration, and set an example. His name is Sylvain Guyez, Pastry Chef at the Ritz Carlton, Pentagon City, and Ronald Mesnier's nephew. The variety and volume of products produced in our small department was amazing but always the highlight of every event.*

5. Question: *How did this experience affect your later professional development?*

Answer: *While working with Chef Guyez I came to love making the pastries and desserts that would astonish the guests and leave them wanting more. I am very grateful that I had this opportunity. I must admit that I also love the ability to do both cooking and pastry.*

6. Question: *Who were your mentors when you were starting out?*

Answer: *Chef Guyez and Chef Gunther Behrendt at The Culinary Institute of America were my mentors.*

7. Question: *What would you list as the greatest rewards of your professional life?*

Answer: *Working with great pastry chefs in some wonderful places and dealing with all the craziness was great. And, now, sharing my knowledge and inspiring students.*

8. Question: *What traits do you consider essential for anyone entering the field?*

Answer: *Anyone entering this field must love to do this. It's hard work and you'll have to deal with an ever-changing work schedule. Remember, you're probably working most weekends when your friends are off.*

9. Question: *If there was one message you would impart to all students in this field what would that be?*

Answer: *Always keep learning and sharing your knowledge. Inspire others; be a leader. Take advantage of the opportunities that come your way. You'll never know where they may lead.*

CHAPTER 12

Chocolate

After reading this chapter, you should be able to:

■ Understand the process of chocolate making.

■ Know the differences between milk, semisweet, and white chocolates.

■ Define how cocoa butter affects the quality of chocolate.

■ Demonstrate what tempering is.

■ List the difficulties of working with chocolate.

■ Demonstrate how to prepare the recipes in this chapter.

Most professional bakers and restauranteurs know that, in order to entice customers to order dessert in their restaurant or bakery, they must place chocolate, in some form, on the menu. One of the most popular foods in the world, chocolate can be soothing to the soul and comforting to the palate. Chocolate desserts range from confections to pastries and are desired for their flavor, richness, texture, and mouthfeel. Eating chocolate is a personal experience: Some prefer the mildness of sweet milk chocolate; others prefer a stronger, more pronounced bitter-sweet flavor.

An entire chapter is being devoted to the subject of chocolate, where it originates and the process of taking it from the cacao bean to the mouth-watering state known as *chocolate*. Several recipes in this book contain chocolate.

The Origins of Chocolate

Chocolate originates from the *Theobroma cacao* tree, also referred to as the cocoa tree. The cacao tree grows in warm, moist, tropical places such as in Africa, South America, Asia, and within the United States, Hawaii. The fruit of the tree grows in the form of pods. Each pod contains a white pulp that surrounds 20 to 40 cacao beans (Figure 12–1). Because of their fragile nature, the pods are harvested by hand. Once harvested, the beans are removed from the pods and allowed to ferment.

Almost every plant bearing fruit or beans has natural yeast on them. This is true for cacao beans as well. These yeast eat the natural sugars on the cacao beans and form carbon dioxide gas and alcohol much like the yeast used in making yeast breads. The fermentation process develops the chocolate flavor and aroma of the beans.

The fermented beans are then laid in the sun or in specially heated rooms to dry, where any excess moisture evaporates and the beans continue to develop their flavor. After drying, the beans are sent to various chocolate manufacturers around the world. There, the beans are roasted in huge ovens and mixed with other beans to get just the right flavor the manufacturer is looking for. This blending of different beans affects the quality of the chocolate produced. The higher the percentage of high-quality beans, the higher the quality of chocolate. Each chocolate manufacturer's chocolate varies in flavor, texture, and melting properties.

After roasting, the beans are crushed to remove their hard outer shell or hull. The material inside the shells are called nibs. The nibs are then smashed and pulverized to form the dark liquid that forms the basis for all rich, dark chocolate. This dark liquid is known as chocolate liquor, although it contains no alcohol, as the name implies. At this stage, some chocolate liquor can be poured into block molds, cooled, and sold as unsweetened chocolate. Unsweetened chocolate is also referred to as baking chocolate, or bitter chocolate. It can also be sold in the form of disks or small blocks.

The remaining chocolate liquor then goes into a huge shell-shaped vat where it is rotated and stirred constantly with paddles or blades, causing heat to be produced from the friction of mixing. This stage is called *conching* (konking), so named for the spiral shell. Conching imparts a velvety smoothness to the chocolate while developing flavor through the evaporation of excess

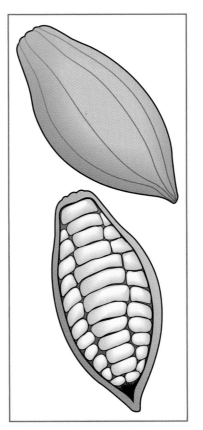

FIGURE 12–1

water and acids. After conching, various types of chocolate can be produced depending on how much cocoa butter and other ingredients are added to the chocolate liquor.

Cocoa Butter and Mouthfeel

Cocoa butter, the saturated fat within chocolate liquor, is responsible for the smooth and silky mouthfeel quality of chocolate. Cocoa butter, a hard and brittle fat at room temperature, contains small amounts of a natural emulsifier known as *lecithin.* Cocoa butter determines the quality of chocolate. The more cocoa butter a chocolate contains, the higher its quality and the better the mouthfeel.

The melting point of cocoa butter at approximately 95°F (35°C) is slightly below human body temperature, which is approximately 98.6°F (37°C). When a piece of chocolate is placed on the tongue, it begins to melt, dissolving into a silky puddle.

Cocoa butter may also be separated during processing from chocolate liquor and cooled into cakes to be sold to chocolate manufacturers for blending into their own signature candies and chocolates. Cocoa butter may also be added to thin down melted chocolate for coating cakes, pastries, and confections.

The Most Common Types of Chocolate

Some of the more common types of chocolate include unsweetened chocolate, milk chocolate, semisweet and bittersweet chocolate, couverture, cocoa powder, white chocolate, chocolate chips, and a variety of imitation chocolate-flavored products.

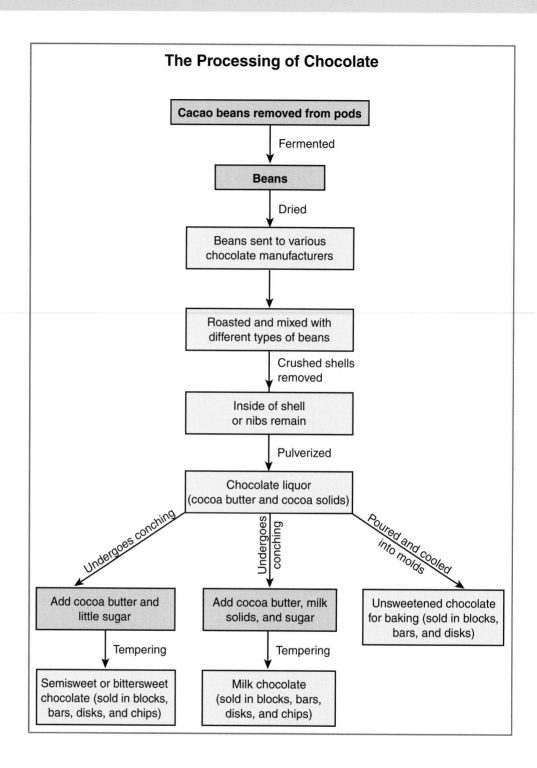

The Processing of Chocolate

Cacao beans removed from pods

↓ Fermented

Beans

↓ Dried

Beans sent to various chocolate manufacturers

↓

Roasted and mixed with different types of beans

↓ Crushed shells removed

Inside of shell or nibs remain

↓ Pulverized

Chocolate liquor (cocoa butter and cocoa solids)

Undergoes conching → Add cocoa butter and little sugar

Undergoes conching → Add cocoa butter, milk solids, and sugar

Poured and cooled into molds → Unsweetened chocolate for baking (sold in blocks, bars, and disks)

Add cocoa butter and little sugar ↓ Tempering → Semisweet or bittersweet chocolate (sold in blocks, bars, disks, and chips)

Add cocoa butter, milk solids, and sugar ↓ Tempering → Milk chocolate (sold in blocks, bars, disks, and chips)

Unsweetened Chocolate

Unsweetened chocolate is chocolate liquor that is sold as solid blocks. By law it must contain at least 50% cocoa butter. Unsweetened chocolate is also referred to as baking or bitter chocolate. It is commonly used in baked goods such as cookies, cakes, and frostings.

Milk Chocolate

Milk chocolate is chocolate liquor that is mixed with cocoa butter, a minimum of 12% milk solids, sugar, and sometimes lecithin (a fatty substance used as an emulsifier to replace some of the cocoa butter). Milk chocolate is not a good substitute for chocolate in baking because the milk solids can burn and the flavor is too mild. However, it can be successfully melted and incorporated into a wide variety of desserts such as mousses, candies, and frostings.

Semisweet and Bittersweet Chocolate

Semisweet chocolate or bittersweet chocolate contains less sugar than milk chocolate and varying amounts of cocoa solids. Many people assume semisweet chocolate is sweet or milder than bittersweet chocolate, but this is not always the case. Individual chocolate manufacturers differ in what constitutes the best tasting chocolate. The amounts of sugar and cocoa solids added to semisweet or bittersweet chocolates vary greatly amongst various manufacturers. For example, one manufacturer's semisweet chocolate may actually contain less sugar and more cocoa solids than another manufacturer's bittersweet chocolate. Semisweet and bittersweet chocolate are sold in blocks, bars, and small disks.

Couverture

Chocolate that contains at least 32% cocoa butter and is of high quality is known as *couverture* (a French word meaning "to cover" or "to coat"). It is used to make high-quality candy bars and to coat candies and create decorations for all sorts of desserts and pastries. Couverture comes in milk, semisweet, bittersweet, and white chocolate with the chocolate liquor, cocoa butter, and sugar contents listed on the label. Couverture differs from other chocolate products that are sold commercially in that it contains a higher percentage of cocoa butter. The extra cocoa butter gives couverture a more fluid consistency when melted that is better suited to dipping candies or coating cakes and other desserts. Couverture is sold in blocks, bars, and small disks that eliminate the need for chopping.

Cocoa Powder

When cocoa butter is pressed out of chocolate liquor under high pressure, the dry, powdery residue that remains during the separation process is cocoa powder (Figure 12–2). The fat content of cocoa powder varies by brand but, in general, it must have a minimum of 10% cocoa butter by law. Low-fat cocoa powders are produced, but require a special process to be made and are not frequently used in the bake shop. Cocoa powder is used in drinks and many chocolate desserts.

There are two types of cocoa powder. The first type is natural cocoa powder, which is the powder formed naturally during the separation of cocoa butter from the chocolate liquor. It is acidic and can therefore react with a chemical leavener like baking soda to neutralize and form carbon dioxide, and help leaven baked goods. The second type is Dutch processed cocoa powder,

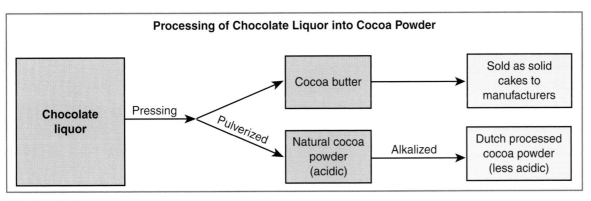

FIGURE 12–2

which has been treated with an alkali (basic solution) to neutralize some of the acidity within the chocolate. It does not react with baking soda. Dutch processed cocoa powder is darker in color and milder in flavor than natural cocoa powder.

White Chocolate

In order to be called chocolate by law, a product needs to contain chocolate liquor. The term *white chocolate* is really a misnomer because it is not chocolate at all. Although it does not contain chocolate liquor, high-quality white chocolate does contain cocoa butter. What is referred to as white chocolate is actually a candy made from cocoa butter, milk solids, sugar, lecithin (a fatty substance used as an emulsifier), and flavorings like vanilla. There are two types of white chocolate: white chocolate with cocoa butter and white chocolate–like coatings that contain little or no cocoa butter. High-quality white chocolate contains a minimum of 32% cocoa butter. It is referred to as *white couverture.* (See Couverture.)

Chocolate Chips

Chocolate sold as "kiss-like" morsels or bits of chocolate come in semisweet, milk, and white chocolate varieties. While some chips do contain 100 percent cocoa butter, others contain fats such as hydrogenated vegetable shortening that are specially formulated to withstand baking temperatures without melting or losing their shape. For this reason, if a recipe calls for high-quality semisweet chocolate, semisweet chocolate chips should not be substituted. The melting properties of chocolate chips differ from a higher quality chocolate and the final product may be altered in texture and quality because of it.

Imitation Chocolate-Flavored Products

Chocolate that contains little or no cocoa butter is of lesser quality and is referred to as compound coating or chocolate-flavored coating. There is no cocoa butter present, so legally these products cannot be called chocolate. The cocoa butter has been replaced with vegetable fats like hydrogenated palm kernel or cottonseed oils. These coating chocolates have a longer shelf life, but lack the smooth, melt-in-your-mouth texture of chocolates containing cocoa butter because they have high melting points. Coating chocolates tend to be used to coat candies. It is best to use high-quality chocolate containing cocoa butter in recipes calling for chocolate.

Imitation White Chocolate-Flavored Products

Low-quality white chocolate or white chocolate-like coating is sold as small white chips or disks and is known as white confectionery coating or compound coating (Figure 12–3). It can also be

---

Difference Between White Chocolate and White Chocolate-like Coatings

High Quality "White Chocolate" (White Couverture) = Cocoa Butter + Lecithin + Sugar + Milk Solids + Flavorings

Poorer Quality White Chocolate-like Coatings (White Confectionary Coating, Compound Coating) = Hydrogenated Vegetable Shortening + Lecithin + Sugar + Milk Solids + Flavorings

FIGURE 12–3

referred to as *summer coating* because summer coating has a higher melting point and is easier to work with in the warmer months. All or part of the cocoa butter in white chocolate–like coatings has been replaced with vegetable fat (e.g., hydrogenated palm kernel or cottonseed oils). These solid vegetable fats have a higher melting point than cocoa butter and tend to leave an unpleasant, waxy film and taste on the tongue. They do not melt as quickly or have the same properties as higher quality white chocolate.

Water and Chocolate

Chocolate reacts poorly when melted together with small quantities of water or when exposed to humid conditions or steam. Even one drop of water in melted chocolate can cause it to tighten into a solid mass and become hard, mottled in color, cakey, or brittle. This phenomenon is known as seizing. That is why it is critical when melting chocolate to have a perfectly clean, dry bowl. Chocolate contains cocoa solids (which are dry) immersed within a rich cocoa butter. Cocoa solids are the material that makes up the cocoa bean and include cocoa butter and cocoa powder. Even one drop of water will cause the dry particles of solids to stick together, absorbing water and swelling, and clump into a chalky-looking, grainy mass. This phenomenon can destroy a recipe. While a few drops of water cause chocolate to tighten up into a solid mass, a larger quantity of water does not cause seizing. Chocolate can be melted over a double boiler or over low heat with water, having no adverse effects. Using a minimum amount of water is crucial. Usually a minimum of 1 tablespoon of water (or other liquid such as coffee, milk, or cream) to 2 ounces of chocolate is a safe combination that will not seize. Many recipes avoid this problem by first melting chocolate by itself or with a fat like butter or oil.

Another way seizing can occur is if a cold substance such as milk, cream, or eggs are added too quickly to warm melted chocolate. The contrast in temperature causes the chocolate to cool down and the cocoa butter to solidify suddenly, resulting in chunky bits of chocolate instead of a smooth mass. Allowing the chocolate to come to room temperature beforehand or warming the cold substance you wish to add can prevent seizing.

> To prevent seizing: When melting chocolate with a water-based liquid, be sure there is a minimum of 1 tablespoon of liquid for every 2 ounces of chocolate.

Properly Melting Chocolate

Melting chocolate properly is essential to the success of a recipe. Chocolate is very sensitive to temperature and can easily burn. A good rule of thumb to remember is to never rush the melting process. As high-quality chocolates melt easily in the mouth at less than 100°F (38°C), it is not necessary to use very high temperatures to melt them.

Semisweet and bittersweet chocolate should be melted at no higher than 120°F (49°C), and milk and white chocolate should never be melted at higher than 115°F (46°C). When melting temperatures exceed the recommended limits, flavors in the chocolate may be compromised. Treat chocolate gently and melt it slowly to ensure the smoothest texture.

There are a few different ways to melt chocolate. First, chocolate can be melted over a double boiler or a hot water bath. The water level in the bottom pot should never touch the bowl or the underside of the top portion of the double boiler. The water is brought to a boil first and then removed from the heat. Chopped chocolate is then placed over the pot of hot water either in a bowl or the top portion of the double boiler.

If left alone, chocolate will melt, even with minimal stirring. Chopping the chocolate beforehand increases its surface area, allowing the heat to come into contact with many surfaces of the chocolate at once. Keep any steam or drops of water from getting inside the bowl or container of chocolate.

Some chefs are successful at melting chocolate slowly and gently to this point, then they get into trouble when lifting the bowl of chocolate off the pot of water and adding it to other ingredients. The bottom of the bowl is guaranteed to be covered in beads of water from the steam produced from the bottom pot of water. Be sure to wipe the bottom of the bowl thoroughly before pouring the chocolate into another container or adding it to other ingredients. This prevents any water drops from getting into the chocolate causing it to seize.

The second way to melt chocolate is to microwave it at a very low power setting. This must be done slowly and carefully because microwaves have "hot spots" and inconsistencies in temperature.

When chopped chocolate will be melted with other ingredients such as butter, oil, or cream, it may be placed directly over low heat in a small pot.

Chocolate should be melted:

- Slowly
- Gently
- Avoiding humidity, steam, or small quantities of water-based liquids to prevent seizing

The Fat Crystals in Cocoa Butter

Working with chocolate can be difficult. Chocolate reacts poorly to humidity, small quantities of water, and adverse temperatures. The reason for this difficulty is due to the temperamental nature of cocoa butter.

The fat in cocoa butter is composed of six different types of fat crystals. Each type of crystal forms and melts at different temperatures. The various melting points of each fat crystal create problems for chocolate, causing the crystals to separate out. This affects the shelf life, the appearance, and the quality of the chocolate. Of the six types of fat crystals, the one that melts at the highest temperature is the one that is considered to be the most stable. It is known as the *beta-type crystal* and has a melting point of 95°F (35°C). When the melted fat crystals in chocolate cool down and begin to set up and solidify, it is the ones with the highest melting point, the beta-crystals, that solidify fastest. As they solidify they create smaller, finer crystals that start a chain reaction that takes over and dominates the chocolate, helping to make it appear shiny and break with a snap. This chain reaction of crystals is similar to the crystallization process in sugar syrups.

Why Chocolate Is Tempered

To create an environment in which beta-crystals dominate the melted chocolate, the chocolate must go through a process known as tempering. The goal of tempering is to take melted chocolate and create the highest percentage of beta-crystals as possible after the chocolate has solidified.

Tempering is a three-stage process wherein the fat crystals in the chocolate are stabilized by melting and then by cooling and rewarming. Properly tempered chocolate can remain at cool room temperature without any loss of quality for up to 1 year or more.

Chocolate that is sold commercially in blocks, bars, and chips has already gone through tempering. Chocolate decorations made from untempered chocolate cannot be kept at room temperature without becoming speckled and mottled in appearance. Instead of drying to a shiny finish, the chocolate looks like a dull mud pie that is crackled and flakes off into pieces. The chocolate is said to be *out of temper*. This is due to various fat crystals within the chocolate that separate out, becoming unstable and rising to the surface. (See Blooming.)

Determining When Tempering Is Necessary

Chocolate that will be used as an ingredient within a recipe such as in a buttercream, mousse, cake, cookie, torte, glaze, ganache, truffle, or fudge does not require tempering. Desserts and chocolate-dipped fruit that will be refrigerated and consumed within a short period of time do not require tempering either.

It is only when the chocolate will be used as a decoration to be placed on a dessert or used to coat candies or cakes that it needs to be tempered. For example, the chocolate for a truffle filling would not need to be tempered, but the chocolate used to dip the truffle and coat the outside would need to be tempered.

If there are time constraints, tempering can be avoided by using compound coatings that contain little or no cocoa butter and therefore do not require tempering. However, their taste is different and may be less desirable than using chocolate containing cocoa butter.

Desserts in which chocolate is an ingredient do not require tempering. Candies dipped in chocolate and other chocolate decorations using melted chocolate require tempering if they are not refrigerated or consumed quickly.

Tempering

Tempering is not an easy process. Even small fluctuations in temperature or humidity in the kitchen will affect whether or not tempering will be successful. Cocoa butter content among different brands of chocolate can also produce inconsistencies that affect a chocolate's ability to be tempered. Chefs who temper chocolate by hand over time become familiar with these fluctuations and can compensate for them.

Even though the directions for tempering seem simple, the art of tempering is indeed quite tricky. Conditions in the kitchen may be perfect for tempering one time and not the next. Once the chocolate is perfectly "in temper," usually within the temperature range of 86° to 91°F (30° to 33°C) for semisweet and bittersweet chocolate and 86° to 89°F (30° to 32°C) for milk and white chocolate, the challenge is in keeping it there. Many warming tools can be used. Anything from a hair blower to a heating pad or hot water bath can be used to maintain a consistent temperature. If the temperature goes above or below the ideal temperature range, the chocolate must be melted down again and the process repeated. Test the temperature of each of these warming tools to make sure which setting will be the correct one to getting the chocolate within the correct temperature range.

Directions for tempering are briefly discussed in the following paragraphs, but be aware that the ability to temper comes over time with much practice.

The art of tempering requires speed because there is only a small window of time between a chocolate's ideal tempering temperature and the chocolate becoming too cool and solidifying. One basic piece of equipment necessary for tempering is a thermometer specialized for chocolate work with a range of 80° to 130°F (26° to 54°C).

Tempering involves three stages: melting, cooling, and rewarming. (See Table 12–1, Correct Temperature Ranges for Each Stage of Tempering.)

1. Melting stage. The first stage in the tempering process begins with melting down the chocolate in a bowl over a pot of water that has been brought to a simmer and removed from the heat. Be sure the bottom of the bowl does not touch the water. The steam rising from the water below is just gentle enough to melt the chocolate in the bowl without scorching it. Melting dissolves all the different types of fat crystals within the chocolate. This presents an "even playing field" to control the process of recrystallizing in a controlled environment. Keep the chocolate at this stage for 15 to 30 minutes, while stirring, to maintain the melt-

Table 12–1 Correct Temperature Ranges for Each Stage of Tempering

TEMPERING STAGE	SEMISWEET AND BITTERSWEET CHOCOLATE	MILK CHOCOLATE	WHITE CHOCOLATE
Melting	Not to exceed 120°F (49°C)	Not to exceed 115°F (46°C)	Not to exceed 115°F (46°C)
Cooling	82°–84°F (27°–29°C)	80°–82°F (26°–28°C)	80°–82°F (26°–28°C)
Rewarming	86°–91°F (30°–33°C)	86°–89°F (30°–32°C)	86°–89°F (30°–32°C)

ing temperature. The melting temperature should not go above 120°F (49°C) for semisweet and bittersweet chocolate or 115°F (46°C) for milk and white chocolate.

2. Cooling stage. The second stage involves cooling the chocolate by removing it from the heat and either stirring it or working a small portion of it on a marble slab. The chocolate is cooled to approximately 82° to 84°F (27° to 29°C) for semisweet or bittersweet chocolate and 80° to 82°F (26° to 28°C) for milk and white chocolate. During this stage the chocolate begins to thicken because the fat crystals are recrystallizing as the temperature drops. Ideally, the fat crystals should be very small and be of the beta type. The presence of beta-crystals starts a chain reaction that creates more and more beta-crystals. This process is slow because it takes time for beta-crystals to form properly.

3. Rewarming stage. Because the cooled chocolate is no longer fluid enough to use for dipping, it is necessary to rewarm it. The rewarming is done very gradually and the chocolate is brought to a specific range of temperatures. For semisweet and bittersweet chocolate, the temperature range is 86° to 91°F (30° to 33°C) and 86° to 89°F (30° to 32°C) for milk and white chocolate. The idea here is not to melt all the fat crystals but to try to maintain only the smallest and finest ones. As the chocolate hardens, the crystals with the highest melting point take over. The fat crystals with the highest melting point (the beta-crystals) will not melt down at these rewarming temperatures and create a chocolate that can be stored at room temperature, extending its shelf life. Even at the ideal temperature range, the chocolate can appear too viscous for dipping. Do not be tempted to increase the temperature to increase the chocolate's fluidity. Chocolate manufacturers may add melted cocoa butter at this stage to thin it down. Note the rewarming stage is not called the remelting stage for a reason: The chocolate is warmed up just enough to melt down the fat crystals with the lower melting points while keeping the crystals with the highest melting points intact.

> The three stages of tempering include melting, cooling, and rewarming.

(*Note:* These temperatures are a general guideline and may vary. Manufacturers of specific brands of chocolate may suggest different temperatures for each stage of tempering.)

Two Methods of Tempering

There are two methods of tempering chocolate: the table method and the seeding or injection method. They are described using semisweet chocolate. Please refer to Table 12-1, Correct Temperature Ranges for Each Stage of Tempering, to temper milk and white chocolate.

Method 1: The Table Method

The table method is so named because part of the melted chocolate is poured onto a table or marble slab to increase its surface area and cool it down. Once cooled to the proper temperature, this chocolate is added back into the remaining melted chocolate. It is then rewarmed to the proper temperature interval. The chocolate must be stirred constantly because it sets very quickly.

PROCEDURE:

1. Chopped chocolate is melted in a bowl over a pot of water that has been brought to a simmer and removed from the heat while being stirred constantly (Figure 12–4). The temperature of the chocolate should not exceed 120°F (49°C).

2. The bowl of chocolate is removed from the heat and the bottom and sides are wiped completely dry, removing all traces of moisture.

3. Pour approximately two thirds of the chocolate onto a clean, dry marble slab or stainless steel surface (Figure 12–5). Using an offset spatula, spread the chocolate out thinly so it increases its surface area, cooling it down. Scrape and mix it back together (Figure 12–6). This process is continued until the chocolate becomes pasty and dull looking. Variations exist as to what fraction of the melted chocolate should be poured onto the work surface.

4. When the chocolate reaches 82° to 84°F (28° to 29°C) quickly scrape it back into the bowl with the remaining melted chocolate (Figure 12–7). Set the bowl of chocolate over a pot of simmering water, off the heat, or on top of a warm heating pad and mix gently until the mixture is rewarmed to approximately 86° to 91°F (30° to 33°C) (Figure 12–8). Test the chocolate by dropping a small amount onto a parchment-lined sheet pan and allow it to harden. It should harden within several minutes. When lifted up, it should break with a snap. The chocolate is tempered and ready to use as long as the temperature range is maintained.

FIGURE 12–4

FIGURE 12–5

FIGURE 12–6

FIGURE 12–7

FIGURE 12–8

Method 2: The Seeding or Injection Method

The seeding or injection method of tempering is somewhat simpler than the table method. Very finely grated or shaved chocolate, also known as the seeding chocolate, is gradually added into the melted chocolate. The shaved chocolate cools down the melted chocolate until the approximate temperature is reached and then rewarmed to the proper temperature interval for tempering. Some methods call for a few larger chunks of chocolate to be used in place of the shaved particles. These larger pieces are allowed to remain in the melted chocolate just until the proper cooling temperature is reached and then removed. The advantage here is that the larger chunks are easier to remove and thus maintain a better control over the temperature. The proper ratio of melted chocolate to grated chocolate is 4 to 1. So for 16 ounces (2⅔ cups; 454 g) of melted chocolate, 4 ounces (⅔ cup; 113 g) of grated chocolate should be used as the seeding chocolate.

PROCEDURE:

1. Chopped chocolate is melted in a bowl over a pot of water that has been brought to a simmer and removed from the heat while being stirred constantly. The temperature of the chocolate should not exceed 120°F (49°C).
2. One fourth of the amount of the same type of chocolate is grated or shaved into very fine particles.
3. The melted chocolate is removed from the pot of water and the bottom of the bowl wiped completely dry. A small quantity of shaved chocolate is then sprinkled and stirred into it (Figure 12–9). After it melts, another small quantity of shaved chocolate is stirred in and allowed to melt. This process is repeated until the correct cooling temperature is reached and the chocolate has thickened. As the correct temperature is reached, the chocolate shavings will take longer to melt. Be sure they are completely melted before rewarming.
4. Place the chocolate over the pot of water and rewarm it (or place the bowl on top of a warm heating pad), monitoring the temperature carefully, until it is within the correct interval for tempering (Figure 12–10). The chocolate is now tempered and ready to use as long as the temperature range is maintained.

Determining When the Chocolate Is Tempered

There are a few ways to know whether the chocolate has been tempered successfully. Some chefs dip a palette knife into the chocolate and then allow it to dry at room temperature. Properly tempered chocolate should set up and dry within several minutes and should be shiny without any streaks.

FIGURE 12–9

FIGURE 12–10

FIGURE 12–11A

FIGURE 12–11B

Or, they may place a small amount of tempered chocolate onto a parchment-lined sheet pan and allow it to harden. If it looks shiny and breaks with a snap after a few minutes, it is properly tempered.

To see the difference between how tempered chocolate and untempered chocolate set up, melt some chocolate and spread it onto a parchment-covered sheet pan next to the tempered chocolate. The tempered chocolate will harden within several minutes, whereas the untempered chocolate stays soft for a much longer period of time. When it finally does harden, the untempered chocolate will look mottled and dull. After a few days, successfully tempered chocolate will continue to be shiny and hard, while the untempered chocolate will look brittle with blotches of white (Figure 12–11A and B).

Commercial Tempering Machines

A large commercial tempering machine is used to temper large quantities of chocolate. A commercial chocolate temperer is thermostatically controlled to keep temperatures exactly where they should be. The chocolate goes through the three stages of tempering within the machine and, once tempered, can remain within the ideal temperature range for long periods of time. Commercial chocolate tempering machines are also available in smaller sizes for smaller operations.

What to Do with Tempered Chocolate

Once the chocolate has been tempered, it can be shaped into various decorations to garnish cakes, tortes, small pastries, pies, tarts, cookies, or individual desserts such as mousses and ice creams. It can also be poured into candy molds or used to coat confections such as truffles, caramels, nuts, fruits, and marzipan.

Tempered chocolate may be poured into a parchment cone or pastry bag and piped into various shapes, designs, and patterns. Once they have hardened, they may be stored in an airtight container at cool room temperature for weeks until needed.

Tempered chocolate can be spread onto a sheet pan covered with parchment paper. Once the chocolate is partially firm, various shapes may be cut out using cookie cutters or a knife.

Tempered chocolate can be spread onto a plastic sheet called *acetate* and, when it becomes tacky, it can be wrapped around a cake or tart until firm. Once firm, the acetate is peeled off, leaving a decorative, hard chocolate coating.

Blooming

The proper temperature to store most chocolate is between 56° and 60°F (13° and 16°C). Tempered chocolate that has been exposed to temperature variations or humidity can develop a whitish-gray, spotty outer coating. These improper storage conditions cause the chocolate to go "out of temper." When this occurs, the chocolate is said to have "bloomed." Blooming does not affect the taste of the chocolate but it is unsightly.

For chocolate that will be melted down and used within a recipe, blooming is not a problem. However, bloom on chocolates that are sold as confections can be a major problem. There are two types of bloom: sugar bloom and fat bloom.

Sugar Bloom

Just as small quantities of water, humidity, or steam can wreak havoc on melted chocolate, humidity can also adversely affect solid chocolate that is being stored on a shelf. Humidity is nothing more than water in the air that can come into contact with the chocolate. Sugar crystals within the chocolate are hygroscopic and absorb moisture from the air, thus providing an incompatibility between the small quantity of water and the chocolate. This causes a recrystallization of sugar on the surface of the chocolate as the water evaporates. The chocolate tastes gritty and looks dull, streaky, and gray. This is known as *sugar bloom.* Chocolate that has been left uncovered in the refrigerator may develop sugar bloom.

(Note: Chocolate with sugar bloom can be used for baking but it should never be retempered for candy making. Because it has absorbed a small quantity of moisture from the sugar, it has the potential to seize up.)

Fat Bloom

Fat bloom occurs when chocolate has been stored over 70°F (21°C) or it has not been properly tempered. The crystals of fat travel to the surface of the chocolate and recrystallize on the outside, forming a whitish coating.

Fat bloom only affects the appearance of the chocolate, not the quality. The chocolate can be used in a recipe or tempered again to restabilize the fat crystals.

Ganache

Ganache is one of the most versatile concoctions in the pastry kitchen. Ganache is a rich combination of cream and chocolate (see Chapters 7 and 8). Depending on the ratio of chocolate to cream used in its preparation, ganache can be used in many different types of desserts. In general, the greater the ratio of chocolate to heavy cream, the firmer the ganache will be when it is chilled because the cocoa butter in the chocolate solidifies. For example, ganache can be used to prepare a fudge sauce using a 1 to 1 ratio of chocolate to cream. Rich chocolate frostings and glazes generally use a ganache made with a 2 to 1 ratio of chocolate to cream. Some recipes for ganache may include other ingredients such as egg yolks, butter, sugar, corn syrup, or cocoa powder to increase the richness and improve texture. Other ingredients such as liqueurs, preserves, and extracts may also be added for flavor.

One popular way that ganache is used is to produce the centers for the rich confection known as *truffles.* Truffles are created from chilled or whipped ganache and shaped into balls. Ganache can be blended with many different ingredients such as coconut cream, coffee, fruit purées, nuts, praline, caramel, or liqueurs and extracts to create a wide variety of flavors. The ganache can also be formed around a single nut or a piece of candied ginger, dried fruit, or a fresh fruit such as a raspberry to give added texture and flavor before being rolled into a ball and chilled.

A ratio of 3 parts chocolate to 1 part cream by weight (a 3:1 ratio) is the average ratio for truffles, although there are many variations. The large proportion of chocolate is to ensure that, once the chocolate is chilled within the cream, it resolidifies and can hold its shape.

Once formed, truffle centers can be frozen, wrapped in an airtight container and stored for several weeks before being thawed in the refrigerator, and then dipped into melted chocolate or rolled in confectioners' sugar, cocoa powder, crushed nuts, toasted coconut, or crushed praline.

Truffles can be made from semisweet, milk, or white chocolate ganache. Some truffle recipes may blend different chocolates together. A nice contrast is created when a white chocolate center is dipped into dark chocolate or vice versa.

Dense versus Light Truffle Centers

Truffle centers can be dense and thick or light and fluffy. Depending on how the ganache filling is treated after preparation will determine whether a truffle is dense or light.

Dense Truffles

To create a dense truffle, the warm ganache filling is simply poured into an ungreased shallow pan with sides or a bowl and chilled until it is firm. A small ice cream scoop ($\frac{1}{2}$ ounce; 15 mL) or melon baller is used to scoop balls of ganache onto a sheet pan lined with parchment paper that has been dusted with either cocoa powder or confectioners' sugar.

The balls are chilled until almost firm and rolled in the palms of the hands until smooth and uniform. They can be rolled into cocoa powder or confectioners' sugar and chilled until ready to serve. Or they can be dipped into melted, tempered couverture and coated with finely chopped nuts, praline, or coconut. Truffles should be stored in an airtight container.

Light Truffles

To create a lighter truffle, the ganache is beaten in an electric mixer using the paddle attachment until smooth and it appears to have lightened in color. The air beaten into the ganache produces a light, almost whipped texture. The ganache should not be overbeaten or it will become grainy and stiff.

The ganache can then be placed into a pastry bag fitted with a large round tip and piped into balls onto a sheet pan lined with parchment paper that has been dusted with cocoa powder or confectioners' sugar. The truffles are then rolled in the palm of the hands until smooth and uniform. Like dense truffles, the light truffles may be rolled in cocoa powder or confectioners' sugar or dipped in melted, tempered couverture.

Molding Chocolate

Tempered chocolate can also be poured into clean, dry molds to form a shell and then filled with various fillings such as ganache or caramel. The filled molds are then covered with tempered chocolate to seal the filling inside. Because chocolate contracts after it hardens, the chocolates are easily popped out of the molds.

Chocolate Molds

High-quality chocolate molds generally are made from polycarbonate. Other materials used to create molds include tin and thin plastic.

Polycarbonate molds are used most often by professional confectioners because of their sturdiness and their ability to form chocolates with a smooth, glossy finish. The molds need to be clean and dry to produce high-quality chocolates. They are generally cleaned by hand (never in a machine) using warm, soapy water. Abrasive materials should never be used or the molds could become scratched. These scratches would, in turn, be transferred to the chocolate.

It is not necessary to clean the molds in soapy water after each use. However, it is necessary to polish or buff the inside of the molds before each use. Buffing is done using cotton wool or balls of cotton to remove dried bits of chocolate.

Procedure to Mold Chocolates

1. Ladle tempered chocolate, preferably couverture, over clean, dry molds and spread the chocolate into each opening or shell using an offset spatula, filling each shell to overflowing.
2. Turn the mold upside down and gently tap to remove any excess chocolate. There should be a thin chocolate coating inside each shell. Scrape the flat side of the mold with an offset spatula to clean the excess chocolate off the surface of the mold. Refrigerate the mold right side up (chocolate facing up) until the chocolate shells have set and hardened.
3. Steps 1 and 2 are repeated to increase the thickness of the shell. Ideally, the shell of chocolate should be no thicker than $\frac{1}{16}$ of an inch (1 mm).
4. Using a pastry bag, carefully fill each chocolate shell with ganache or other filling to slightly less than full. The filling must not come up to the edge of the chocolate shell where a layer of melted chocolate will be poured to seal the filling into the shell. If any shell is filled too high, the shell will not seal and filling will leak out. The temperature of the filling should not exceed 70°F (21°C) or the chocolate shell could melt.
5. Using a ladle, gently pour tempered chocolate over the filling to cover and seal each filled shell. If the filling is liquidy and soft, fill a pastry bag with tempered chocolate and pipe it over the filled shells until the shells are completely sealed.
6. Using an offset spatula, gently scrape the excess chocolate off into a bowl. Allow the chocolates to harden in a refrigerator.
7. Once hardened, the shells will look slightly contracted and will have pulled away from the sides of the mold. Place a sheet pan lined with parchment paper onto a work surface. After tapping the mold onto the work surface, turn the mold upside down over the prepared sheet pan and the chocolates should fall out. If some chocolates have remained, tap the mold again to loosen them and turn the mold over to release the chocolates.

RECIPES

Coconut Truffles (This chapter, page 377)
Espresso Hazelnut Truffles (This chapter, page 374)
Jeweled Semisweet Chocolate Drops (This chapter, page 373)

Jeweled Semisweet Chocolate Drops

Makes approximately 80 to 100 chocolate drops

Lessons demonstrated in this recipe:

- How to use tempered chocolate to prepare a simple candy.
- Decorating the chocolate drops with nuts, brittle, dried fruits, and ginger adds flavor and texture.

MEASUREMENTS			INGREDIENTS
U.S.		**METRIC**	
1 pound	2⅔ cups	454 g	semisweet chocolate, tempered
As desired		As desired	assortment of coarsely chopped nut brittle, toasted almonds, hazelnuts, cashews, dried cranberries, cherries, raisins, chopped candied ginger, chopped peppermint candies, and non-pareils

1. Cover a sheet pan with parchment paper. Have assorted nuts, brittle, dried fruit, and candied ginger ready.

2. Place the tempered chocolate into a disposable plastic pastry bag. Using kitchen scissors, snip the end of the bag to create a ¼-inch (6-mm) hole.

3. Pipe the chocolate into small, dime-sized round drops onto the prepared sheet pan. The drops will spread out slightly.

4. Decorate each drop with a few of the candies, nuts, or dried fruits (Figure 12–12).

5. Allow the drops to harden at cool, room temperature. The drops can be stored in an airtight container at room temperature for at least 2 weeks.

Variation: Semisweet chocolate can be replaced with an equal amount of white chocolate, milk chocolate, or bittersweet couverture.

FIGURE 12–12

Jeweled Semisweet Chocolate Drops

ESPRESSO HAZELNUT TRUFFLES

Makes approximately 60 to 70 truffles (½ ounce; 15 g each)

Lessons demonstrated in this recipe:

- How to prepare truffles using ganache, a combination of cream and chocolate.
- A ratio of 3 parts chocolate to 1 part cream (a 3:1 ratio) allows the truffle centers to become firm when chilled.
- Beating air into the ganache creates a truffle that is lighter in texture.
- Truffles are dipped chilled, so they retain their shape even in the warm, melted chocolate.
- The truffles are dipped into tempered couverture because it has a higher percentage of cocoa butter and a more fluid consistency when melted. If tempering is too time consuming, melted compound coating can be used.

MEASUREMENTS			INGREDIENTS
U.S.		**METRIC**	
8 fluid ounces	1 cup	240 mL	heavy cream
	2 teaspoons	2 g	instant espresso or instant coffee powder
2 ounces	4 tablespoons	55 g	unsalted butter
	1 tablespoon	5 mL	light corn syrup
1½ pounds	4 cups	680 g	semisweet chocolate, coarsely chopped
1½ fluid ounces	3 tablespoons	45 mL	hazelnut liqueur or hazelnut syrup
			sifted unsweetened cocoa powder for dusting
			60 to 70 hazelnuts, skinned and toasted
DIPPING CHOCOLATE:			
1½ pounds	4 cups	680 g	tempered semisweet couverture
or			
1½ pounds	4 cups	680 g	semisweet compound coating, coarsely chopped and melted over a double boiler
			finely crushed Hazelnut Praline (Chapter 7, page 218) placed in a pie plate

1. In a heavy saucepan, bring the cream, coffee, butter, and corn syrup to a boil over medium-high heat.

2. Remove pan from the heat and add the chopped chocolate and the liqueur or syrup. Whisk the mixture gently until the chocolate is melted (Figure 12–13).

FIGURE 12–13

3. Pour the mixture into a bowl and chill it in the refrigerator, stirring often, until it thickens and can hold its shape when dropped from a spoon. If time is limited, the mixture can also be placed in the freezer for 20 minutes, stirring frequently. The mixture can also be stirred over an ice water bath for 15 to 20 minutes (Figure 12–14). Be sure to completely scrape down to the bottom of the bowl with a rubber spatula where the mixture will harden first.

4. If an ice water bath is used, wipe any water off the bottom of the bowl using a kitchen towel (Figure 12–15). Using a rubber spatula, scoop the mixture into the bowl of an electric mixer and, using the paddle attachment, beat the mixture until it becomes smooth and appears somewhat lighter in color (Figure 12–16). This should take between 10 and 15 seconds at medium speed. Do not overbeat the ganache or it will become grainy and stiff.

5. Immediately scoop the mixture into a pastry bag fitted with a large round tip and pipe out balls the size of large grapes (approximately $\frac{1}{2}$ ounce; 15 g) onto a parchment-lined sheet pan that has been dusted with cocoa powder (Figure 12–17).

(*Note:* The truffles can be frozen in an airtight container for several weeks. Thaw them in the refrigerator until they are ready to be dipped.)

6. Make an indentation into each ball and push one hazelnut into the center (Figure 12–18). Roll the truffles in the palms of your hands (using extra cocoa powder if necessary to prevent them from sticking) until they are smooth balls and place them back onto the same sheet pan (Figure 12–19). Refrigerate the balls until they are very firm.

FIGURE 12–14

FIGURE 12–15

FIGURE 12–16

FIGURE 12–17

FIGURE 12–18

FIGURE 12–19

(*Note:* The dipping step can be omitted completely and the truffles can instead be rolled in confectioners' sugar or unsweetened cocoa powder.)

> **TIP** If any of the chocolate coating has cracked, the truffles can be re-dipped into the chocolate.

(*Note:* If the truffle centers are too cold, the layer of chocolate on the newly dipped truffles may crack. Repeat the dipping process again to ensure no further cracking occurs once the chocolate has hardened.)

7. **Roll each truffle between the palms of your hands to wipe off any excess cocoa powder. Use one of the following methods to dip the truffles.**

DIPPING METHOD 1

Place one chocolate truffle at a time into tempered couverture. Using a truffle dipping fork or spoon (a special handled fork or hollowed spoon), pick up the truffle (Figure 12–20A). Gently knock off the excess chocolate. Place the truffle onto a sheet pan covered with parchment paper. Sprinkle each truffle with hazelnut praline. Alternatively after dipping, the truffle can be rolled into the praline to cover it completely. Toasted, chopped nuts can be used instead of the praline.

DIPPING METHOD 2

Using gloved hands, scoop a small amount of chocolate into the palm of one hand and roll each truffle in the chocolate (Figure 12–20B). Place each truffle, knocking off excess chocolate, onto a parchment-lined sheet pan. Sprinkle hazelnut praline over each truffle. Allow the chocolate on the truffles to harden.

Alternatively, place some praline into a small bowl and roll the dipped truffle in it to completely cover it. Set the praline-covered truffle onto a sheet pan covered with parchment paper (Figure 12–21).

FIGURE 12–20A

FIGURE 12–20B

FIGURE 12–21

COCONUT TRUFFLES

Makes 100 to 110 truffles
($\frac{1}{2}$ ounce; 15 g each)

Lessons demonstrated in this recipe:

- How to prepare truffles using a coconut-flavored white chocolate ganache.
- High-quality white chocolate containing cocoa butter will yield a truffle with a smooth, rich, creamy mouthfeel.
- A denser center is created by pouring the ganache mixture into a shallow container with sides and scooping out truffles using an ice cream scooper or melon baller.

MEASUREMENTS			INGREDIENTS
U.S.		**METRIC**	
8 fluid ounces	1 cup	240 mL	heavy cream
2 ounces	4 tablespoons	55 g	unsalted butter
2¾ ounces	¼ cup	80 g	cream of coconut
1½ pounds	4 cups	680 g	high-quality white chocolate, chopped
	1 teaspoon	6 g	finely grated lime zest
1¼ ounces	½ cup	35 g	sweetened, shredded coconut, toasted and finely chopped
1 fluid ounce	2 tablespoons	30 mL	coconut rum
	2 teaspoons	10 mL	coconut extract
			sifted confectioners' sugar
DIPPING CHOCOLATE:			
1½ pounds	4 cups	680 g	tempered semisweet couverture
or			
1½ pounds	4 cups	680 g	semisweet compound coating, coarsely chopped and melted over a double boiler
1¼ ounces	½ cup	35 g	sweetened, shredded coconut, toasted
			sifted unsweetened cocoa powder

1. In a heavy saucepan, bring the cream, butter, and the cream of coconut to a boil.

2. Remove the mixture from the heat and add the white chocolate (Figure 12–22). Whisk gently until the chocolate is melted. Add the lime zest, coconut, coconut rum, and coconut extract (Figure 12–23).

FIGURE 12–22

FIGURE 12–23

3. Pour the mixture into an ungreased shallow pan with sides. Chill the ganache until the mixture is firm.

4. Using a melon baller or a small ice cream scoop (½ ounce; 15 mL), form the ganache into small balls and place them into a bowl with sifted confectioners' sugar (Figure 12–24). Coat each ball in sugar and place it back on the sheet pan.

5. Chill the truffles for approximately 10 to 15 minutes or until they are firm enough to roll. Roll the truffles between the palms of your hands into smooth balls (Figure 12–25). Chill the truffles until they are quite firm.

6. On a sheet pan covered with parchment paper, lightly combine the toasted coconut and enough cocoa powder to coat the coconut and form a powdery mixture. Set aside. Dip the chilled truffle centers into tempered semisweet chocolate using Dipping Method 1 or 2 in the previous recipe for Espresso Hazelnut Truffles.

7. Roll each truffle in the coconut cocoa mixture and set the truffle on a sheet pan lined with parchment paper (Figure 12–26).

8. Chill the truffles until firm. When the truffles are cut in half, they will resemble miniature coconut halves.

TIP If the truffles are not to be dipped right away, they can be kept for several weeks in an airtight container in the freezer. Thaw the truffles in the refrigerator before dipping.

(*Note:* The truffles can simply be rolled in confectioners' sugar or unsweetened cocoa powder and left undipped.)

Look for other recipes containing chocolate throughout this textbook.

FIGURE 12–24

FIGURE 12–25

FIGURE 12–26

Chocolate Truffles

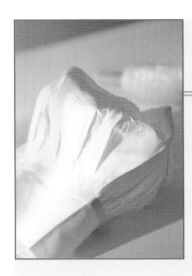

Professional Profile

BIOGRAPHICAL INFORMATION

Ronald Stapleton
Owner
Cocoa Passion
McLean, VA

1. Question: *When did you realize that you wanted to pursue a career in baking and pastry?*

Answer: *When I was 4 or 5 years old I would help my dad, who was a baker, make various twists and shapes from his bread dough at home. This was the beginning of my interest. As I got older I would make cakes or desserts for family dinners on Sundays. First I worked with package mixes and then when I was about 9 or 10, I developed the confidence to start making cakes from scratch.*

2. Question: *Was there a person or event that influenced you to go into this line of work?*

Answer: *The person who influenced me the most was my brother. He bought me my first culinary book,* The International Confectioner. *He continued to encourage me. I also purchased a book,* The Great Chefs of France *that had big impact on me. I was mesmerized by a picture of a frozen strawberry cake that had fine chocolate line work piped on top. It was so pretty that it showed me that cooking can also be an expression of art.*

3. Question: *What did you find most challenging when you first began working in baking and pastry?*

Answer: *The only aspect that presented a real challenge was the attitude of some of the chefs. Every other aspect of the trade was a joy.*

4. Question: *Where and when was your first practical experience in a professional baking setting?*

Answer: *On the advice of my school counselor in England I got a job in a hotel when I was about 13. I worked during school holidays and during weekends. It was the polite nature and caring encouragement of the pastry chef there that stimulated my interest in chocolate and sugar work. I remember spending spare time listening to top 10 music and trying to roll that perfect chocolate cigarette!*

5. Question: *How did this experience affect your later professional development?*

Answer: *Seeing the wonderful creations of the pastry chef and then looking at pictures in the publications he had, gave me the incentive to carry on with my endeavor to become a pastry chef. It encouraged me to go farther.*

6. Question: *Who were your mentors when you were starting out?*

Answer: *My parents were my early mentors, especially my mother who always encouraged me to pursue my dreams. My brother and many teachers were also influential, but there are a few instructors who stand out such as Willy Pfund, a 5 foot, 1 inch "Sugar Emperor." What he couldn't do in sugar just couldn't be done. Another important teacher was Chef Roger Taylor who taught me how to use the love I had for this profession.*

7. Question: *What would you list as the greatest rewards of your professional life?*

Answer: *It is watching students develop skills that they thought would be impossible and seeing the sheer delight on their faces when they achieve their goals. It is also a great reward to be able to practice a skill and art form that I love.*

8. Question: *What traits do you consider essential for anyone entering the field?*

Answer: *A person must have patience, be "people friendly," and have the ability to work hard and creatively.*

9. Question: *If there was one message you would impart to all students in this field what would that be?*

Answer: *Follow your aspirations and never give up when the going gets tough. Most of all, enjoy. If you don't enjoy your work you will never develop satisfaction.*

CHAPTER 13

Healthy
Baking

After reading this chapter, you should be able to:

- Modify some commonly used ingredients in recipes to healthier alternatives.

- Recognize the difference between more healthy and less healthy fats.

- List the three criteria for fat substitution.

- Identify the differences between sugar substitutes.

- Successfully substitute healthier ingredients for less healthy ingredients in recipes.

It is common knowledge that overindulging

in pastries and desserts laden with fats, sugars, and refined grains can take a toll on health. With worries over increased incidence of obesity, heart disease, diabetes, and cancer, it is important for the pastry chef to create healthier alternatives for consumers with special dietary needs.

Consuming desserts does not have to be taboo if the baker knows how to modify ingredients to create healthier baked goods. Even rich desserts can be incorporated into a healthy diet when consumed in moderation.

Because so many recipes can have so many variations, trial and error through experimentation is the best way to discover which healthier ingredients provide the best taste and texture when compared with the original. Once a recipe is tested, adjustments can be made. The challenge is to create a product that is just as delicious and eye appealing as the richer version. However, not all rich recipes for baked goods can be successfully modified.

This chapter discusses how to modify recipes using healthy ingredients to replace some of the fat, sugar, and refined grains in the original recipe. The ultimate goal of this process is to create healthier alternatives for baked goods. Ideally, the final product should produce a similar baked good as the original recipe and have a higher nutritional value.

How to Modify Ingredients Containing Refined Grains with Healthier Alternatives to Enrich Baked Goods

Refined grains such as white flour provide baked goods with structure through gluten development. Most white flour is refined, which means it has had the outer part of the grain or the hull removed. Although many flours are supplemented with vitamins and minerals, fiber is lacking.

Enriching baked goods by modifying refined grains with healthier alternatives is not difficult to do. With some experimentation, it is easy to replace a portion of a refined flour with a flour or grain containing more fiber. Because flour from a wheat plant provides structure for baked goods, it is important to make substitutions carefully. For example, replacing all-purpose white flour with whole wheat flour or ground almonds would provide no gluten development and therefore would be a poor substitution. On the other hand, replacing only a portion of the all-purpose white flour with whole wheat flour or ground almonds would work well. Individual tastes should be a guide as to exact amounts of replacements and can vary greatly.

The following is a list of some substitutions to modify refined grains with healthier alternatives:

■ Replace one fourth the amount of white flour with an equal amount of whole wheat flour to improve the fiber content of the baked good. If more fiber is desired and the taste has not been altered too much, up to one fourth more whole wheat flour can be used to replace an equal amount of white flour.

- In recipes calling for pastry flour such as pie crusts and shortcakes, whole wheat pastry flour can be substituted for white pastry flour, partially or as a whole substitution. The color of the finished baked good will be slightly darker.
- Soy flour can also be used to replace one fourth the amount of white flour. It provides increased nutrition, but no gluten-forming proteins with which to build structure.
- Finely ground oatmeal (ground in a food processor) can be used to replace one fourth the amount of white flour. It provides texture and soluble fiber, but no gluten-forming proteins with which to build structure.
- Flax seed meal (finely ground flax seeds) can be added to most batters and doughs. It provides increased fiber content and omega-3 fatty acids (a heart-healthy fat).
- Raw wheat germ or oat or wheat bran can also be used to replace one fourth of the white flour.
- Ground or chopped nuts can be used in moderation to provide fiber, healthy fats, and a crunch in many baked goods.

Fats—Why They Are Necessary

Entirely eliminating fat in baked goods is not an option because fats provide many desirable qualities desirable such as tenderness, flakiness, and flavor. Replacing all fat will produce an undesirable product that no one will want to eat. Replacing less healthy fats with some healthier alternatives needs to be done with much thought. To do this properly, understanding the role fats play in baked goods is crucial. A wide variety of fats are available to the baker. Some of these fats are less healthy than others. For simplicity, fats may be broken down into two main categories: saturated and unsaturated based on the predominate type of fat. (See Table 13–1, Saturated versus Unsaturated Fats; and Table 13–2, Fats in Order of Healthfulness.)

Saturated Fats

With few exceptions, fats that are considered less healthy tend to be solid at room temperature. Fats directly derived from animals, like lard and butter, are known as saturated fats. Most saturated fats

Table 13–1 Saturated versus Unsaturated Fats

SATURATED FATS	UNSATURATED FATS
Tend to be solid at room temperature. Exceptions include tropical oils such as coconut oil and palm kernel oil	Tend to be liquid at room temperature. Exceptions include partially hydrogenated margarines and shortenings
Saturated fats are associated with an increased risk of heart disease and certain cancers	Monounsaturated fats and specific polyunsaturated fats such as omega-3 fatty acids are associated with reducing heart disease. Transfats, although unsaturated, are associated with an increased risk of heart disease and certain cancers
Include lard, butter, cocoa butter, tropical oils, and fats derived from animals (e.g., high-fat dairy products)	Include canola oil, olive oil, peanut oil, nuts (e.g., walnuts and almonds), and seeds (e.g., flaxseed)

Table 13-2 Fats in Order of Healthfulness

MONOUN- SATURATED FATS	ARE PREFERRED OVER	POLYUN- SATURATED FATS	ARE PREFERRED OVER	SATURATED FATS	ARE PREFERRED OVER	TRANS- FATS
■ Canola oil ■ Olive oil ■ Peanut oil		■ Corn oil ■ Safflower oil ■ Sunflower oil		■ Butter ■ Lard ■ Cocoa butter ■ Coconut and palm kernel oils		■ Solid vegetable shortening ■ Margarine

are derived from animal products, although a few are derived from plants, including coconut, palm kernel oils, and cocoa butter (the fat from cocoa beans). All these fats except cocoa butter (see The Healthy Benefits of Chocolate) tend to raise the "bad" cholesterol in the body known as LDLs (low-density lipoproteins) and can increase the risk of getting heart disease and certain cancers.

All fats are composed of a carbon chain. Each carbon is bonded to hydrogen atoms. The chemical structure of a saturated fat is one in which as many hydrogen atoms as possible are bonded to carbon atoms with single bonds. (See The Chemical Structure of Different Types of Fats.) The more single bonds that exist, the more saturated the fat is, and the more solid it is at room temperature. Fats that are solid at room temperature include lard, butter, and cocoa butter.

Unsaturated Fats

Unsaturated fats contain at least two carbon atoms that are double bonded to each other. Therefore, fewer hydrogen atoms are able to bond to the carbons that the molecule contains. Because the molecules are not as saturated as they could be, these fats are referred to as unsaturated fats. These fats can be either monounsaturated or polyunsaturated. Monounsaturated fats have one double bond (*mono* meaning "one"), and polyunsaturated fats have more than one double bond (*poly* meaning "many"). (See The Chemical Structure of Different Types of Fats.)

Unsaturated fats tend to be liquid at room temperature and originate from plants, not animals. Fats, especially monounsaturated fats, are associated with reducing LDLs, while raising the good cholesterol (HDLs or high-density lipoproteins). Monounsaturated fats include olive and canola oils. Polyunsaturated fats include corn, safflower, and soybean oils.

There is a special group of polyunsaturated fats known as *omega-3 fatty acids*. Their health benefits include lowering the risk of heart disease and strokes by keeping blood from thickening and forming clots in blood vessels. Omega-3 fatty acids are found in certain oils, such as canola and olive, nuts, and seeds, and are particularly well known for their presence in various types of fish.

TRANS-FATS

Because liquid fats such as vegetable and canola oils can become rancid quickly, food manufacturers discovered a way to make these oils last longer. Increasing the shelf life of oils involves a process known as *hydrogenation*. Hydrogenation involves chemically infusing the oil with hydrogen atoms, breaking the double bonds already present in the fat and bonding hydrogen atoms to the newly vacated spots on the carbon. This process also alters the fat physically; as the oil is hydrogenated, it gradually becomes more solid, altering its formerly liquid state. Liquid fats that have been chemically and physically altered by having hydrogen added to them are called *hydrogenated fats*. These fats can be partially or fully hydrogenated.

Those fats that are partially hydrogenated are referred to as trans-fats because of their chemical structure. (See The Chemical Structure of Different Types of Fats.) Trans-fats take on a chemi-

The Chemical Structure of Different Types of Fats

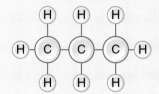

Saturated FAT

Where carbons bond to as many hydrogens as possible as single bonds

Trans-FAT

Where hydrogens sit across the double carbon-to-carbon bond

Unsaturated FATS

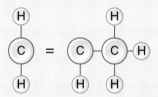

Monounsaturated FAT

One carbon-to-carbon bond is a double bond

Polyunsaturated FAT

More than one double carbon-to-carbon bond

cal structure such that hydrogen atoms sit diagonally across the double bond. Although not all partially hydrogenated fats are trans-fats, most tend to be. Trans-fats, also referred to as *trans-fatty acids*, are solid or partially solid at room temperature. These types of fats are in many processed food products such as crackers, cakes, cookies, cereals, margarines, and vegetable shortenings. Trans-fats are extremely unhealthy. They have been associated with an increased risk of heart disease and certain cancers. The amount of trans-fats in the diet should be kept to a minimum. They decrease the good cholesterol and increase the bad cholesterol. These fats have been mentioned in this book as solid vegetable shortenings that can leave a waxy, unpleasant film on the tongue when eaten. To identify whether or not a product contains trans-fats, it is important to read the nutrition facts and ingredient list on the package. Ingredients are listed in order of magnitude. For example, if a hydrogenated fat is listed at the top of the list, then it is one of the most prominent ingredients in the product. Trans-fats will be listed as partially hydrogenated oil or a combination of oils. Some of the oils used in these products include corn, soybean, sunflower, coconut, and palm kernel oils.

Those fats that are fully hydrogenated become saturated fats because each carbon is bonded to as many hydrogens as possible and the bonds are single.

TRANS-FAT–FREE SOLID MARGARINES AND LOW-FAT BUTTERS

Many margarines on the market are partially hydrogenated and would not make a healthy alternative to butter; however, there are some margarines and low-fat butters that are manufactured using no trans-fats. These solid margarines and butters are good substitutes for regular butter and shortening, especially in recipes for pie crusts, cakes, cookies, or muffins that use the creaming method of mixing. To identify which baking fats are trans-fat–free, it is necessary to read the nutrition facts and the list of ingredients provided on the package.

FRACTIONATED OILS

There are products on the market containing fats referred to as *fractionated oils*. The process of fractionation is a healthy alternative to hydrogenating oils. It too involves the changing of liquid oils into solids, and it is more expensive than hydrogenation. The process begins by heating an oil and then cooling it. As the oil cools down, it forms a thicker, less liquid layer on the top, with a thinner, more liquid layer on top of that. The thinner layer is removed. As each layer cools, it is skimmed off. This separation of each of the fractions (or layers) of oil by melting point eventually leads to the identification of the specific fraction of oil that is needed for a specific baked good. Fractionation uses no chemicals and produces fats that are healthier, and much more easily digested than hydrogenated oils.

> Saturated fats tend to be solid at room temperature, with a few exceptions. Unsaturated fats tend to be liquid at room temperature.

Substituting Healthier Fats for Less Healthy Fats

Although it is generally not difficult to substitute a healthier fat for a portion of a less healthy one in most recipes, replacing less healthy fats with healthier alternatives can become challenging. The end product will be slightly altered, but if the proper substitution is made, it should taste as good as the original. Quality and taste should never be compromised; compromising will render the final product undesirable.

It is important to realize that simply replacing all the butter in a recipe with canola oil will not reduce calories. What it will do is replace all the saturated fat with a healthier unsaturated one. On the other hand, replacing half of the butter with apple sauce or fruit purée will also decrease the fat by half. However, the texture of the final product may be different from the original.

To understand how to correctly replace one fat for another, the baker needs to first read the recipe and then ask three questions:

- What role does the fat play in the recipe?
- What form of fat is used (solid or liquid)?
- What is the mixing method used in the recipe?

> Three Key Points of Fat Substitution
>
> - The role of fat
> - What form the fat takes
> - The mixing method used

The Role of Fat

The baker is truly in the "fat business" because the majority of pastries and desserts rely on some form of fat to contribute to flavor, texture, and richness. Understanding the role that fat plays in a particular recipe is crucial to replacing it with a healthier alternative. (See Table 13-3, Healthier Fat Substitutes.)

Fats play many roles in baking. They can act as tenderizers, preventing gluten from forming, aerators (to hold air within batter, which helps leavening), and separators to keep layers of dough apart, causing flakiness. They also determine how much spread a baked good will have depending upon the melting point of the fat.

Because of the important role that fats play, much thought must go into how to replace them with healthier versions. Many low-fat baked goods are very high in sugar. Because sugar acts as a tenderizer, much like fat, replacing more sugar for less fat is a substitute that commonly works.

Table 13–3 **Healthier Fat Substitutes**

WHEN FATS ACT AS:	SUBSTITUTE:	WHY?
Tenderizers	Healthier oils or fruit purées which contain sugar for tenderness	The fats in healthier oils and the sugar in fruit purées will coat gluten strands and tenderize
Aerators	A solid low-fat margarine or butter containing no trans-fat, or half of the original butter plus half light butter or fruit purée; another option would be to fold beaten egg whites into the batter	Trans-fat free margarines and low-fat butters still hold some air to help leaven. Beaten egg whites folded into the batter will also leaven
Creators of flakiness	Solid low-fat margarine or butter containing zero trans-fats cut into cubes and frozen for approximately 1 hour	Freezing a trans-fat free margarine or low-fat butter will increase the melting point, creating some flakiness

In general, to substitute another ingredient for the fat within a recipe, the baker must determine why the fat is being used and what its ultimate contribution will be to the finished baked good.

What Form the Fat Takes

Recipes call for the fat to be in a certain form. For instance, does the recipe call for the fat to be melted or just softened? Is the fat supposed to be chilled and cut into small pieces? Knowing the form of the fat ultimately helps the baker determine which fat substitute will or will not work.

For example, if the fat in a recipe calls for melted butter, substituting a healthier oil, such as canola, would be easy to do. If the taste of butter is important to maintain the integrity of the recipe, instead of replacing all of the butter with oil, replacing half of the butter with oil would reduce the saturated fat content by half while still maintaining a desirable buttery taste. A melted trans-fat free, low-fat butter would also make a good substitute for the melted butter.

If the fat is used to create flakiness, a solid fat would be best. Substituting a portion of a healthier solid fat, like a low-fat margarine or butter with zero trans-fats would work well. Because melting point is critical to creating flaky layers, freezing the healthier fat would increase flakiness by delaying its melting in the oven.

The Mixing Method Used

The mixing method used in a recipe can determine the type of healthier fat that can be substituted for a less healthy one. For example, if a recipe uses the creaming method and calls for butter to be creamed with sugar, a healthier fat in the solid form like a low-fat margarine or butter with no trans-fats would be best as a substitute. Another option is to use half the amount of solid fat called for in the recipe so it can be creamed, and then substitute an equal amount of apple sauce for the remaining solid fat which lowers the fat content considerably while maintaining the original mixing method. If a recipe uses the one-bowl method and calls for melted butter, a liquid, healthier fat like canola or nut oil would make a good substitution. A melted trans-fat free, low-fat butter would also work.

In general, up to half of the fat can be replaced with a healthier fat replacement in most recipes calling for a solid fat. However, not all recipes are conducive to modifying the fat-based ingredients.

Some Healthier Fat Substitutions

The following are some healthier fat substitutions:

- Low-fat, trans-fat–free, solid margarine or butter makes a healthier substitution for regular butter.
- Low-fat or fat-free versions of cream cheese, sour cream, yogurt, cottage cheese, and ricotta cheese make healthy substitutions for the fully fatted versions of these.
- Canola, olive, sunflower, walnut, and other flavorful oils make good substitution for melted fats, although they do not decrease calories.
- Puréed tofu makes a healthy substitution for heavy creams and eggs.
- Puréed fruits and vegetables can make healthy fat substitutions by providing natural sweetness, which helps tenderize and moisten baked goods. Up to half of the fat can be replaced with one of the following in recipes calling for solid fat:
  - Apple or pear purée
  - Apple butter
  - Apple sauce
  - Puréed cooked white beans
  - Puréed sweet potatoes
  - Puréed carrots
  - Puréed prunes
  - Mashed bananas

Replacing Sugar with Sugar Substitutes

The role sugar plays in baking is important because it has certain properties that are necessary to achieve desired characteristics in baked goods. Sugar (chemically known as sucrose) is a carbohydrate—a nutrient needed for energy. It provides 4 calories per gram. Sugar tenderizes by interfering with gluten formation, it moistens and extends shelf life due to its hygroscopic nature, it aerates fats, stabilizes egg foams, browns, and causes spread.

Many commercial low-fat baked goods contain large quantities of sugar. Because fat and sugar are the two ingredients most commonly used for tenderizing baked goods, sugar is the ingredient that is typically increased if the fat is decreased. The overabundance of sugar in baked goods has created some health concerns associated with obesity and diabetes. Excess sugar that is not metabolized immediately is stored as fat in the body.

Sugar substitutes or replacements for sucrose may be used with a varying degree of success in baked goods. This is because all sugar substitutes are not equal. There are many substitutes for sucrose on the market: fructose, honey, maple syrup, and date sugar, to name a few. Although seeming more natural and less processed than white granulated sugar, these substitutes do little to decrease the carbohydrate level of baked goods.

Unlike natural sugar from the sugar cane plant, beets, or honey, there are substitutes for sugar that are artificial or manmade. These artificial sweeteners do not raise blood sugar levels. They were created for diabetics who need to control their carbohydrate intake. There are a number of artificial sweeteners on the market and they differ greatly (in their chemical structures, in their physical properties, and in how they are metabolized in the body).

Sugar substitutes such as artificial sweeteners do not have the same properties as real sugar because many of them are not carbohydrates. All sugar substitutes sweeten, but that is where their similarities end. Not all sugar substitutes can be creamed with fat, provide tenderness or moistness to a baked good, or caramelize to create browning.

Manufacturers of sugar substitutes are recognizing that their products do not have the same properties as sucrose in the bake shop. Because of the problems that bakers have experienced in using these artificial sweeteners, many manufacturers are putting combinations or blends of their products with sucrose on the market. The addition of some sucrose restores the desired characteristics to the sweetener so it can be more successfully used in baking.

Although there are several artificial sweeteners on the market, only four of them are discussed: saccharin, aspartame, sucralose, and sugar alcohols. (See Table 13–4, Comparison of Sugar, Artificial Sweeteners, and Sugar Alcohols.)

Saccharin

Saccharin is widely used to sweeten beverages and can be sprinkled over foods. It is not a good substitute for white granulated sugar in baking in that it has none of the chemical properties of sucrose such as the ability to be creamed with fat, and it lacks the ability to help baked goods to brown. Although it contains zero calories, there are health concerns associated with it when ingested in large quantities. One major disadvantage of saccharin is its bitter aftertaste.

Aspartame

Aspartame is not a carbohydrate like sucrose. It is actually the combination of two amino acids—aspartic acid and phenylalanine. It, too, sweetens but possesses none of the same chemical properties of sucrose and is not a good substitute. One disadvantage of baking with aspartame is that it loses its sweetness when heated. Aspartame contains 4 calories per gram (just like sucrose), but it is so intensely sweet that only a small amount needs to be used, so in effect, it is noncaloric.

Sucralose

Sucralose comes the closest in chemical makeup to sucrose because it is made from sucrose with only one modification. There is a chlorine atom attached to the end of the sucrose molecule. It can be substituted in a 1 to 1 ratio for sugar in most recipes. Unlike aspartame, sucralose can withstand high temperatures without losing its sweetening abilities. There are some drawbacks to using sucralose. Sucralose does not brown like sucrose and it is unable to aerate fats when creaming. This can cause undesirable changes in texture in baked goods. It is 600 times sweeter than sucrose and contains zero calories. Because it is not metabolized like sucrose, a true carbohydrate, blood sugar is not raised.

Sugar Alcohols

The last type of sugar substitute discussed is part of a category of artificial sweeteners known as *sugar alcohols*. Sugar alcohols are composed of an alcohol attached to a sugar molecule. Sugar and other natural sweeteners are carbohydrates. Sugar alcohols are not true carbohydrates but are modified versions of them.

Table 13–4 Comparison of Sugar, Artificial Sweeteners, and Sugar Alcohols

SUGAR	ARTIFICIAL SWEETENERS	SUGAR ALCOHOLS
A carbohydrate providing 4 calories per gram	Contain no carbohydrates and provide little or no calories in that very little is needed to provide the same sweetness as sugar	Modified versions of carbohydrates so that they provide some calories
Raises blood sugar levels	Blood sugar levels are not raised	Blood sugar levels are affected to a lesser degree than sugar
Overconsumption of sugar causes the excess to be stored as fat		Overconsumption may cause digestive upset

(*Note:* All sugar alcohols end in "-OL". This makes it easy to identify them in a product's ingredient list.)

Many types of sugar alcohols are used frequently in candies and gum because they do not promote tooth decay. Frequently used sugar alcohols include xylitol, mannitol, and sorbitol. Many commercial bakeries specializing in sugar-free pastries and other baked goods use sugar alcohols instead of sucrose. Although sugar alcohols do not raise blood sugar levels as much as sucrose, they do contain calories. Diabetics should check with their health professional to be sure they are able to eat them. One disadvantage to consuming sugar alcohols is that, when consumed in large quantities, they can cause digestive problems. Because they are so slowly absorbed from the small intestine, it fills with fluid, causing gas and diarrhea.

Because sugar alcohols are modified carbohydrates, they tend to have more of the same properties as sucrose and they generally do not give a strange aftertaste when eaten.

MALTITOL

One sugar alcohol that works well as a sugar substitute in baking is maltitol. It is derived from maltose, a natural sugar. Maltitol behaves like sucrose in that it is able to be used in baked goods without any change in texture to the batter or dough. It even works well when it is creamed with fat for cakes and desserts using the creaming method. Baked goods using maltitol also brown nicely. Maltitol provides 3 calories per gram and it is used in a 1 to 1 substitution by volume with sucrose. For equal sweetness, it can be used in a 1 to 1 ratio by weight considering that only a slight difference in weight exists between sucrose and maltitol and the fact that maltitol is only 90% as sweet as sucrose. It is available commercially in granular or in liquid form. The recipes in this chapter that use maltitol use it in the granular form.

Easy Substitutions

To lower the fat content of a baked good, make the following substitutions in the recipe in a 1 to 1 ratio:

- Substituting fat-free yogurt or sour cream in a cake or cookie dough works well to replace their fully fatted counterparts.
- Fat-free cream cheese tends to impart a rubbery, plastic texture so a low-fat version would work best.
- Replacing healthier oils for melted butter works well in most recipes. If more of a buttery flavor is desired, substitute just half the butter with oil.
- Many low-fat recipes substitute egg whites for whole eggs because only the yolk contains fat. However, egg whites tend to be drying agents and healthier products that use just whites tend to be rubbery in texture. For recipes where there are other tenderizing ingredients, egg whites can be used successfully. For every whole egg, try substituting two egg whites. One or two yolks can be added back in to help the texture, but recipes will vary. In general, if a recipe does not contain an inordinate amount of eggs, leave them in the recipe unless cholesterol is a health issue.
- Pasteurized egg substitutes are generally fat and cholesterol free and can be used instead of whole eggs. Read the manufacturer's conversion of how much egg substitute is equivalent to one whole egg.
- Replace fat-free cow's milk or soy milk for whole cow's milk.
- When making a pie or tart crust, substitute half the fat for a solid, low-fat trans-fat–free margarine or butter. Cube it, wrap it in plastic, and freeze it for at least 1 hour. Cutting the fat in while it is partially frozen will keep it from melting too quickly in the oven and thus maintains the flakiness desired. If using partially frozen margarine or butter, it should not harden to the point where it cannot be cut into the dry ingredients.

- Try substituting puréed fruit or vegetables for some of the fat for a cake using the creaming method. Substitute half of the fat with the same weight of applesauce, apple butter, fruit purée (sold as a fat replacement), or sweet potato purée. It is crucial that the puréed fruit or vegetables be strained so that there are no stringy fibers. Therefore, strained baby vegetables work extremely well, although they provide less fiber than a fresh, puréed fruit or vegetable.
- To sweeten desserts only, replace sugar with an equal amount of an artificial sweetener such as sucralose or a lesser amount (or to taste) of aspartame.
- For baked goods in which it is desirable that the sugar substitute has some of the same properties of sugar, try a granular sugar alcohol such as maltitol used in a 1 to 1 substitution for sugar.
- To substitute unsweetened cocoa powder for unsweetened chocolate, replace 1 ounce (30 g) of unsweetened chocolate with ½ ounce (3 tablespoons; 15 g) unsweetened cocoa powder plus ½ fluid ounce (1 tablespoon; 15 mL) canola oil *or* ½ ounce (1 tablespoon; 15 g) low-fat, trans-fat–free solid margarine or butter.
- To substitute unsweetened cocoa powder for semisweet chocolate, replace 6 ounces (1 cup; 170 g) semisweet chocolate with 1 ounce (6 tablespoons; 30 g) unsweetened cocoa powder plus 3½ ounces (7 tablespoons; 100 g) granulated sugar plus 2 fluid ounces (¼ cup; 60 mL) canola oil *or* 2 ounces (4 tablespoons; 60 g) low-fat, trans-fat–free solid margarine or butter.

How to Modify a Recipe's Ingredients with Healthier Alternatives

The first step in effectively changing a recipe's ingredients to be healthier is to scan the list of ingredients to see which ones could be altered without creating major changes to the final baked good.

The conservative approach is best when modifying desserts to be healthier. Try not to substitute all the fat in a recipe for low-fat or no-fat versions right away. Experimenting through trial and error is crucial to success and it may be necessary to prepare the recipe many times, using varying amounts of ingredients in order to achieve the desired end product. Making a product that neither resembles nor tastes like the original is a waste of time and ingredients because no one will want to eat it. For some recipes, completely substituting healthier fats or whole grains will work nicely. For others, only a small amount of fat or whole grains may be able to be replaced without losing desirable taste and texture.

The appropriate substitute is important. For example, applesauce would not be an appropriate substitute for fat in a flaky pie crust recipe. Applesauce is sweet and will tenderize and moisten up to a point, but no matter what, it will never create flakiness in a pie crust. A better choice would be to replace half the original type of fat with a healthier solid fat to ensure that the crust is flaky. The recipe for Fudge Swirl Sour Cream Pound Cake (Chapter 7) is reprinted below. The original ingredients are printed on the left while healthier ingredient alternatives are printed on the right. One or more of the alternatives may be used to replace the original ingredients depending on how healthy the baker wishes to make the pound cake.

How to Substitute Unsweetened Cocoa Powder for Chocolate

1. To replace unsweetened chocolate: 1 ounce (30 g) unsweetened chocolate = ½ ounce (3 tablespoons; 15 g) unsweetened cocoa powder + ½ fluid ounce (1 tablespoon; 15 mL) canola oil *or* ½ ounce (1 tablespoon; 15 g) low-fat, trans-fat–free solid margarine or butter.

2. To replace semisweet chocolate: 6 ounces (1 cup; 170 g) semisweet chocolate = 1 ounce (6 tablespoons; 30 g) unsweetened cocoa powder plus 3½ ounces (7 tablespoons; 100 g) granulated sugar plus 2 fluid ounces (¼ cup; 60 mL) canola oil *or* 2 ounces (4 tablespoons; 60 g) low-fat, trans-fat–free solid margarine or butter.

FUDGE SWIRL SOUR CREAM POUND CAKE

The nutrition facts here are for the *original* Fudge Swirl Sour Cream Pound Cake. Table 13–5 lists the ingredients for the Original Fudge Swirl Sour Cream Pound Cake and offers healthier substitutions.

Nutrition Facts		
Fudge Swirl Sour Cream Pound Cake		
Serving Size 2-ounce slice		
Amount Per Serving		
Calories 328		Calories from Fat 124
		% Daily Value*
Total Fat	14.29 g	22%
Saturated Fat	8.45 g	42%
Cholesterol	86.74 mg	29%
Sodium	146.40 mg	6%
Total Carbohydrate	45.77 g	15%
Dietary Fiber	0.87 g	3%
Sugars	25.71 g	
Protein	5.08 g	10%
Vitamin A: 9%	Vitamin C 0%	
Calcium 4%	Iron 8%	

*Percent Daily Values are based on a 2,000 calorie diet. Your daily values may be higher or lower depending on your calorie needs.

Table 13–5 Fudge Swirl Sour Cream Pound Cake—Original Ingredients and Healthier Substitutions

ORIGINAL	HEALTHIER SUBSTITUTIONS
4 ounces (½ cup; 115 g) unsalted butter,	4 ounces (½ cup; 115 g) unsalted trans-fat free softened margarine, softened **or** 2 ounces (¼ cup; 55 g) unsalted butter and 2 ounces, (¼ cup; 55 g) applesauce
4 ounces (115 g) cream cheese, softened	4 ounces (115 g) low-fat cream cheese, softened **or** 4½ ounces (½ cup; 130 g) applesauce **or** half sugar and half maltitol
14½ ounces (2 cups; 400 g), granulated sugar	14 ounces (2 cups; 400 g) granulated maltitol
12½ ounces (3 cups; 405 g) all-purpose flour	9½ ounces (2 cups; 270 g) all-purpose flour and 3¼ ounces (1 cup; 95 g) soy flour
1 teaspoon (4 g) baking powder	1 teaspoon (4 g) baking powder
½ teaspoon (2 g) baking soda	½ teaspoon (2 g) baking soda
½ teaspoon (3 g) salt	½ teaspoon (3 g) salt
4 large eggs (190 g)	Pasteurized egg substitute equivalent to 4 large eggs (190 g)
2 teaspoons (10 mL) vanilla extract	2 teaspoons (10 mL) vanilla extract
8 ounces (1 cup; 225 g) sour cream	8 ounces (1 cup; 225 g) fat-free sour cream **or** no-fat plain or vanilla yogurt
3½ ounces (7 tablespoons; 100 g) Rich Chocolate Sauce (See Chapter 11)	3½ ounces (7 tablespoons; 100 g) reduced-fat "ganache" fudge sauce (recipe in this chapter) **or** 2 ounces (⅓ cup; 55 g) semisweet mini chocolate chips

Because the original recipe uses the creaming method, a solid fat is needed. The healthier version also requires a solid fat. The following are a few options for replacing some of the fat:

■ Replacing the butter, a saturated fat, with a lower fat, trans-fat–free margarine or butter will reduce the saturated fat significantly but not necessarily the calories.

■ To cut calories and saturated fat, half the amount of regular butter can be combined with an equivalent amount of applesauce.

■ To cut calories and saturated fat, eliminate the cream cheese completely and replace it with an equal amount of applesauce by weight. (If choosing this alternative, use the 4 ounces [½ cup; 115 g] trans-fat free, low-fat–solid margarine or butter to replace the unsalted butter.)

■ Low-fat cream cheese can replace its fully fatted counterpart.

This recipe requires an acidic dairy product such as sour cream to combine with the baking soda to provide leavening. Here are two healthier options:

■ Fat-free sour cream

■ Plain or vanilla fat-free yogurt

Reduced-fat "ganache" is used to replace the ganache in the original recipe. The original ganache recipe contains heavy cream and chocolate whereas the low-fat version contains unsweetened cocoa powder, fat-free milk, and only a small amount of chocolate; this decreases the saturated fat significantly. A small amount of mini chocolate chips can be sprinkled on top of the cake instead of using ganache in the middle of the cake.

Maltitol, a sugar alcohol, can be used successfully to replace all or half of the sugar in this recipe. Sugar-free chocolate can also be substituted for the chocolate in the reduced-fat ganache recipe.

Replacing one third of the amount of refined flour with soy flour increases the nutritional value of the finished cake. Another option might be to replace the soy flour with whole wheat flour.

The following are substitutions that would not work:

■ Vegetable purées or a greater amount of applesauce than is listed earlier. This would not allow enough creaming to occur, compromising the cake's volume and texture and would cause the batter to darken too much.

■ Replacing a liquid fat such as oil for the butter would not allow any creaming to occur and the texture of the cake would be heavy.

Antioxidants and Health

Antioxidants are chemical substances found mostly in plants. Antioxidants help cells within the human body repair themselves after being exposed to damaging substances or from the environment. They are associated with reducing the risk of heart disease and certain cancers. These chemical substances exist in many plant foods such as in fruits, vegetables, soy products, beans, chocolate, seeds, nuts, oils, grains, and wine. Eating foods that are rich in antioxidants is a great way to maintain a healthy diet.

The Health Benefits of Chocolate

Although chocolate does contain saturated fat in the form of cocoa butter, it does not raise blood cholesterol levels to the same extent as other saturated fats. Chocolate contains some health benefits if eaten in moderation. Chocolate contains *cocoa solids nonfat*, a term used to describe the nonfat solids within the cocoa bean. Chocolate that contains the highest levels of cocoa solids nonfat contains the highest amounts of antioxidants. Cocoa solids nonfat contain chemicals known as polyphenolic compounds. Polyphenolic compounds are substances found in plants that act as powerful antioxidants. These compounds also enhance the flavor of chocolate and the color of the cocoa bean. The more cocoa solids nonfat that are present in the chocolate, the stronger the chocolate flavor will be. That is why these benefits are seen only in the darkest or the most bittersweet types of chocolate.

RECIPES

HEALTHY CHOCOLATE CHIP POUND CAKE

Makes approximately 16 2-ounce servings

Lessons demonstrated in this recipe:

- How to modify the Fudge Swirl Sour Cream Pound Cake (Chapter 7) to create a healthier dessert while maintaining the original recipe's taste and texture.
- Using low-fat butter, low-fat cream cheese, and fat-free sour cream or yogurt cuts down on the saturated fat while still maintaining the creaming method of mixing.
- Substituting soy flour for some of the all-purpose flour boosts the nutritional value of the cake.
- Replacing the soy flour with whole wheat flour is another option to add fiber.
- Using pasteurized egg substitute instead of eggs reduces the cholesterol and fat content.
- Substituting a small amount of semisweet chocolate for the ganache makes this cake healthier, yet still gives in to chocolate cravings.

MEASUREMENTS		METRIC	BAKER'S %	INGREDIENTS
U.S.				
2 ounces	¼ cup	55 g	15%	low-fat, trans-fat free margarine or butter, softened *or* 1 ounce (2 tablespoons; 30 g) unsalted butter, softened and 1 fluid ounce (2 tablespoons; 30 mL) canola oil
6 ounces		170 g	47%	low-fat cream cheese, softened
14¼ ounces	2 cups	400 g	112%	granulated sugar
8 ounces	2 cups	270 g	74%	all-purpose flour
9½ ounces	1 cup	95 g	26%	soy flour
	1 teaspoon	4 g	1.1%	baking powder
	½ teaspoon	2 g	0.5%	baking soda
	½ teaspoon	3 g	0.8%	salt
3 each		140 g	39%	large eggs or pasteurized egg substitute
1 each		28 g	8%	large egg white
	2 teaspoons	10 mL	2.7%	vanilla extract
8 ounces	1 cup	225 g	62%	fat-free sour cream or fat-free vanilla or plain yogurt
2 ounces	⅓ cup	60 g	16%	mini semisweet chocolate chips or chopped semisweet chocolate
			404.1%	Total Healthy Chocolate Chip Pound Cake percentage

1. Preheat oven to 350°F (175°C). Spray a 10-inch (25-cm) false-bottom tube pan with cooking spray and set aside.

2. In the bowl of an electric mixer, using the paddle attachment, cream the butter (or the butter and oil), cream cheese, and sugar until light in color and fluffy. This can take up to 5 to 6 minutes. Stop the mixer and scrape down the bowl with a rubber spatula occasionally.

3. In a small bowl, whisk together the flours, baking powder, baking soda, and salt. Set aside.

4. In a small bowl, whisk together the eggs or egg substitute, egg white, and the vanilla extract.

5. On low speed, add the egg mixture into the creamed butter in thirds waiting for the mixture to blend together uniformly before adding more egg (Figure 13–1).

6. On low speed, add one third of the flour mixture into the eggs and butter. Blend until combined and add one half of the yogurt or sour cream.

7. Add another one third of the flour mixture, blending well, followed by the remaining yogurt or sour cream. Stop the machine and scrape down the sides of the bowl.

8. Add the remaining one third of the flour mixture and mix until well combined (Figure 13–2). Remove the bowl from the mixer.

9. Using a rubber spatula, scrape around the bottom and sides of the bowl to make sure the mixture is smooth and well combined.

10. Pour the batter into the prepared pan and smooth it with a rubber spatula.

11. Sprinkle the chocolate chips evenly over the batter (Figure 13–3).

12. Bake for 50 to 60 minutes or until a cake tester placed into the center of the cake comes out clean.

13. Cool thoroughly and remove from the pan.

FIGURE 13–1

FIGURE 13–2

FIGURE 13–3

Nutrition Facts
Healthy Chocolate Chip Pound Cake

Serving Size: 2-ounce slice

Amount Per Serving		
Calories 294	Calories from Fat 103	
		% Daily Value*
Total Fat	11.75 g	18%
Saturated Fat	4.91 g	25%
Cholesterol	56.78 mg	19%
Sodium	180.30 mg	8%
Total Carbohydrate	41.27 g	14%
Dietary Fiber	1.22 g	5%
Sugars	26.24 g	
Protein	7.38 g	15%
Vitamin A 5%	Vitamin C	0%
Calcium 7%	Iron	8%

*Percent Daily Values are based on a 2,000 calorie diet.
Your daily values may be higher or lower depending on
your calorie needs.

Healthy Chocolate Chip Pound Cake

HEALTHY LEMON POUND CAKE

(Another healthy variation to the Fudge Swirl Sour Cream Pound Cake)

Makes approximately 16 2-ounce servings

Lessons demonstrated in this recipe:

- Having the options of substituting the butter with trans-fat free, low-fat butter or half butter and half canola oil still allows the creaming method to be used with healthier fats.
- Puréed low-fat cottage cheese replaces some of the fat, adding a creamy texture to the cake.
- Adding a small amount of soy flour increases the nutritional value of the pound cake.

MEASUREMENTS				INGREDIENTS
U.S.		METRIC	BAKER'S %	
4 ounces	½ cup	115 g	32%	low-fat cottage cheese (not fat free)
14½ ounces	2 cups	400 g	112%	granulated sugar
4 ounces	½ cup	115 g	32%	low-fat, trans-fat–free butter, softened *or* 2 ounces (¼ cup; 55 g) unsalted butter and 2 fluid ounces (¼ cup; 60 mL) canola oil
9½ ounces	2 cups	225 g	74%	all-purpose flour
3¼ ounces	1 cup	95 g	26%	soy flour
	1 teaspoon	4 g	1.1%	baking powder
	½ teaspoon	2 g	0.5%	baking soda
	½ teaspoon	3 g	0.8%	salt
3 each		140 g	39%	large eggs or pasteurized egg substitute
	2 teaspoons	10 mL	2.7%	vanilla extract
8 ounces	1 cup	225 g	62%	fat-free lemon yogurt
	2 teaspoons	10 mL	2.7%	lemon extract
	1 tablespoon	18 g	4.9%	grated lemon zest
			389.7%	Total Healthy Lemon Pound Cake percentage

1. Preheat oven to 350°F (175°C). Spray a 10-inch (25-cm) false-bottom tube pan with cooking spray and set aside.

2. Pureé the cottage cheese and half of the sugar in a food processor until the mixture is smooth. Set aside.

3. In the bowl of an electric mixer, using the paddle attachment, cream the butter (or the butter and oil), the remaining sugar, and the pureéed cottage cheese and sugar mixture until light in color and fluffy. This can take up to 5 to 6 minutes. Stop the mixer and scrape the bowl with a rubber spatula occasionally.

4. In a small bowl, whisk together the flours, baking powder, baking soda, and salt. Set aside.

5. In a small bowl, whisk together the eggs or egg substitute and the vanilla extract.

6. On low speed, add the egg mixture into the creamed butter in thirds waiting for the mixture to blend together uniformly before adding more egg.

7. On low speed, add one third of the flour mixture into the eggs and butter. Blend until combined and add one half of the lemon yogurt.

8. Add another one third of the flour mixture, blending well, followed by the remaining lemon yogurt. Add the lemon extract and the lemon zest. Stop the machine and scrape down the sides of the bowl.

9. Add the remaining one third of the flour mixture and mix until well combined. Remove the bowl from the mixer.

10. Using a rubber spatula, scrape around the bottom and sides of the bowl to make sure the mixture is smooth and well combined.

11. Pour the batter into the prepared pan and smooth it with a rubber spatula.

12. Bake for 50 to 53 minutes or until a cake tester placed into the center of the cake comes out clean.

13. Cool thoroughly and remove from the pan. Dust the cake with confectioners' sugar, if desired.

Nutrition Facts
Healthy Lemon Pound Cake

Servings: 16
Serving Size: 2-ounce slice

Calories 269	Calories from Fat 81	
		% Daily Value*
Total Fat	9.14 g	14%
Saturated Fat	3.33 g	17%
Cholesterol	51.95 mg	17%
Sodium	160.70 mg	7%
Total Carbohydrate	40.29 g	13%
Dietary Fiber	0.99 g	4%
Sugars	26.17 g	
Protein	7.31 g	15%

Vitamin A	4%	Vitamin C	0%
Calcium	5%	Iron	7%

*Percent Daily Values are based on a 2,000 calorie diet. Your daily values may be higher or lower depending on your calorie needs.

Healthy Lemon Pound Cake

HEALTHIER FUDGE BROWNIES

Makes approximately 40 2-ounce brownies

Lessons demonstrated in this recipe:

- ■ How to prepare a healthier fudge brownie by substituting a healthier fat such as canola oil for the butter.
- ■ Using cocoa powder instead of unsweetened chocolate reduces the fat.
- ■ Instant coffee is added to intensify the chocolate flavor.
- ■ Whole eggs are replaced with fat-free egg whites and a no-fat, no-cholesterol, pasteurized egg substitute.
- ■ Adding baby sweet potatoes, carrots, or apple sauce provides natural sweetness and moistness and replaces half of the fat.
- ■ A small amount of semisweet chocolate is mixed into the batter to provide intensity of chocolate flavor.

MEASUREMENTS				INGREDIENTS
U.S.		**METRIC**	**BAKER'S %**	
$^1/_4$ ounce	2 tablespoons	10 g	3.5%	instant coffee powder
	2 teaspoons	10 g	3.2%	hot water
6 fluid ounces	$^3/_4$ cup	150 mL	48%	canola oil
6 ounces	$^1/_2$ cup + 1 tablespoon	170 g	55%	puréed sweet potato or carrots from processed baby food, *or* unsweetened applesauce
$4^3/_4$ ounces	1 cup	135 g	44%	all-purpose flour
$6^1/_4$ ounces	$1^1/_4$ cups	155 g	56%	bread flour
	2 teaspoons	8 g	2.6%	baking powder
$3^3/_4$ ounces	$1^1/_4$ cups	110 g	35%	unsweetened cocoa powder
	$^1/_2$ teaspoon	3 g	1%	salt
7 ounces	1 cup	200 g	127%	light brown sugar (packed, if measuring by volume)
1 pound + $1^3/_4$ ounces	$2^1/_2$ cups	500 g	161%	granulated sugar
4 each		110 g	36%	large egg whites
$3^1/_4$ fluid ounces		95 mL	31%	pasteurized egg substitute (equivalent to 2 large eggs)
	$1^1/_2$ teaspoons	7.5 mL	2.4%	vanilla extract
$7^1/_2$ ounces	$1^1/_4$ cups	215 g	69%	semisweet chocolate, coarsely chopped
$4^1/_2$ ounces	$1^1/_4$ cups	125 g	40%	walnuts, chopped
			714.4%	Total Healthier Fudge Brownies percentage

1. Preheat the oven to 350°F (175°C). Spray one 11- by 15-inch (27.5- by 37.5-cm) pan or two 8-inch (20-cm) square pans with nonstick cooking spray. Alternatively, line pans with aluminum foil and spray with nonstick cooking spray. Set aside.

2. Whisk together the coffee, the oil, and puréed sweet potato, carrots, or applesauce. Set aside.

3. In a mixing bowl, whisk the two flours, the baking powder, the cocoa powder, and the salt. Set aside.

FIGURE 13–4

FIGURE 13–5

4. In the bowl of an electric mixer, using the paddle attachment, beat the sugars on low speed and gradually add the egg whites, pasteurized eggs, coffee mixture, and vanilla. Beat at medium speed until well blended.

5. Blend in the flour mixture on low speed and mix just until blended. Remove the bowl from the mixer and mix in the chopped chocolate using a spatula (Figure 13–4). Do not overmix.

6. Pour the batter into the prepared pan (or pans) and spread evenly. Scatter the walnuts over the batter (Figure 13–5).

7. Bake for 28 to 30 minutes until the brownie is set and firm on top but still slightly soft. A knife inserted into the center will not come out completely clean.

8. Cool completely and chill for 1 hour in the refrigerator to make cutting into bars easier.

Nutrition Facts
Healthier Fudge Brownies
Serving Size: 2-ounce brownie

Amount Per Serving		
Calories 210	Calories from Fat 73	
		% Daily Value*
Total Fat	8.49 g	13%
Saturated Fat	1.69 g	8%
Cholesterol	0.03 mg	0%
Sodium	39.49 mg	2%
Total Carbohydrate	33.60 g	11%
Dietary Fiber	1.66 g	7%
Sugars	22.79 g	
Protein	2.80 g	6%
Vitamin A 0%	Vitamin C 0%	
Calcium 2%	Iron 6%	

*Percent Daily Values are based on a 2,000 calorie diet. Your daily values may be higher or lower depending on your calorie needs.

Healthier Fudge Brownies

HEALTHY FLATBREAD

Makes approximately
24 ½-ounce pieces

Lessons demonstrated in this recipe:

■ How to prepare a healthy snack cracker that uses no trans-fats or saturated fats.

■ Flax seeds contain phytoestrogens and omega-3 fatty acids, which are believed to interfere with cancer growth, lower blood cholesterol, and act as antioxidants.

■ Grinding a portion of the flax seeds ensures that their healthy qualities will be easily absorbed by the body.

■ A small portion of whole wheat flour is used to add fiber.

■ Olive oil, a healthy monounsaturated fat, is used instead of butter or shortening.

MEASUREMENTS				INGREDIENTS
U.S.		**METRIC**	**BAKER'S %**	
2 ounces	⅓ cup	60 g	30%	flax seeds ground to a powder in a coffee grinder
4¾ ounces	1 cup	135 g	67%	all-purpose flour
2¼ ounces	½ cup	65 g	33%	whole wheat flour
	½ teaspoon	1 g	0.5%	garlic powder
	½ teaspoon	1 g	0.5%	onion powder
	½ teaspoon	2 g	1%	baking powder
	¾ teaspoon	4 g	2%	salt
	¼ teaspoon	1 g	0.5%	coarsely ground black pepper
1½ ounces	¼ cup	45g	23%	flax seeds, left whole
1 fluid ounce	2 tablespoons	30 mL	15%	extra virgin olive oil
4 fluid ounces	½ cup	120 mL	60%	skim milk
			232.5%	Total Healthy Flatbread percentage

1. Place all the dry ingredients from the ground flax seeds to the whole flax seeds in the bowl of an electric mixer and, using the paddle attachment, blend the mixture well on low speed.

2. With the machine still on low speed, add the olive oil and blend until combined (Figure 13–6). Slowly add the milk until a dough forms (Figure 13–7).

3. Remove the dough from the bowl and wrap it in plastic wrap. Chill the dough for 15 to 30 minutes.

FIGURE 13–6

FIGURE 13–7

FIGURE 13–8

4. Preheat the oven to 325°F (165°C). Divide the dough into quarters. On a lightly floured surface, roll one fourth of the dough into a rough rectangle as thin as possible. Using a pizza cutter, slice the dough into lengths 4 to 5 inches (10 to 12.5 cm) long by $\frac{1}{2}$ to 1 inch (1.25 to 2.5 cm) wide strips and place them on a baking sheet covered with parchment paper (Figure 13–8). The dough shapes do not need to be perfect rectangles. Alternatively, any shape can be cut out from the dough such as stars, circles, squares, triangles, or rectangles.

5. Bake the dough for 20 to 22 minutes or until golden brown and crispy. Cool on racks. The crackers can be stored in an airtight container at room temperature for 2 week13or frozen in an airtight container for 4 to 6 months. Thaw the crackers at room temperature and serve. If the crackers have become soft, place them in a 325°F (165°C) oven for 10 minutes to crisp them.

Nutrition Facts
Healthy Flatbread
Serving Size: $\frac{1}{2}$-ounce piece

Calories 59	Calories from Fat 23	
		%Daily Value*
Total Fat	2.66 g	4%
Saturated Fat	0.31 g	2%
Cholesterol	0.10 mg	0%
Sodium	77.02 mg	3%
Total Carbohydrate	7.22 g	2%
Dietary Fiber	1.37 g	5%
Sugars	0.36 g	
Protein	1.75 g	3%
Vitamin A 0%	Vitamin C 0%	
Calcium 2%	Iron 3%	

*Percent Daily Values are based on a 2,000 calorie diet. Your daily values may be higher or lower depending on your calorie needs.

Healthy Flatbread and Cinnamon Sugar Flatbread

CINNAMON SUGAR FLATBREAD (VARIATION OF HEALTHY FLATBREAD)

Makes approximately 24 ½-ounce pieces

Lessons demonstrated in this recipe:

- How to prepare a sweet variation of the healthy flat bread recipe.
- A small portion of soy flour is used to provide phytoestrogens, which are associated with cancer prevention.
- Soy flour also contains isoflavones (antioxidants), which lower LDLs—the bad cholesterol.

MEASUREMENTS				INGREDIENTS
U.S.		METRIC	BAKER'S %	
4¾ ounces	1 cup	135 g	82%	all-purpose flour
1 ounce	¼ cup	30 g	18%	soy flour
	1 tablespoon	15 g	9%	granulated sugar
	½ teaspoon	2 g	1.2%	baking powder
	½ teaspoon	3 g	1.8%	salt
1 fluid ounce	2 tablespoons	30 mL	18%	canola oil
4 fluid ounces	½ cup	120 mL	73%	skim milk
1 each		28 g	17%	egg white, beaten (to be used as an egg wash)
3½ ounces	½ cup	100 g	61%	granulated sugar mixed with ½ teaspoon (1 g) ground cinnamon
			281%	Total Cinnamon Sugar Flatbread percentage

1. Place all the dry ingredients (from the all-purpose flour to the salt) in the bowl of an electric mixer and, using the paddle attachment, blend the mixture well on low speed.

2. With the machine still on low speed, add the canola oil and blend until combined. Slowly add the milk until a dough forms.

3. Follow the same procedure for steps 3 and 4 of the Healthy Flatbread, including preheating the oven to 325°F (165°C). After cutting the pieces of dough and placing them on the sheet pan, brush each piece of dough with some egg wash using a pastry brush. Sprinkle each piece of dough with some cinnamon sugar (Figure 13–9).

FIGURE 13–9

4. Bake the dough for 20 to 22 minutes or until golden brown and crispy. Cool on racks. The crackers can be stored in an airtight container at room temperature for 2 weeks or frozen in an airtight container for 4 to 6 months. Thaw the crackers at room temperature and serve. If the crackers have become soft, place them in a 325°F (165°C) oven for 10 minutes to crisp them.

Nutrition Facts
Cinnamon Sugar Flatbread
Serving Size: $\frac{1}{2}$-ounce piece

Amount Per Serving		
Calories 105	Calories from Fat 22	
		% Daily Value*
Total Fat	2.52 g	4%
Saturated Fat	0.21 g	1%
Cholesterol	0.20 mg	0%
Sodium	106.47 mg	4%
Total Carbohydrate	18.01 g	6%
Dietary Fiber	0.56 g	2%
Sugars	10.37 g	
Protein	2.85 g	6%

Vitamin A	0%	Vitamin C	0%
Calcium	2%	Iron	3%

*Percent Daily Values are based on a 2,000 calorie diet. Your daily values may be higher or lower depending on your calorie needs.

HEALTHY BLUEBERRY CINNAMON MUFFINS

Makes 12 4-ounce muffins

Lessons demonstrated in this recipe:

- How to prepare a healthy, nutritional muffin using the creaming method.
- Using a small amount of a trans-fat free margarine or butter allows the sugar to be creamed to aid leavening and obtain a lighter, cake-like texture.
- Using some applesauce replaces some fat while adding moistness.
- Adding some whole wheat flour adds texture and fiber.
- Blueberries contain a pigment known as anthrocyanin, which contains healthy antioxidants associated with lowering the risk of certain cancers.

MEASUREMENTS				INGREDIENTS
U.S.		**METRIC**	**BAKER'S %**	
7 ounces	1½ cups	200 g	75%	all-purpose flour
2¼ ounces	½ cup	65 g	24%	whole wheat flour
	¾ teaspoon	1 g	0.4%	ground cinnamon
	2 teaspoons	8 g	3%	baking powder
	½ teaspoon	3 g	1.1%	salt
2 ounces	4 tablespoons	60 g	21%	low-fat (trans-fat–free) margarine or butter
7 ounces	1 cup	200 g	79%	granulated sugar or maltitol granules
3 ounces	¼ cup	8 g	32%	applesauce
2 each		94 g	35%	large eggs or pasteurized egg substitute equivalent to 2 large eggs
4 fluid ounces	½ cup	120 mL	45%	soy milk or skim cow's milk
	1 teaspoon	5 mL	1.9%	vanilla extract
1 pound	2½ cups	455 g	171%	fresh or frozen blueberries (not thawed)
	1 teaspoon	1 g	0.4%	all-purpose flour
As needed		As needed		extra granulated sugar for sprinkling
			488.8%	Total Healthy Blueberry Cinnamon Muffins percentage

1. Preheat the oven to 375°F (190°C). Spray a 12-cup muffin tin with nonstick cooking spray and set it aside.

2. Whisk the flours, cinnamon, baking powder, and salt in a mixing bowl until they are well combined. Set aside.

3. In the bowl of an electric mixer, using the paddle attachment, cream the low-fat margarine or butter with the sugar or maltitol on medium speed until the mixture looks light in texture. On low speed, add the applesauce and blend well.

4. Gradually add the eggs or egg substitute and continue to mix on low speed until the mixture is well combined.

5. On low speed, add one third of the dry ingredients and blend well.

6. Add one half of the milk and blend. Add another one third of the dry ingredients. Blend only until combined.

7. Add the remaining milk and the vanilla extract, stopping the machine to scrape down the sides of the bowl periodically.

8. Add the remaining one third of the dry ingredients and blend until just combined.

9. In another bowl, combine the blueberries and flour until they are well coated.

10. Add the blueberries to the batter and gently fold them in using a rubber spatula.

11. Fill each muffin cup with batter. The cups should be approximately four fifths full (Figure 13–10).

12. Sprinkle some sugar over each muffin and bake for 20 to 30 minutes or until the muffins are golden brown or a knife or cake tester inserted into the middle of a muffin comes out clean. Serve warm or at room temperature.

FIGURE 13–10

Nutrition Facts
Healthy Blueberry Cinnamon Muffins
Serving Size: 4-ounce muffin

Amount Per Serving		
Calories 222	Calories from Fat 35	
		% Daily Value*
Total Fat	4 g	6%
Saturated Fat	1.95 g	10%
Cholesterol	40.26 mg	13%
Sodium	117.06 mg	5%
Total Carbohydrate	42.66 g	14%
Dietary Fiber	1.89 g	8%
Sugars	21.93 g	
Protein	4.50 g	9%
Vitamin A 4%	Vitamin C 6%	
Calcium 2%	Iron 9%	

*Percent Daily Values are based on a 2,000 calorie diet. Your daily values may be higher or lower depending on your calorie needs.

Healthy Blueberry Cinnamon Muffins

LOW-FAT CHOCOLATE RASPBERRY MOUSSE

Makes 20 4-ounce servings

Lessons demonstrated in this recipe:

- How to prepare a healthy mousse using puréed tofu and no heavy cream.
- This would make a wonderful dessert for anyone who is lactose intolerant or cannot handle fat from dairy products.
- Tofu, a soy protein, contains antioxidants that are associated with reducing the risk of heart disease and certain cancers.
- Tofu is a "flavor follower" and takes on any flavor that surrounds it.
- Using some high-quality chocolate plus cocoa powder intensifies the chocolate flavor.
- High-quality dark chocolate such as semisweet or bittersweet contains healthy antioxidants.

MEASUREMENTS			INGREDIENTS
U.S.		METRIC	
1 pound + 8½ ounces	3 cups	685 g	reduced-fat silken tofu
5¼ ounces	1⅓ cups	150 g	confectioners' sugar
6½ ounces	½ cup	180 g	seedless raspberry preserves
3 ounces	⅔ cup	85 g	Dutch processed cocoa powder
4 fluid ounces	½ cup	120 mL	piping hot brewed coffee
8 ounces	1⅓ cups	230 g	high-quality semisweet or bittersweet chocolate, finely chopped and melted over a double boiler
	2 teaspoons	10 mL	vanilla extract

1. In a food processor, purée the tofu, the confectioners' sugar, and the preserves until the mixture is completely smooth. Stop the machine and scrape down the sides periodically with a rubber spatula.

2. In a mixing bowl, whisk together the cocoa powder and the hot coffee.

3. Whisk the melted chocolate and the vanilla into the cocoa and coffee mixture. Continue to whisk the mixture until the chocolate and cocoa are smooth.

4. Place the chocolate mixture into the food processor with the puréed tofu and process until the mousse is completely smooth (Figure 13–11).

FIGURE 13–11

TIP   Low-fat chocolate raspberry mousse may also be used as a filling between cake layers

5. Divide the mousse into stemmed glasses and chill for at least 1 to 2 hours or overnight. The mousse will thicken over time. Alternatively, place the mousse in a large bowl, covered, and chill it for 1 to 2 hours or overnight. Once thickened, it can then be placed in a pastry bag fitted with a large star tip and piped decoratively into stemmed glasses or individual prebaked tart shells (Figure 13–12).

6. Garnish with fresh raspberries just before serving.

FIGURE 13–12

Nutrition Facts
Low-Fat Chocolate Raspberry Mousse

Serving Size: 4 ounces

Amount Per Serving		
Calories 133	Calories from Fat 35	
		% Daily Value*
Total Fat	4.27 g	7%
Saturated Fat	2.39 g	12%
Cholesterol	0 mg	0%
Sodium	35.50 mg	1%
Total Carbohydrate	24.36 g	8%
Dietary Fiber	2.04 g	8%
Sugars	8.61 g	
Protein	3.51 g	7%
Vitamin A 0%	Vitamin C 1%	
Calcium 2%	Iron 7%	

*Percent Daily Values are based on a 2,000 calorie diet. Your daily values may be higher or lower depending on your calorie needs.

Low-Fat Chocolate Raspberry Mousse

REDUCED-FAT "GANACHE"

Makes approximately 8
1-ounce servings

Lessons demonstrated in this recipe:

- The combination of low-fat cocoa powder and a small amount of high-quality chocolate produces a deep chocolate flavor.
- Replacing heavy cream with fat-free milk and corn syrup greatly reduces the fat content.
- Adding instant coffee crystals intensifies the chocolate flavor.
- A dark chocolate such as semisweet or bittersweet contains healthy antioxidants.

MEASUREMENTS			INGREDIENTS
U.S.		METRIC	
$\frac{3}{4}$ ounce	$\frac{1}{4}$ cup	25 g	unsweetened cocoa powder
$1\frac{3}{4}$ ounces	$\frac{1}{4}$ cup	55 g	superfine sugar
2 fluid ounces	$\frac{1}{4}$ cup	60 mL	light corn syrup
2 fluid ounces	$\frac{1}{4}$ cup	60 mL	skim milk
1 fluid ounce	2 tablespoons	30 mL	water
	$\frac{1}{2}$ teaspoon	1 g	instant espresso or coffee crystals
$\frac{1}{4}$ ounce	$\frac{1}{2}$ tablespoon	5 g	unsalted butter
4 ounces		115 g	high-quality semisweet or bittersweet chocolate chips, finely chopped

1. Sift the cocoa and sugar into a small, heavy saucepan.

2. Add the corn syrup, milk, water, coffee, and butter. Bring the mixture to a rolling boil, whisking constantly. Remove the mixture from the heat.

3. Add the chopped chocolate and whisk until smooth (Figure 13–13). Use as a dessert sauce while still warm or at room temperature.

FIGURE 13–13

Nutrition Facts
Reduced-Fat Ganache

Serving Size: 1 ounce

Amount Per Serving		
Calories 136	Calories from Fat 43	
		% Daily Value*
Total Fat	5.35 g	8%
Saturated Fat	3.20 g	16%
Cholesterol	2.06 mg	1%
Sodium	17.62 mg	1%
Total Carbohydrate	25 g	8%
Dietary Fiber	1.72 g	7%
Sugars	14.60 g	
Protein	1.39 g	3%
Vitamin A 1%	Vitamin C	0%
Calcium 2%	Iron	5%

*Percent Daily Values are based on a 2,000 calorie diet.
Your daily values may be higher or lower depending on
your calorie needs.

Reduced-Fat Ganache

Professional Profile

BIOGRAPHICAL INFORMATION

Mark Hodgson
Pastry Chef Instructor
City College of San Francisco
San Francisco, CA

1. Question: *When did you realize that you wanted to pursue a career in baking and pastry?*

Answer: *While growing up I baked Lucy's Lemon Squares (from a Peanuts cookbook), bagels from scratch, char sui bao (barbecued pork buns), cheese-cakes of every flavor, mocha tortes, and English muffins over and over again. I derived great pleasure in creating something at home that other people would normally have to purchase in a store. Although I considered going to culinary school after high school it wasn't until I was in graduate school that I decided to pursue my passion for food as a profession.*

2. Question: *Was there a person or event that influenced you to go into this line of work?*

Answer: *After preparing elaborate holiday meals on a table that stretched from one end of the room to the other in my tiny apartment, my friends and family began to suggest that I should consider cooking or baking professionally. They saw how much pleasure cooking brought to me.*

3. Question: *What did you find most challenging when you first began working in baking and pastry?*

Answer: *I didn't go to a culinary school so initially working in a professional kitchen overwhelmed me. The words people spoke were an unknown jumble; a foreign language in which I had no sense of the meanings. Terms such as ganache, frangipane, crème anglaise, tuiles, joconde, genoise, or tammis, chinoise, mandoline, rondeau were all completely new to me. Likewise, the intensity of working in a professional kitchen where the expectations were very high and the work is very fast paced proved challenging.*

4. Question: *Where and when was your first experience in a professional baking setting?*

Answer: *I began my professional pastry career in the Ritz-Carlton Hotel in San Francisco in 1994 as the fourth cook/apprentice in the pastry shop. I still remember making huge batches of chocolate chip cookie dough in a 60-quart mixer and then scooping the dough with an ice cream scoop into four full sheet pans for what seemed like hours (with blisters forming). Coworkers made it clear that I was going a bit slow and that I would have to pick up the pace.*

5. Question: *How did this first experience affect your later professional development?*

Answer: *I was determined that I would overcome my weakness. Little by little I improved my speed, my organization, my knowledge of kitchen and pastry terminology and, most importantly, I learned how to successfully make all of the pastries and baked goods we produced. I thrived at the Ritz. In an industry where employee turnover is quite high I was an exception; I worked there for 11 years.*

6. Question: *Who were your mentors when you were starting out?*

Answer: *The pastry chefs that have had the most significant impact on my career are Paul Masse, because of his endless patience and depth of experience and knowledge; Kim O'Flaherty, because of her dedication and passion for the art of pastry; and Michel Willaume, because of his pure talent and originality. All three chefs inspired and challenged me in different ways.*

7. Question: *What would you list as the greatest rewards in your professional life?*

Answer: *Recognition from guests and coworkers for creating spectacular presentations and delicious desserts are the greatest rewards for me as a pastry chef. As a pastry instructor my daily reward comes from seeing the expressions on students' faces when they make beautiful and delicious creations. I derive the greatest pleasure of all from watching the students progress and improve from the beginning to the end of the program and knowing that I played some small part in that process.*

8. Question: *What traits do you consider essential for anyone entering the field?*

Answer: *To succeed in the pastry industry you must develop or possess a thick skin, determination, and "stick-to-itiveness." You should also enjoy the precision, consistency, attention to detail, and organization that pastry requires. Having a sweet tooth is also helpful.*

9. Question: *If there was one message you would impart to all students in this field what would that be?*

Answer: *You need to be a bit obsessive about getting things right and not let mistakes get you down. Remember your mistakes; they are invaluable lessons on what not to do the next time!*

CHAPTER 14

Troubleshooting

After reading this chapter, you should be able to:

- Correct common mishaps or prevent them from happening the next time.

- Use common sense when baking, making sure to understand the methods, tools, and equipment needed.

Alas, bakers and pastry chefs are only human and sometimes

those wonderful sweet pastries, breads, and desserts do not come out as anticipated. Most often there is a reason why mishaps occur and why the baked good can be "saved"—or not. Lessons should always be learned from mistakes that are made.

If the recipe describes how a particular batter or dough should feel or how a method of mixing should look, use your sense of touch and sight to verify these descriptions. If your recipe does not come close to how the ingredients should behave, maybe something went wrong.

Many times a correction can take place if the chef is aware of what each step should look like. The earlier an error is detected, the fewer ingredients will be wasted. If a correction cannot be made, throwing out a minimum of ingredients is better than having to dispose of the entire finished product.

The lesson to be learned from this chapter is this: Pay attention to mistakes, mishaps, and maloccurrences. They do happen. They can teach us a great deal of information that can be applied the next time the recipe is made.

Many of the "great" pastry chefs of the world have become "great" because they learned from their mistakes.

This chapter is titled "Troubleshooting." The term *troubleshooting* refers to locating and repairing a breakdown found in any type of work. In this case, the various mistakes that can occur in a bake shop and how to rectify them are discussed. There may be more than one reason why a poor outcome has occurred. It is the baker's job to review preparation techniques to determine which ones to correct.

What Can Go Wrong?

There are so many variables in baking; sometimes it is a daunting task to figure out what went wrong. The following is a general list to help you to determine where a mishap could have occurred:

- Directions were misread, misinterpreted, or not followed properly.
- Measuring of ingredients was inaccurate.
- The designated mixing method was not followed, for example, undermixed or overmixed.
- Inappropriate ingredients were substituted for the original.
- Flavorings were too little or too much.
- The wrong size pans were used.
- The pans were not prepped properly.
- The oven was not working properly.
- The baked good did not come out of the pan.

Directions Were Misread, Misinterpreted, or Not Followed Properly

The most common mistake bakers make is the simplest one to correct. Problems can arise if recipe directions are not read properly, not understood by the reader, and then not followed properly. Read the directions thoroughly *at least twice* to make sure it is clear what ingredients, tools, and equipment are needed

Measuring of Ingredients Was Inaccurate

There are no shortcuts to measuring. Proper measuring is crucial to the success of a recipe. Choose to measure by weight or volume. If measuring by volume, knowing how to properly measure liquids versus solids is critical to achieving a successful outcome.

The Designated Mixing Method Was Not Followed

There are different methods of mixing for different types of baked goods. If the proper mixing methods are not followed, the final product will be altered in texture, volume, and taste.

Inappropriate Ingredients Were Substituted for the Original

Substituting one ingredient for another may not work unless the substitutions are very close. For example, if you ran out of sour cream for a cake but plain yogurt was available, it would make a good substitution. However, substituting pastry flour for bread flour in a recipe would not yield the same result.

Flavorings Were Too Little or Too Much

Everyone has different tastes, and spices and flavorings may need to be adjusted. A recipe may call for 1 teaspoon (5 mL) vanilla extract or 1 teaspoon (5 g) salt, but the final product may taste bland to you. Punching up the flavor by adding more vanilla extract or salt would increase the flavor of the final product. This sounds so simple, but the correct flavoring is most important. Yeast breads can be bland if they are underfermented or too little salt is added.

The Wrong Size Pans Were Used

Many times the wrong size pans are used. Baking a cake batter in an 8-inch (20-cm) pan when the recipe calls for a 10-inch (25-cm) pan will result in the batter rising up and overflowing in the oven. Or, in the reverse, pouring batter into too big a pan may result in a thin, dense cake with little moistness.

The Pans Were Not Prepped Properly

No matter how hard you try, a cake will not come out of a pan that was not prepped properly, whether by greasing or by greasing, parchment papering, greasing again, and flouring. Sometimes it is important to know when greasing a pan is not appropriate. If a recipe states that a cake pan should not be greased and it was, the cake may have poor volume because the batter could not climb up the sides of the pan.

The Oven Was Not Working Properly

This is a common problem. Many ovens vary in temperature because they need to be recalibrated. Place an oven thermometer on the middle rack toward the middle of the oven to determine whether the temperature that the oven is set for is in fact the true temperature. This is a smart investment because many baked goods can be overbaked or underbaked if an oven's temperature is incorrect, which can affect the product's texture and volume.

The Baked Good Did Not Come Out of the Pan

Nothing is more frustrating than spending time baking a wonderful product and then it does not release from the pan. Remove cakes while they are still barely warm but not cold. Wait a few minutes before removing dropped and rolled cookies from the pan to allow them to firm up. It is important to follow each specific recipe's directions on how to remove the baked good.

The remaining portion of this chapter is devoted to listing—in chapter order—the many common problems that could occur while baking and the reasons for each. When appropriate, solutions are given. Sometimes there may be several explanations why a mishap has occurred. The baker should be able to determine exactly what went wrong and how to correct it.

Custards

WHAT IF . . .?	REASON	SOLUTION
Custard comes out lumpy and curdled.	Eggs were not tempered properly.	■ Add hot milk or cream in a slow stream to eggs while whisking constantly.
	Custard was not strained into a bowl over an ice water bath and carryover cooking caused eggs to curdle.	■ Strain the custard to remove the lumps.
	Custard was overcooked and the temperature went over 185°F (85°C), causing the egg proteins to lump together and curdle.	■ If there is only a slight amount of curdling, process the custard in a blender, then strain the custard to remove lumps.
A pastry cream thickens nicely, but when left overnight in a refrigerator, it thins out.	The alpha-amylase, an enzyme in the egg yolks, dissolved the starch so the custard thinned out.	■ Be sure the custard comes to a boil to destroy alpha-amylase.
A skin forms on a custard left in the refrigerator.	Casein, a protein in milk, dries out when exposed to air.	■ Place plastic wrap directly onto the surface of the custard or cover it with some melted butter, while still hot, to prevent any exposure to air.
After stirring in flavorings, the custard thins out.	The custard was overstirred, causing the starch granules to break down.	■ Blend in flavorings gently, never vigorously, while the custard is still warm. Once it has set, try not to overstir it.
	Too much sugar or acidic ingredients were used, causing the starch to break down and release liquid, thinning the sauce.	■ Be sure to add citrus juices after the full thickening power of the starch has occurred.

WHAT IF . . .?	REASON	SOLUTION
After combining egg yolks and sugar together, small, hard, yellow clumps develop.	This is a common occurrence when eggs, especially yolks, and granulated sugar are placed together in a bowl and allowed to stand unmixed for several minutes. The yellow clumps are actually proteins in the egg that have dried out and joined together. The sugar's hygroscopic properties draw water from within the egg toward the sugar, leaving the egg yolk dry, appearing curdled and "cooked."	■ Keep the yolks and granulated sugar separate until ready to begin the recipe, then continuously whisk them together before tempering them with hot milk.

Yeast Breads

WHAT IF . . .?	REASON
Bread does not rise well.	■ Yeast was not alive. ■ Salt was added directly to the yeast. ■ Flour had too little protein, causing little gluten to develop. The less gluten developed, the less carbon dioxide can be trapped within the dough. ■ Dough was overmixed or undermixed. ■ Oven temperature was too high and formed a crust before the bread had time to rise.
Bread splits open.	■ Dough was overmixed and gluten was weakened. ■ Dough was not fermented or proofed long enough. ■ After shaping, seam was placed on top. ■ Dough was not formed well during makeup. ■ No cuts were made in the dough to allow the dough to expand. ■ Not enough steam formed to keep the top crust moist, allowing it to expand. ■ Oven temperature was too high and expanding gases split the bread.
Bread is too dense with small holes.	■ Not enough liquid was used. ■ Too much flour was used. ■ Not enough yeast was used. ■ Too much salt was used, which slowed fermentation. ■ Dough was not fermented or proofed long enough.

WHAT IF . . .?	REASON
Bread is too coarse with large holes.	■ Too much liquid was used. ■ Too much yeast was used. ■ Dough was not properly mixed. ■ Dough was proofed for too long.
Bread breaks apart and crumbles.	■ Flour used had too little protein. ■ Not enough salt was used to strengthen the gluten. ■ Dough was overproofed.

Pies and Tarts

WHAT IF . . .?	REASON
Crust is too tough.	■ Flour had too much protein. ■ Too little fat was used. Solution: ■ More fat prevents gluten from forming and creating a tough crust. ■ Too much liquid was used, which developed gluten. ■ Dough was overmixed.
Crust falls apart easily.	■ Dough was dry and crumbly because not enough liquid was used. ■ Too much fat was used, causing little gluten to form, making the dough too crumbly. ■ Too little fat was used, which prevented the dough from holding together. ■ Fat was cut in too completely, preventing any gluten from forming and creating too much tenderness and little structure. ■ Flour had too little protein so dough had little structure to hold together. ■ Dough was not mixed properly.
Crust is undercooked on bottom.	■ Filling had too much liquid. ■ Filling was hot when placed in pie shell. ■ Used wrong dough. Mealy dough for bottom crust prevents sogginess. ■ Crust was underbaked. ■ Oven temperature was too low. ■ Crust was not baked on lowest rack in oven or directly on the bottom, which would place the crust closest to the heat.
Crust shrinks down from sides.	■ Too little fat was used. Solution: ■ More fat prevents gluten from being developed and decreases the likelihood that the crust will shrink. ■ Flour had too much protein.

WHAT IF . . .?	REASON
	■ Dough had too much water, which developed too much gluten.
	■ Dough was overhandled and too much gluten formed.
	■ Dough was not rested long enough and gluten did not relax.
	■ Dough was stretched to fit into the pan and sprang back during baking.
	■ Pie weights, such as raw beans, were not used.
	■ Dough was not docked or stippled before baking.
	■ Rolled-out pie shell was not chilled before baking.
Crust is not flaky enough.	■ Too little fat was used.
	■ Incorrect fat was used, which melted too quickly in oven.
	■ Fat needed to be colder or frozen for a short time to keep it from melting too quickly.
	■ Dough was overhandled or overheated, and the fat melted into the dough.
	■ Dough was overmixed and fat was blended in too well, resulting in a tender crust.
	Solution:
	■ Increase the amount of solid vegetable shortening because it has a higher melting point and decrease the amount of butter.
Dough is too elastic to roll out.	Too much gluten formed from one of the following:
	■ Flour had too much protein.
	■ Too much liquid was used, which developed gluten.
	■ Dough was not rested.
	Solution:
	■ Add a small amount of an acid (e.g., citrus juice or vinegar) to decrease or break up gluten formation.
	■ Refrigerate the dough for a short time to relax gluten.

Quick Breads

WHAT IF . . .?	REASON
Quick bread is tough.	■ Dough or batter was overmixed, causing too much gluten to form.
	■ Flour used had too much protein.
	■ Bread spent too long in the oven and was overbaked.

WHAT IF . . .?	REASON
Tunnels or knobby shapes appear in and on top of quick bread.	■ Batter was overmixed, causing too much gluten to form. In the oven, trapped gases are unable to escape through the thick network of gluten in the dough and explode in the batter while baking, forming large holes and tunnels. ■ Chemical leaveners were not evenly distributed in batter, causing pockets of gases to form and creating holes. Solution: 　■ Use flours with low protein. 　■ Thoroughly blend dry ingredients with chemical leavening agents before adding wet ingredients.
Batter does not rise well.	■ Not enough leavening was used or chemical leaveners were old. ■ If creaming method was used, fat and sugar were creamed improperly. ■ Oven temperature was too low.
Muffin is not cake-like.	■ Not enough air was beaten into the fat and sugar. Solution: 　■ Creaming method should be used instead of the muffin method.
Muffins do not rise above the edges of the muffin pan.	■ Paper liners were used. Solution: 　■ Grease muffin pans and omit liners for higher muffins.

Cakes

WHAT IF . . .?	REASON
Cake does not rise well.	■ Not enough leavening was used or chemical leaveners were old. ■ The oven was not hot enough, causing the cake to fall. ■ The oven was too hot and set the cake before it had a chance to rise. ■ The fat and sugar were not creamed for long enough.
Cake overflows in oven.	■ Cake pan was overfilled or too small. Solution: 　■ Fill pans no more than one half to two thirds full.
Cake is too dark.	■ Oven temperature was too hot. ■ Too much baking soda was used. Because acidic batter does not brown easily, too much baking soda neutralizes the batter's acidity, causing it to brown more intensely. ■ Too much sugar in the batter caramelized and burnt. ■ Cake was overbaked.

WHAT IF . . .?	REASON
Cake falls apart and is crumbly.	Too little gluten was formed because: ■ A flour with too little protein was used, there was too much fat or sugar, or there was too little liquid. ■ Batter was improperly mixed. ■ Cake was removed from the pan while still too warm.
Cake is tough.	■ Too much gluten was formed because a flour with too much protein was used, too little sugar or fat was used, or batter was overmixed or improperly mixed. ■ Too much egg protein.
Cake is heavy and dense.	■ There was not enough leavening. ■ Oven was not hot enough. ■ Batter was improperly mixed. ■ Too much sugar or fat was used.
Cake has large holes or tunnels throughout.	■ Batter was overmixed, forming too much gluten because trapped gases are unable to escape and consequently explode in the batter during baking. ■ Chemical leaveners were not evenly distributed in batter, causing pockets of gases to form and creating holes in the cake. ■ Oven temperature was too high.
Cake does not come out of pan.	■ Pans were improperly prepared before baking. ■ Cake was not removed while still warm.
Egg-foam cake collapses.	■ Excess air was incorporated into the egg foam. ■ Too little air was incorporated into the egg foam.

Frostings

WHAT IF . . .?	REASON	SOLUTION
Fudge frosting sets up too fast before it is spread on the cake.	■ Not enough corn syrup or cream of tartar was used to prevent crystallization. ■ Frosting was overbeaten and sugar crystals solidified suddenly, causing the fudge to stiffen. ■ The fudge was cooled too rapidly and beaten while still too warm.	■ Place the fudge over a double boiler or in a bowl set over a hot water bath and allow it to melt down. Cool it down until it is of a spreadable consistency.
For buttercreams using hot sugar syrups		
There are lumps in the buttercream.	■ The sugar syrup formed lumps or crystals on the sides of the pan during cooking, which were poured into the eggs during beating.	■ Do not stir the sugar syrup while it is cooking. Wash down any sugar crystals that appear on the sides of the pan with a pastry brush dipped in water.

WHAT IF . . .?	REASON	SOLUTION
	▦ The sugar syrup was stirred, precipitating crystals out of solution. ▦ The sugar syrup was poured directly onto the whip while beating it into the yolks.	▦ Pour the sugar syrup down the sides of the bowl, not directly onto the whip.
Buttercream looks broken and curdled.	▦ Butter was too cold.	▦ Be sure butter is softened by leaving it at room temperature or microwaving it on low power until softened but not melted. ▦ To save a broken buttercream, beat in a small amount of melted butter until it comes together.
Buttercream seems to melt and not thicken up.	▦ Butter melted because it was added before the beaten sugar and egg mixture was cool enough.	▦ Beat sugar syrup and egg mixture until the bottom of the bowl feels cool to the touch before adding the butter. ▦ Whisk thin buttercream over an ice water bath to firm up, and then beat it to smooth it out.
Buttercream is grainy. After freezing, buttercream is too stiff to spread.	▦ Buttercream was overbeaten. ▦ Buttercream was not softened first.	▦ Do not overbeat buttercream. ▦ Defrost buttercream in the refrigerator and when ready to use leave it at room temperature until softened. ▦ Place stiff buttercream in a bowl and set over a hot water bath. Stir briskly until it smooths out and softens a bit.

Cookies

There are so many variations of cookie recipes that certain characteristics may be desirable for some and viewed as mistakes by others.

WHAT IF . . .?	REASON	SOLUTION
Cookies are too pale.	■ They were underbaked. ■ Oven temperature was not high enough. ■ Batter or dough was too acidic; and acidic batters do not brown well. ■ More sugar was needed, which contributes to browning.	■ Add some baking soda to neutralize any acidity. ■ Corn syrup browns at a lower temperature than granulated sugar, so substitute 1 to 2 tablespoons ($\frac{1}{2}$ to 1 ounce; 15 to 30 g) corn syrup in place of some of the granulated sugar to increase browning.
Cookies are too dark.	■ They were overbaked. ■ Oven temperature was too high. ■ Too much sugar was used. ■ Too much baking soda was used. ■ Sheet pan was too close to the heat.	 ■ Because acidic doughs do not brown as easily, decrease baking soda and increase acidity level. ■ Place cookie dough on two sheet pans put together or place sheet pans on a higher rack in the oven.
Cookies are too hard.	■ Dough was overmixed, forming gluten. ■ Not enough liquid was used. ■ Flour had too much protein, so too much gluten formed. ■ Not enough fat was used. ■ Cookies were overbaked. ■ Oven temperature was too hot.	
Cookies are too crumbly.	Too little gluten was formed because: ■ Flour had too little protein. ■ Too much fat was used. ■ Too much sugar was used. ■ Not enough eggs were used.	 ■ Use a flour with a higher protein level. ■ Add some water to the batter to encourage gluten to form, making a cookie that holds together with a better structure. ■ Decrease the fat, sugar, or both. ■ Add more egg protein to hold the batter together.

WHAT IF . . .?	REASON	SOLUTION
Cookies are too puffy.	■ A reduced-fat or shortening was used causing water in dough to form steam and puff ■ Flour had too little protein, leaving water in the dough to form steam and puff.	■ Use a flour with a higher protein content, which will bind with water in the dough, forming less steam and puffiness.
Cookies spread out and flatten.	■ Oven temperature was too low and dough had more time to spread before it set. ■ Too much sugar was used. ■ Too much baking soda which caused gluten to weaken. ■ Too little flour was used. ■ A fat such as butter that melts at a lower temperature was used. ■ A liquid fat such as oil was used.	■ Use a flour with a higher protein content. ■ Replace some of the butter or oil with shortening. ■ Use less baking soda. ■ Dough or batter was too warm. ■ Bake at a higher temperature so cookies will set faster ■ Chill dough before baking. ■ Sheet pans were overgreased. ■ Use ungreased or parchment-covered sheet pans.
Cookies do not spread out enough.	■ Fat had too high a melting point. ■ Dough was too thick. ■ Not enough sugar was used. ■ Sheet pans were not greased.	■ Use a different fat with a lower melting point such as butter or oil. ■ Use less flour. ■ Use a lower protein flour. ■ Add more liquids such as water, milk, or cream. ■ Use more baking soda. ■ Use more sugar to increase spread. ■ Used greased pans to make dough spread out.

Working with Sugar

WHAT IF . . .?	REASON	SOLUTION
During the preparation of caramel, the sugar syrup suddenly solidifies.	The mixture crystallized suddenly for the following reasons: ■ Undissolved sugar crystals left on the sides of the pan caused a chain reaction to occur suddenly causing the syrup to solidify. ■ The pan was shaken and a chain reaction of crystallization occurred.	■ Wash down the sides of the pan with a pastry brush dipped in water, and never place a spoon that has been in the mixture back in the pan without washing it first. ■ Adding a small amount of an acid (e.g., citrus juice, vinegar, or cream of tartar) in the beginning helps prevent crystallization.

WHAT IF . . .?	REASON	SOLUTION
	■ A spoon was dipped in the syrup, removed, and then placed back in the syrup, unwashed.	■ Using a small amount of corn syrup helps prevent crystallization.
The sugar syrup boils over.	■ The pan used was too small, not allowing enough room for the mixture to boil and bubble up.	
The sugar syrup never hardened when it cooled.	■ The sugar syrup did not reach the correct temperature.	■ Use a candy thermometer clamped to the side of the pan to determine the correct temperature of the syrup.

Frozen Desserts

WHAT IF . . .?	REASON
Ice cream is grainy or icy.	■ Too little sugar was used in the base.
	■ Sugar in the base was not dissolved.
	■ Base was not chilled thoroughly.
	■ Base froze either too slowly or too quickly, creating large ice crystals.
	■ After processing, while ice cream was stored in the freezer, the temperature fluctuated, causing large ice crystals to form.
Base never freezes or takes a long time to process.	■ Base was not chilled thoroughly.
	■ Too little salt was used for the brine, or the ice cream machine was not cold enough or not working properly.
Frozen dessert is too slushy and soft.	■ Too much sugar or alcohol was used in the base, preventing the liquid ingredients from freezing to a more solid state.
	For ice cream:
	■ Custard was undercooked and did not thicken properly.
	■ Base was not processed long enough.
	■ Base was not cold enough during processing.
Frozen dessert lacks flavor.	■ The base needed to be more boldly flavored because cold temperatures dull flavors.

Chocolate

WHAT IF . . .?	REASON	SOLUTION
While melting chocolate over a double boiler, the chocolate seizes.	■ A drop or two of moisture caused the chocolate to clump.	■ Gently whisk in a small amount of fat, such as vegetable shortening or oil, to smooth it out. ■ Add some water or other liquid to help smooth out the chocolate. This should be done only if thinning out the chocolate will not disrupt the recipe's balance. A few drops of water or moisture can cause seizing, whereas a greater amount of water will not. Use at least 1 tablespoon (15 mL) of liquid for every 2 ounces (60 g) of chocolate.
While folding cold whipped cream into warm melted chocolate, seizing occurs and small chunks of chocolate are distributed throughout.	■ Adding a cold ingredient to warm chocolate will immediately set the chocolate because the crystals of cocoa butter in the chocolate recrystallize instantly with the sudden cold temperature.	■ Allow the melted chocolate to cool down to room temperature before adding it to cold ingredients. To add it to cold whipped cream, vigorously whisk one quarter of the cream into the chocolate to lighten it up and get it acclimated to the cool cream. Then gradually fold in the remaining whipped cream with a rubber spatula. Or, one fourth of the heavy cream (unwhipped) can be blended with the melted chocolate and then folded into the remaining cream that has already been beaten to soft peaks.
Ganache becomes grainy and hard while being beaten.	■ The ganache was over-beaten or rapidly chilled, causing fat crystals in the chocolate to suddenly recrystallize and solidify.	■ Rewarm the ganache over a hot water bath until it is liquefied. Allow it to cool and thicken before beating it again for a lesser amount of time.

WHAT IF . . .?	REASON	SOLUTION
Chocolate melts too quickly over direct heat and burns.	■ Chocolate was melted over too high a heat.	■ Melt chocolate slowly and gradually in a bowl set over a pot of water that has been brought to a simmer and removed from the heat. The steam from the water is enough to melt the chocolate. The bottom of the bowl holding the chocolate should never touch the water. Never melt chocolate over boiling water or direct heat.
Ganache looks separated and curdled.	■ Ganache—which is an emulsion of drops of fat from cream and fat crystals of cocoa butter that are suspended in the water-based liquids from the cream—has separated into its fat and water-based components, which causes it to have a curdled appearance.	■ To bring the ganache back to a smooth texture, warm the ganache in a bowl set over a pot of warm, not simmering water, just to make it barely warm. Remove the bowl of ganache and set aside. Warm up to 2 fluid ounces ($\frac{1}{4}$ cup; 60 mL) heavy cream in a bowl set over the same pot of warm water until it is the same temperature as the ganache. Pour a small amount of the cream into a bowl. Slowly whisk the barely warm ganache into the bowl and it should smooth out. Keep adding small amounts of warm cream if necessary.
While pouring melted chocolate out of the double boiler or a bowl set over a hot water bath into other ingredients, it seizes up suddenly.	■ Moisture or drops of condensation from the bottom of the bowl or double boiler fell into the other ingredients while pouring and solidified the chocolate.	■ After melting chocolate, always use a kitchen towel to dry the bottom and sides of the pot or bowl before pouring chocolate into other ingredients.
After dipping truffles into tempered chocolate, a grayish-white coating develops on the surface.	Fat bloom appeared because: ■ Chocolate used to coat truffles was not properly tempered. ■ Chocolate truffles were not stored at the proper temperature.	■ Properly temper chocolate or use a compound coating for dipping. ■ Store chocolate at cool room temperature of 56° to 60°F (13° to 16°C) so the fat crystals in chocolate do not melt and rise to the surface.

Baking Tools and Equipment Defined

Specific tools and equipment that a baker uses to make the baking process easier are defined in the following paragraphs.

Baguette Pan

Also known as a French Bread Pan (Figure A–1), the baguette pan is a long metal pan formed into half cylinders, which are joined together, side by side. Frequently, small holes are placed in the metal to allow for better air circulation and a crispier crust.

Bain Marie

Bain marie is a French term for a hot water bath. It can be used in two ways. As a double boiler, with a bowl placed over a pot of simmering water, it can melt chocolate or warm other delicate ingredients (see Double Boiler). It can also be used as a technique to ensure a gentle, even baking for custard desserts like crème brûlée, crème caramel, or cheesecakes. Ramekins full of crème brûlée custard or a springform pan filled with cheesecake batter can be placed in a larger rectangular pan, which is then filled half way with hot water. The water surrounds the custard providing it with a consistent temperature that prevents the eggs in the custard from curdling (Figure A–2A and B).

Baker's Peel

A baker's peel is used to transfer bread or pizza dough in or out of an oven when it will not be baked directly on a sheet pan. It has a long handle and flat shovel-like blade. Peels can be made of wood, stainless steel, or a combination of the two (Figure A–3).

Banneton

A banneton is a woven basket made of coiled reed or willow in various shapes and sizes that is sometimes lined with cloth (Figure A–4). The basket is dredged in flour and rustic and hearth type bread doughs are allowed to rise in it. The weave of the basket imprints an attractive pattern onto the dough before it is baked. The dough is removed from the basket and placed on a baker's peel, to be then placed on baking tiles in the oven.

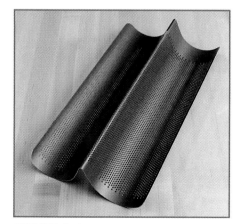

FIGURE A–1

FIGURE A–2A

FIGURE A–2B

FIGURE A–3

FIGURE A–4

FIGURE A–5

Bench Scraper

The bench scraper, also known as a dough scraper, is a small tool consisting of a rectangular blade attached to a wooden or plastic handle (Figure A–5). It is used to cut and scale pieces of dough and to clean work surfaces by scraping it against a table to loosen pieces of dough or flour.

Bowl Scraper

A bowl scraper is a small, flexible piece of plastic used to scrape around the inside of a mixing bowl to loosen doughs or stiff batters for easier removal (Figure A–6).

Coupler

A coupler is a plastic cone-shaped tube that is used to allow various pastry tips to fit onto a pastry bag to facilitate the piping of frostings and batters (Figure A–7). The coupler allows tips to be changed during decorating without having to change pastry bags.

Double Boiler

In a double boiler, a saucepan nestles halfway down into another saucepan (Figure A–8). The bottom saucepan is filled with just enough water so that the bottom of the top saucepan does not actually touch the water. The water is brought to a simmer, removed from the heat, and then the food to be gently heated is placed into the top saucepan. This is one way to make a bain marie.

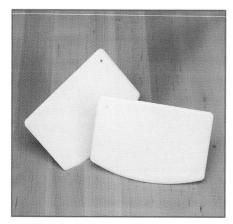

FIGURE A–6

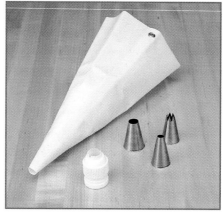

FIGURE A–7

FIGURE A–8

Electric Mixer

An electric mixer is an invaluable piece of equipment used by professional bakers to mix various batters and doughs of all types (Figure A–9). Commercial mixers come in various sizes and are described by the capacity of the mixing bowl and the power of the motor used.

FIGURE A–9

The vertical or planetary mixer is the most common type of mixer and comes in sizes that range from smaller portable models known as tabletop mixers to larger floor models (or stationary models) that are bolted to the floor or wall. The one feature that all vertical mixers have is that as the mixing bowl remains fixed in one place, the actual tool that mixes fits into an attachment arm that rotates 360 degrees to reach all areas of the inside of the mixing bowl. This rotation, on its own axis, resembles the motion a planet takes around the sun—hence the name planetary mixer. The main mixing attachments include a paddle used for mixing cake batters, cookie doughs, and quick breads; a whip used for mixing egg foams, meringues, and whipped cream; and a dough hook used to mix and knead bread doughs.

Another feature all vertical mixers have is that the mixing bowl is easily removed for cleaning.

The spiral mixer is a specialized mixer with a spiral-shaped rotating arm used exclusively for heavy bread and bagel doughs. There are no interchangeable attachments and the mixing bowls can be fixed or removable.

The spiral mixer differs from the vertical mixer in that both the bowl and the spiral-shaped rotating arm rotate at the same time providing extra power to develop firm doughs quickly and easily. The mixing bowl and rotating arm are powered by separate motors.

False Bottom Tart Pan

The tart pan is used for making a fluted pastry crust that can be filled with sweet or savory fillings; it has a removable bottom for easy removal (Figure A–10). Tart pans are round, square, or rectangular. The round tart pans have varying diameters.

Grater

A rectangular strip or box of metal with sharp holes of varying sizes (Figure A–11). Foods are passed up and down the sharp holes to allow small slivers to fall through to the other side. Graters can be used to remove the outer peel from citrus fruits or to shred cheese, vegetables, or chocolate.

Loaf Pan

A rectangular pan used to bake breads and cakes (Figure A–12); available in various sizes.

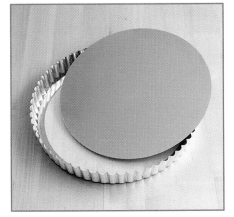

FIGURE A–10

FIGURE A–11

FIGURE A–12

Metal Cake Ring

Metal cake rings have round sides like cake pans but with no top or bottom (Figure A–13). They come in various diameters and heights and are used to mold layers of cake with fillings that need to firm up and set before standing on their own. Cake rings can also be used as cake pans; these are placed on aluminum foil on a sheet pan and filled with batter.

Microplane Zester

Also known as a rasp, this is a long, narrow, rectangular strip of metal with raised, sharp cuts, sometimes attached to a handle, used for grating hard cheese, chocolate, and citrus peels (Figure A–14). It is so named after the tool used by a carpenter.

Muffin Tin

A muffin tin, or muffin pan, consists of round metal impressions in which muffin batter can be baked to form small cakes or muffins (Figure A–15). Muffin tins come in professional (holding 2 to 4 dozen muffins in standard, full, or half sheet pan sizes), miniature (12 muffins), standard (6 muffins), and jumbo (6 muffins) sizes.

Offset Spatula

Also referred to as a palette knife or cake spatula, an offset spatula has a wide, dull blade with a slight bend just before the handle and is used to remove cookies and small pastries from a sheet pan (Figure A–16). If the spatula is elongated and the tip rounded, it can be used also to frost cakes or cookies and spread fillings.

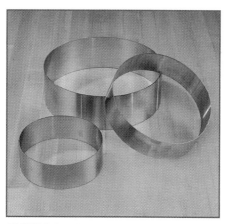

FIGURE A–13

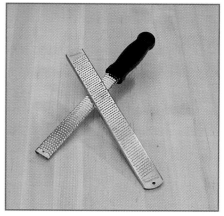

FIGURE A–14

FIGURE A–15

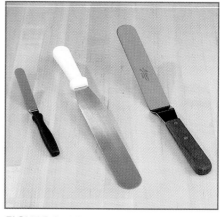

FIGURE A–16

Parchment Paper

Parchment paper is specially treated paper used to line cake and sheet pans to prevent foods from sticking (Figure A–17). Parchment, also known as baking paper, comes in rolls that can be cut or in precut rectangular sheets 16 ³⁄₈ by 24 ³⁄₈ inches (41.5 by 62 cm), which fit perfectly into a full sheet pan. It will not burn in a hot oven and can also be used to make parchment cones for piping chocolate and thin icings.

FIGURE A–17

Pastry Bag

A pastry bag is a cone-shaped hollow bag made of various materials with a narrow opening at one end and a large opening at the other. It is used to fill and pipe out decorations onto a cake or batters for cookies or candies through a pastry tip fitted onto the narrow end of the bag. (See Figure A–7.)

Pastry Brush

A brush resembling a paintbrush that comes in various sizes and is used to lightly cover foods with glazes, butter, water, or egg washes (Figure A–18).

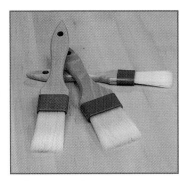

FIGURE A–18

Pastry Tip

A pastry tip is a hollow metal cone with varying cuts that is fitted onto a pastry bag such that when frostings or batters are piped out, various designs and shapes are formed. (See Figure A–7.)

Popover Pan

A popover pan is a baking pan similar to a muffin pan but with deeper, narrower impressions to bake popovers (puffy, eggy, custard-like muffins) (Figure A–19). The impressions are further apart than those of a muffin pan so as to accommodate the popovers' high rising.

Ramekin

A ramekin is a small baking dish usually made of ceramic or heat-resistant glass used to bake individual soufflés, custards, and cakes. Ramekins come in various sizes (Figure A–20).

FIGURE A–19

FIGURE A–20

Sheet Pans

Sheet pans are rectangular metal bakings pans. They come in two sizes: full and half (Figure A–21). The full sheet pan measures 18 by 26 inches (45 by 66 cm) and has sides that are 1 inch (2.5 cm) high. The half sheet pan measures 13 by 18 inches (32.5 by 46 cm). They are so named because two half sheet pans put together would equal one full sheet pan. Full sheet pans are used mostly in commercial kitchens because they are too large to fit into a standard sized noncommercial oven. Most half sheet pans will fit into a noncommercial oven.

Sieve or Strainer

A sieve is a small tool used to separate out finer particles from coarser ones (Figure A–22). It is a metal bowl with screen-like openings with a handle. It can be used to sift out lumps from dry ingredients like flour or confectioners' sugar, or it can be used to strain lumps from desserts like pastry cream or custard sauces.

Sifter

A sifter is similar to a sieve with one difference (Figure A–23). A sifter is used for dry ingredients only. A sifter looks like a metal cup with a screen bottom and rotating metal wires or screens inside to help separate lumps in dry ingredients like flour, confectioners' sugar, and cocoa powder.

Silicone Baking Mats

Silicone baking mats, as excellent nonstick surfaces, have revolutionized baking. Silicone baking mats are reusable, flexible, plastic rectangular sheets coated with silicone that are able to withstand extreme temperatures, both cold and hot (Figure A–24). They are placed into full sheet or half sheet pans and used instead of parchment paper. Resembling rubber placemats, they require no greasing or flouring and only need to be wiped clean with a damp sponge or cloth. If a silicone baking mat does not fit perfectly into a sheet pan or cookie sheet of a different size, the baking sheet can be flipped over so the bottom is facing up and the silicone mat placed on top.

Silicone Baking Pan

This baking pan is made from flexible silicone that can withstand a wide range of temperatures and has a permanent nonstick surface, much like a silicone baking mat (Figure A–25). They come in a wide variety of shapes and sizes and need to be baked on a rigid surface like a half or full sheet pan.

FIGURE A–21

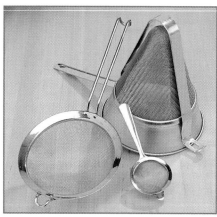

FIGURE A–22

FIGURE A–23

FIGURE A–24

FIGURE A–25

FIGURE A–26

Springform Pan

A springform pan is a round pan with a removable bottom (Figure A–26). It resembles a cake pan but has higher sides for baking cheesecakes and other cakes. The sides of the pan can be detached from the bottom for ease of removal. It is available in various sizes.

Squirt Bottle

Plastic bottles with a narrow opening that are used to pipe sauces on to plates for decoration (Figure A–27).

Tube Pan

This deep, round cake pan with a hollow tube in its center is used to bake angel food, sponge, and pound cakes (Figure A–28). Some tube pans have a removable bottom and metal tabs on the top for balancing the cake as it cools upside down.

Zester

A tool used to cut long, narrow strips from citrus fruits consisting of five angled, round metal holes attached to a handle (Figure A–29). It is similar to a grater.

FIGURE A–27

FIGURE A–28

FIGURE A–29

Procedure to Make Parchment Cake Pan Liners for Round Cake Pans

When baking round cake layers, it is imperative for a pastry chef to know how to line the cake pans with circles made of parchment paper for easy removal. It may appear that the easiest way to make a parchment circle fitted to the size of a round cake pan would be to trace the outline of the cake pan with a pencil after placing it on a piece of parchment paper. However, it is actually quicker and more accurate to make a series of bends and folds. The folded parchment resembles a cone and, after putting the point of the cone into the center of the cake pan, the excess is trimmed off. The cone is unfolded to produce a circle that fits into the pan. Many times in a professional kitchen, scissors may not be available, and this method works well with or without scissors because ripping the paper at the appropriate place is easy to do. Try to become accomplished in this method such that it can be done in a matter of seconds. Whether or not scissors are used does not affect the final outcome (the easy removal of the cake from the pan).

FIGURE B–1

Procedure:

1. Cut a piece of parchment paper large enough so that the diameter of the cake pan fits easily across the top of it (Figure B–1).
2. Fold the parchment paper in half (Figure B–2).
3. Fold the halved paper into quarters (Figure B–3).
4. Now place the quartered paper such that it resembles a book that opens from the left to the right with the spine of the book on the right side (Figure B–4).
5. Bend the top right corner of the paper down to form a triangular shape with the spine of the book bent over to meet the bottom edge (Figure B–5).
6. Bend the right side down again to meet the bottom edge, making a triangular shape similar to a paper cone or a paper airplane (Figure B–6).
7. Place the point of the paper cone in the middle of the cake pan and measure just where the wider end meets the edge of the pan (Figure B–7). Mark the distance with your fingers (Figure B–8).
8. Using scissors or by hand, cut or tear the marked area off (Figure B–9). Open the cone and a circular shape should be just the right size to fit into the cake pan (Figure B–10). Refold and trim the edges if the circle is still too wide.

FIGURE B–2

FIGURE B–3

How to Line Sheet Pans with Parchment Paper

Parchment paper is available commercially that is a perfect fit for a standard full sheet pan (18 by 26 inches or 46 by 66 cm). Cutting the parchment in half crosswise is a perfect fit for a half sheet pan (13 by 18 inches or 33 by 46 cm).

For a rectangular pan or cake pan or sheet pan of a different size, place the pan on top of a piece of parchment paper and trace the pan with a pencil. Cut the rectangle out and lay it into the pan. The bottom of the pan is the only area that needs to be covered with parchment paper.

FIGURE B–4

FIGURE B–5

FIGURE B–6

FIGURE B–7

FIGURE B–8

FIGURE B–9

FIGURE B–10

Proper Preparation of Cake Pans before Baking

It is important to properly prepare cake pans before baking to ensure that the cake can easily be removed. There is nothing more frustrating than to spend a great deal of time preparing a cake batter and then find the cake has stuck to the pan. Follow each recipe's directions on how to prepare the pans. For example, some egg foam cake batters are baked in an ungreased dry cake pan.

Seven Steps to Follow to Ensure a Cake's Removal:
Before Baking:

1. Make sure the cake pan is clean and dry.
2. Lightly grease the bottom and sides of the cake pan with nonstick cooking spray, or apply softened butter or vegetable shortening with a pastry brush (Figure B–11A and B). This helps anchor the parchment paper while the batter is poured into the pan.
3. Place parchment paper, cut to fit, into the bottom of the pan (Figure B–12). The parchment should not go up the sides of the pan.
4. Lightly grease the parchment paper again (as in step 2) (Figure B–13). This ensures that some particles of flour will stick to the bottom of the pan to ease removal later on.

FIGURE B–11A

FIGURE B–11B

FIGURE B–12

FIGURE B–13

FIGURE B–14

FIGURE B–15

(*Note*: Some chefs make a paste of equal parts of flour and fat [butter or vegetable shortening] for white and yellow cakes and cocoa powder and fat for chocolate cakes to brush into the cake pans, combining both greasing and flouring into one step.)

(*Note:* Some chefs flip the cake onto small, round cooling racks without a cardboard circle. After the cake is completely cooled, they place it onto a cake circle. Indentations from the cake rack may be left on the cake.)

5. Place some flour into the pan and rotate it while knocking it against a work surface to evenly distribute the flour to every area of the pan (both bottom and sides) (Figure B–14). Knock the excess flour out (Figure B–15). Be sure to knock out the majority of the flour. There should only be a light coating where the particles have stuck to the spray or butter. If done properly, there should be little or no flour residue when the cake is baked and the parchment is removed. Resist any temptation to touch or handle the inside of the prepared pan.

6. Do not overfill cake pans with batter. Most pans should be filled one half to two thirds full depending on the type of cake being prepared.

After Baking:

7. After baking, allow the cake to cool in the cake pan on a cake rack until the pan feels slightly warm to the touch and can easily be handled. It should not be completely cold. Have ready a cardboard cake circle that measures the exact diameter of the cake. Gently cut around the edges of the cake with a sharp knife to loosen it. Jostle the cake gently as if it were going to be flipped out of the pan. Now rotate the pan and gently jostle it again until it becomes dislodged from the bottom and sides of the pan. Place the cake circle on top of the cake and flip the pan over so the cake circle is under the pan. Do not allow the cake to flip out onto your hand. Some cakes are so tender they crumble and break apart. Gently, remove the cake pan and slowly peel the parchment paper away from the cake. If the cake will be wrapped in plastic and frozen, the parchment may be left on to help prevent any plastic wrap from sticking to the surface. Allow the cake to completely cool on the cardboard circle placed on a cooling rack.

Weights and Measurements

Weights and Measurements

This section has been written to give you instant access to a table of weights and measurements when you are entering recipes. The weights and measurements are only approximate: you can never be precise for many reasons:

1. The moisture contents of products vary constantly.
2. Sizes of individual pieces or particles will vary from container to container.
3. The exact weight of a gallon or a pound of product is seldom a convenient round number.
4. It would be impractical to say that a pint of water is 1 9/10 cups. It is simpler just to say 2 cups.
5. Products containing moisture become lighter as they dry out.
6. Wet products containing sugar become heavier when the moisture evaporates and they become thicker.
7. A cup of flour could weigh 4 ounces. If you sift it, it may weigh less.
8. Any measurement such as a "Level" teaspoon or "Level" cupful is seldom exactly accurate.

Gram Weight Conversion Table

OUNCES	GRAMS	POUNDS	GRAMS
1	28.35	1	453.6
2	56.70	2	907.2
3	85.05	2.5	1134.0
4	113.40	3	1136.8
5	141.75	4	1814.4
6	170.10	5	2268.0
7	198.45	6	2721.6
8	226.80	7	3175.2
9	255.15	8	3628.8
10	283.50	9	4082.4
11	311.85	10	4536.0
12	340.20	15	6804.0
13	368.55	20	9072.0
14	396.90	25	11340.0
15	425.25	30	13608.0
16	453.60	35	15876.0

(*Note:* When you know the weight in ounces, multiply by 28.35 to find the grams. Divide the grams by 453.6 to find the pounds.)

Metric Size Fluid Equivalents

METRIC	U.S. FL OZ	3/4 OZ	1 OZ	1-1/8 OZ	1-1/4 OZ	1-1/2 OZ	CLOSEST PREVIOUS CONTAINER U.S. OZ
1.75 liter	59.2	78.9	59.2	52.6	47.4	39.5	1/2 gal = 64 oz
1.0 liter	33.8	45.1	33.8	30.0	27.0	22.5	qt = 32 oz
750 milliliters	25.4	33.9	25.4	22.6	20.3	16.9	5th = 25.6 oz
500 milliliters	16.9	22.5	16.9	15.0	13.5	11.3	pt = 16 oz
200 milliliters	6.8	9.1	6.8	6.0	5.4	4.5	1/2 pt = 8 oz
50 milliliters	1.7						Miniature = 1.6 oz

Equivalent Measures for Fluids

(*Note:* This table gives measurement/weight equivalencies for water. Other liquids may vary.)

3 teaspoons	=	1 tablespoon
16 tablespoons	=	1 cup
28.35 grams	=	1 ounce
1 cup	=	$\frac{1}{2}$ pint
2 cups	=	1 pint
2 pints	=	1 quart
4 quarts	=	1 gallon
768 teaspoons	=	1 gallon
1 lb (water)*	=	16 fluid ounces
1 lb (water)*	=	1 fluid pint
2 lb (water)*	=	1 fluid quart

Volume Conversions for Recipe Writing

	TEASPOON	TABLESPOON	QUART	CUP
1 teaspoon	1.0	0.333333	0.0052083	0.020833
1 tablespoon	3.0	1.0	0.015625	0.062500
1 cup	48.0	16.0	.25	1.0
1 pint	96.0	32.0	.50	2.0
1 quart	192.0	64.0	1.0	4.0
1 gallon	768.0	256.0	4.0	16.0

Pound and Ounce to Gram

OUNCES	GRAMS
1	28.35
5	141.75
10	283.50
12	340.20
16	453.60

POUNDS	GRAMS
1	453.60
5	2268
10	4536
25	11340
50	22680

Teaspoons and Tablespoons

TEASPOONS	TABLESPOONS
3 = 0.5 ounce	1 = 3 teaspoons
6 = 1 ounce	2 = 1 ounce
48 = 1 cup	4 = 0.25 cup
96 = 1 pint	8 = 0.5 cup
192 = 1 quart	16 = 1 cup
960 = 5 quarts	128 = $\frac{1}{2}$ gallon
768 = 1 gallon	256 = 1 gallon

Decimal Equivalents of Common Fractions

FRACTION	ROUNDED TO 3 PLACES	ROUNDED TO 2 PLACES
$\frac{5}{6}$	0.833	0.83
$\frac{4}{5}$	0.8	0.8
$\frac{3}{4}$	0.75	0.75
$\frac{2}{3}$	0.667	0.67
$\frac{5}{8}$	0.625	0.63
$\frac{3}{5}$	0.6	0.6
$\frac{1}{2}$	0.5	0.5
$\frac{1}{3}$	0.333	0.33
$\frac{1}{4}$	0.25	0.25
$\frac{1}{5}$	0.2	0.2
$\frac{1}{6}$	0.167	0.17
$\frac{1}{8}$	0.125	0.13
$\frac{1}{10}$	0.1	0.1
$\frac{1}{12}$	0.083	0.08
$\frac{1}{16}$	0.063	0.06
$\frac{1}{25}$	0.04	0.04

Fahrenheit to Celsius Conversion (Approximate)

FAHRENHEIT	CELSIUS
32°F	0°C
50°F	10°C
68°F	20°C
86°F	30°C
100°F	40°C
115°F	45°C
120°F	50°C
130°F	55°C
140°F	60°C
160°F	70°C
170°F	75°C
180°F	80°C
185°F	85°C
195°F	90°C
200°F	95°C
212°F	100°C
230°F	110°C
250°F	120°C
265°F	130°C
285°F	140°C
300°F	150°C
325°F	165°C
350°F	175°C
360°F	180°C
375°F	190°C
400°F	200°C
425°F	220°C
450°F	230°C
485°F	250°C
500°F	260°C
575°F	300°C

Sizes and Capacities of Scoops

NUMBER ON SCOOP	LEVEL MEASURE
6	$\frac{2}{3}$ cup
8	$\frac{1}{2}$ cup
10	$\frac{3}{8}$ cup
12	$\frac{1}{3}$ cup
16	$\frac{1}{4}$ cup
20	$3\frac{1}{3}$ tablespoons
24	$2\frac{2}{3}$ tablespoons
30	2 tablespoons
40	$1\frac{2}{3}$ tablespoons
50	$3\frac{3}{4}$ teaspoons
60	$3\frac{1}{4}$ teaspoons
70	$2\frac{3}{4}$ teaspoons
100	2 teaspoons

Sizes and Capacities of Measuring/Serving Spoons

SIZE OF MEASURING/ SERVING SPOON	APPROXIMATE MEASURE
2 ounces	$\frac{1}{4}$ cup
3 ounces	$\frac{3}{8}$ cup
4 ounces	$\frac{1}{2}$ cup
6 ounces	$\frac{3}{4}$ cup
8 ounces	1 cup

Canned Goods

Common Can Sizes and Their Appropriate Contents

CAN SIZE	PRINCIPAL PRODUCTS
No. 5 squat	
75 oz. squat	
No. 10	Institution size—fruits, vegetables, and some other foods
No. 3 cyl	Institution size—condensed soups, some vegetables. Meat and poultry products. Economy family size—fruit and vegetable juices
No. 2 1/2	Family size—fruits, some vegetables
No. 2	Family size—juices, ready-to-serve soups, and some fruits
No. 303	Small cans—fruits, vegetables, some meat and poultry products, and ready-to-serve soups
No. 300	Small cans—some fruits and meat products
No. 2 (vacuum)	Principally for vacuum pack corn
No. 1 (picnic)	Small cans—condensed soups, some fruits, vegetables, meat, and fish
8 oz.	Small cans—ready-to-serve soups, fruits, and vegetables
6 oz.	

Common Can Sizes and Their Approximate Weights and Volumes

Volumes represent total water capacity of the can. Actual volume of the pack would depend upon the contents and the head space from the fluid level to top of can.

AVERAGE CAN SIZE	AVERAGE VOLUME	CANS PER FLUID	APPROX CUPS	CASE	WEIGHT
75 squat				6	4 lb 11 oz
No. 5	56			6	4 lb 2 oz
No. 10	105.1	99 to 117	12 to 13	6	6 lb 9 oz
No. 3 cyl	49.6	51 or 46	$5\frac{3}{4}$	12	46 fl oz
No. 2 1/2	28.55	27 to 29	$3\frac{1}{2}$	24	12 oz
No. 2	19.7	20 or 18	$2\frac{1}{2}$	24	1 lb 13 oz
No. 303	16.2	16 or 17	2	24 or 36	1 lb
No. 300	14.6	14 or 16	$1\frac{3}{4}$	24	$15\frac{1}{2}$ oz
No. 2 (vacuum)		12 fl oz	$1\frac{1}{2}$	24	
No. 1 (picnic)		$10\frac{1}{2}$ fl oz	$1\frac{1}{4}$	48	
8 oz	8.3	8 fl oz	1	48 or 72	8 oz
6 oz	5.8	6 fl oz	$\frac{3}{4}$	48	6 oz

acetate   A clear, thin, flexible plastic sold in rolls, sheets, and strips used in the molding of chocolate, protecting the sides of a cake, or coating the sides of a cake with chocolate.

Acetobacillus   A species of bacteria that exist in a sour dough starter. These bacteria give off acetic acid, providing a slightly tangy taste to the finished bread.

acid   A substance that tastes sour like lemon juice and has a pH of less than 7.0.

active dry yeast   Fresh yeast that has been dehydrated. It is more concentrated than fresh yeast and has a longer shelf life. See *yeast.*

agar   A gelatin-like stabilizer and thickener that is derived from a type of seaweed known as red algae. Also referred to as *agar agar.*

agar agar   See *agar.*

air cell   The pocket of air that forms at the larger end of an egg.

all-purpose flour   Flour made from a combination of hard and soft wheats so as to be suited for all purposes. Protein levels may vary depending on where it is milled.

almond paste   Almonds and sugar ground into a fine paste.

alpha-amylase   An enzyme in raw egg yolks that feeds on starch, causing it to break down.

amino acids   The building blocks of protein.

ammonium carbonate   Also known as ammonium bicarbonate; a chemical leavener that, in the presence of moisture and heat, reacts to produce ammonia, carbon dioxide gas, and water.

angel food cake   An egg foam cake that uses only egg whites and is virtually fat free.

artificial sweeteners   Substitutes for sugar (sucrose) that are artificial or man-made in the laboratory. They do not tend to raise blood sugar levels. Artificial sweeteners may not have all of the properties of sugar and may not be suitable for baking.

artisan breads   Breads that are prepared by bakers who manipulate the dough with their hands with great care and skill using traditional methods.

ascorbic acid   Also known as vitamin C. It is added to flour by the miller to improve gluten quality.

autolyse   A short rest given to a yeast dough before kneading has begun. This rest helps gluten develop properly.

baguette pan   Also known as a *French bread pan*; a long metal pan formed into half cylinders that are joined together side by side. Frequently small holes are placed in the metal to allow for better air circulation and a crisper crust.

bain marie (ban mah-ree)   A French term for a hot water bath that can be used either to warm or melt ingredients or to surround custards in the oven to ensure even cooking.

When used as a double boiler, with a bowl placed over a pot of simmering water, it can melt chocolate or warm other delicate ingredients. It can also be used to ensure a more gentle, even cooking for custard desserts like crème brûlée, crème caramel, or cheesecakes. Ramekins full of crème brûlée custard or a springform pan filled with cheesecake batter can be placed in a larger rectangular pan, which is then filled halfway with hot water. The water surrounds the custard, providing it with a consistent temperature, which prevents the eggs in the custard from curdling.

baked custard   A mixture of eggs and milk or cream, sometimes with additional ingredients, which is poured into a container and baked in the oven until thickened.

baker's peel   A flat shovel-like blade with a long handle used to transfer bread or pizza dough into or out of an oven when it will not be baked directly on a sheet pan. Peels can be made of wood, stainless steel, or a combination of the two.

baker's percentages   A system used by professional bakers (especially bread bakers) in large commercial operations that involves percentages to express formulas in a simple way.

baking   The act of placing a food such as a dough or other unbaked pastry in the oven where dry heat cooks the food. The term *baking* is generally applied to cakes, pies, yeast breads, cookies, and quick breads. Baking helps proteins and starches to set and doughs to rise.

baking chocolate   Chocolate liquor in a solid state; also referred to as *unsweetened chocolate* or bitter chocolate.

baking powder   A chemical leavener that contains sodium bicarbonate (baking soda) and at least one acid that is used to help baked goods such as quick breads and cakes to rise. See *double-acting baking powder* or *single-acting baking powder.*

baking soda   A chemical leavener known as sodium bicarbonate that, when combined with an acid in the presence of moisture, forms carbon dioxide gas and is used to help baked goods to rise.

banneton   A woven basket made of coiled reed or willow of various shapes and sizes, sometimes lined with cloth. Rustic and hearth-type bread doughs are allowed to rise in them, imparting an attractive pattern onto the dough before it is baked.

bar cookie   The category of cookie preparation wherein a stiff dough is shaped into long bars or logs, baked, and sliced.

barm   An English whole wheat sourdough starter made from wild yeast.

base   A substance with a pH of greater than 7.0 that neutralizes acid to produce a salt, for example, baking soda. Also known as an alkali.

bench scraper   A small tool consisting of a rectangular blade attached to a wooden or plastic handle. It is used to cut and scale pieces of dough and to clean work surfaces by scraping it against a table to loosen pieces of dough or flour. Also known as a *dough scraper.*

benching   A stage in yeast dough production in which the dough is scaled into pieces, covered, and allowed to rest for a short period of time before being shaped. See *resting.*

biga (BEE-gah)   An Italian word for a thick sponge starter.

biscuit method   A mixing method for quick breads that resembles the method to make pâte brisée or flaky pie crust by first cutting cold fat into dry ingredients and then adding liquid ingredients.

blind baking   When a pie or tart shell is baked with nothing in it. The shell is lined with parchment paper and pie weights or dried beans to keep its shape during baking. Used most often for pies whose filling is prepared separately and requires no further cooking.

bleaching   The process whereby newly milled flour is exposed to a bleach such as chlorine gas or benzoyl peroxide to whiten it.

bloom   (1) When tempered chocolate that has been exposed to temperature variations and/or humidity, it develops a whitish-gray spotty outer coating. (2) When a cold liquid is added to powdered gelatin, the absorbed liquid causes it to soften and swell so that it appears to be blooming. (3) Also refers to the Bloom rating, a system used to show how firm or strong a specific type of gelatin is.

bowl scraper   A small flexible piece of plastic used to scrape around the inside of a mixing bowl to loosen doughs or stiff batters for easier removal.

bran   The hard outside covering of the wheat kernel. Also known as the *hull.*

bread flour   The general term for flour that is milled from wheats having a higher protein content and generally used for bread making.

breaking   In the milling of flour, when special machines crack or break open wheat kernels to separate them into their component parts.

buttercream   A frosting consisting of butter or vegetable shortening, granulated sugar, corn syrup, or confectioners' sugar, and whole eggs, yolks, or egg whites. There are different types of buttercreams, including both uncooked (e.g., simple buttercream) and cooked variations (e.g., French, Italian, and pastry cream-based buttercreams).

buttermilk   Buttermilk traditionally referred to the liquid left over after cream was made into butter. Presently, the buttermilk on the market is cultured and refers to skim or low-fat cow's milk that is treated with harmless bacteria, giving it a thick consistency and a sour taste. It has a milk fat content between 0.5 and 3 percent. Because of its acidic nature, it reacts well with baking soda, neutralizing it to form carbon dioxide gas to leaven cakes and quick breads. Because of its acidity and low fat content, buttermilk has a longer shelf life than regular cow's milk. Also known as cultured buttermilk.

cacao beans (kah-KAH-oh)   The fruit of the *Theobroma cacao* tree from which chocolate is derived. Also referred to as cocoa beans.

cake   A sweet, tender, moist baked pastry that is sometimes filled and frosted.

cake flour   A flour milled from soft wheats that is typically bleached and used for only the most tender cakes and pastries.

caramelized sugar   Sugar that is cooked to within the temperature interval of 320° to 350°F (160° to 177°C), which causes the sugar to develop a brown color and a rich intense flavor.

cardamom   A fragrant pod related to the ginger family used in Middle Eastern and Indian dishes. It gives a pleasant, pungent aroma to Danish dough.

carrageenan   A type of seaweed from Ireland that is similar to agar that is used to thicken foods containing dairy products.

carryover cooking   Refers to the process in which cooking continues for a short period of time, even though the heat source has been removed.

casein (kay seen)   A protein in milk and other dairy products, which when exposed to air, forms a crusty skin.

chalaza (kuh-LAY-zah)   The white stringy material that anchors the yolk in the center of the egg.

chef   A chef is the beginning stages of a sourdough starter. It is also known as a *seed culture.* After a period of time, the starter becomes healthy enough to bake bread.

chemical leaveners   Refer to *baking powder, baking soda,* and *ammonium carbonate,* which are chemicals that react with liquid ingredients upon mixing and the heat of the oven to produce carbon dioxide gas that leavens baked goods.

chiffon cake   A type of egg foam cake containing a liquid fat and a chemical leavener. These cakes tend to be moister than typical sponge cakes and are baked in a tube pan.

chocolate confectionery frostings   A category of chocolate frosting based on two types of chocolate confections: fudge and truffles.

chocolate liquor   The dark liquidy paste created when chocolate nibs are crushed. Once cooled into bricks or disks, it is known as unsweetened chocolate, baking chocolate, or bitter chocolate.

choux paste   See *éclair paste.*

churn-frozen desserts   Frozen desserts that are churned or stirred as they freeze. Churn-frozen desserts include ice cream, sorbets, and sherbets.

cinnamon   A spice originating from the inner bark of an evergreen laurel tree, cinnamon has a sweet, spicy aroma. It is ground or sold as curls of bark called sticks or quills. It is one of the most popular spices in a baker's kitchen. It is used extensively in pies, cakes, cookies, and fruit desserts.

classifying   The final stage of flour milling wherein the particles of flour are categorized by size.

clear flour   The particles of flour from the outermost layers of the endosperm. It is the flour that remains after patent flour is removed.

clove   A spice originating from the dried, unopened flower buds of the tropical evergreen tree called the clove tree. The buds are sold whole or ground into a deep mahogany powder. Clove is pungent, yet sweet, which is ideal in various cakes, cookies, pies, and other desserts.

coagulation   When proteins are heated and moisture gets trapped between each protein coil, the protein forms a network that produces a thick, gel-like structure. An example of coagulation is when egg proteins thicken a custard.

cocoa butter   The saturated fat that is naturally present in chocolate liquor that gives chocolate its characteristic velvety texture.

cocoa powder   The dry powdery residue that remains when cocoa butter is removed from chocolate liquor.

coconut   Coconut is the fruit grown on tropical palm trees. It consists of a hard, fibrous, outer brown shell, which when cracked open, yields a white, hard flesh with a center filled with a milky liquid called coconut water or coconut milk. Coconut is sold whole in the shell, shredded and sweetened or unsweetened, flaked, grated, and ground.

cold-water test   A test to determine whether a sugar syrup is done by dropping some syrup into a glass of cold water to see how easily it can be gathered into a ball.

compound coating   Chocolate that contains little or no cocoa butter and is used to coat candies. It has a longer shelf life but tends to be of lesser quality.

conching (KONCH-eng)   Part of the chocolate making process when chocolate liquor is rotated and stirred with blades to develop flavor and texture.

conditioning   A general term used for when a miller adds certain chemicals such as ascorbic acid or diastase to newly milled flour to help it produce better gluten, provide the best food for yeast, and overall improve the qualities of the finished baked good.

cookie   A diverse group of small, sweet cakes or pastries that are described and categorized by how the dough is prepared for baking.

couche   A piece of heavy canvas or linen in which a yeast bread can be nestled in order to hold its shape during the proofing process.

coulis   An uncooked fruit sauce of fresh or frozen puréed fruit that is sweetened and strained.

coupler   A coupler is a plastic cone-shaped tube that is used to allow various pastry tips to fit onto a pastry bag to facilitate the piping of frostings and batters. The coupler allows tips to be changed during decorating without having to change pastry bags.

couverture   High-quality chocolate made with at least 32% cocoa butter that is used in baked goods or to make candy bars, or to coat candies and create decorations for all sorts of pastries. Couverture comes in milk, semisweet, or white varieties.

cream cheese   Cream cheese is a soft, spreadable, unaged cheese that is cultured with bacteria to give it a slight tang. It is used in cheesecakes, cookies, pastry doughs, and pie crusts.

cream of coconut   Cream of coconut is also referred to as coconut cream. Cream of coconut is a thick liquid that is intensely rich in coconut flavor and is made from the liquid that rises to the top of the coconut milk, sugar, and other thickening agents. It is used in many desserts and alcoholic beverages.

cream of tartar   Chemically known as potassium hydrogen tartrate, an acidic salt of tartaric acid, cream of tartar is formed during the wine making process and is deposited on the inside of wine barrels. It is used to make meringues more stable and to help prevent candies from crystallizing.

creaming   See *creaming method.*

creaming method   A method of mixing in which granulated or brown sugar is mixed with a softened, solid fat using the paddle attachment of an electric mixer until it is light and fluffy. Air is incorporated into the fat and is instrumental in aiding the leavening process.

crème anglaise (krehm ahn-GLEHZ)   A sweet, French stirred custard sauce used as a dessert sauce or as a base for frozen desserts. It is made from egg yolks, sugar, milk, half-and-half, or heavy cream; and various flavorings. A typical crème anglaise is flavored with vanilla and is referred to as a vanilla custard sauce.

croissant cutter   A tool used to cut croissants from croissant dough that resembles a short rolling pin with a cut out triangular shape between the handles.

croissant dough   A laminated yeast dough that is formed when butter is encased in a base dough containing yeast, then rolled and folded repeatedly to make multiple thin layers. Traditionally the dough is cut into triangles, shaped into crescents and baked. The finished rolls are known as *croissants*.

crystallization   When particles of a pure substance such as sugar (sucrose) form a repeated shape and are packed closely together to form crystals.

curdling   When egg proteins clump together because they are heated for too long and at too high a temperature.

cutting   The technique used to combine fat and dry ingredients until the pieces of fat have been reduced to a desired size. This is accomplished using a pastry blender, food processor, or an electric mixer. This mixing technique is used in the flaky pie crust method of preparing crusts for pies and tarts and in the biscuit method of mixing to prepare quick breads.

Danish dough   The dough from which Danish pastry is made. A rich yeast dough in which fat is enclosed and then rolled and folded repeatedly to make multiple thin layers. It is baked into flaky breakfast pastries and coffee cakes.

degassing   See *punching*.

denature   When proteins such as eggs are heated, beaten, or acidified, causing the protein strands or coils to straighten out or break apart.

desem   A natural starter using only whole wheat flour that produces very dense bread with little acidity.

detrempe (day-trup-eh)   The French term for a base dough used in laminated doughs.

diastase   An enzyme that acts as a catalyst to help starches within flour break down into sugars.

docking   Piercing the bottom and sides of a raw pie or tart shell with a fork or special instrument to prevent the pastry from puffing up and shrinking in the oven. Also referred to as *stippling*.

double-acting baking powder   A chemical leavener that requires both moisture and heat to produce carbon dioxide gas that is used to leaven baked goods.

dough hook   A tool on an electric mixer shaped like a hook that is used to mix yeast doughs. The dough hook helps to simulate the kneading process.

dough scraper   See *bench scraper*.

drop cookie   The category of cookie preparation wherein a dough is dropped from a spoon onto a sheet pan and then baked.

dry method   A method of preparing caramelized sugar by heating sugar in a heavy pan without any water. An acid may be added to prevent crystallization.

Dutch processed cocoa powder   Cocoa powder that has been treated with an alkali to neutralize its acidity.

éclair paste   Also known as *choux paste* or *pâte à choux*. A steam-leavened specialty dough used to prepare cream puffs and éclairs.

egg foam   When air is beaten into whole eggs or egg whites forming a foam that can leaven baked goods.

egg-foam cakes   The category of cakes that uses air beaten into eggs (an egg foam) to leaven them.

egg-foam frostings   Also known as *boiled frostings*, egg-foam frostings consist of Italian and Swiss meringues. Gelatin may also be added.

electric mixer   A piece of equipment used frequently in baking consisting of a bowl fitted to a motor on which three standard attachments are included: a paddle for mixing, a whip for beating, and a dough hook for kneading. Other attachments are available. Electric mixers can be small enough to fit on a work table or large enough to be permanently attached to the floor.

emulsified shortening   A solid fat containing emulsifiers that are able to hold a large amount of liquid to keep cake batters uniformly combined and emulsified. It is generally used for high ratio cakes.

emulsifying agent   A food (e.g., egg yolks) or a food additive that allows two immiscible liquids to stay uniformly mixed together without separating.

emulsion   A uniform mixture of two unmixable substances such as fat and water-based ingredients to create a homogeneous mixture that will not separate.

endosperm   The largest portion of a wheat kernel located under the bran layer. It is used to make white flour.

enrichment   The process of adding certain vitamins and minerals back into the flour that were lost during the milling process.

enzyme   A protein that speeds up a chemical reaction. An example of an enzyme is *alpha-amylase*, which breaks down starch into sugar.

evaporated milk   Unsweetened whole milk from which 60 percent of the water has been evaporated. It contains at least 7.5 percent milkfat and is sold in cans. It is used in confections, frostings, and baked goods.

evaporated skim milk   Unsweetened skim milk from which 60 percent of the water has been evaporated. It is the same as evaporated milk except skim milk is used. It contains less than 0.5 percent milkfat.

false-bottom tart pan   A tart pan used for making a fluted pastry crust that can be filled with sweet or savory fillings that has a removable bottom for easy removal. Tart pans have different shapes, such as round, square, or rectangular. The round tart pans have varying diameters.

false-bottom tart ring base   A thin metal circle used as the removable bottom of a round false-bottom tart pan and which can be used to help separate two cake layers.

fast-rising dry yeast   See *instant active dry yeast.*

fat bloom   When crystals of fat travel to the surface of chocolate and recrystallize on the outside to form a whitish coating.

fermentation   (1) The process of yeast eating sugar and converting it to carbon dioxide gas and alcohol. (2) In yeast dough production, the first rise of a yeast dough in which carbon dioxide gases are produced and become trapped in a network of gluten.

Flexipan   The brand name for a type of baking pan made from flexible silicone that can withstand a wide range of temperatures and has a permanent, nonstick surface, much like a silicone baking mat. Flexipans come in a wide variety of shapes and sizes and need to be baked on a rigid surface such as a half or full-size sheet pan.

flour   Grains or nuts ground or milled into various degrees of fineness to create a meal or powder.

folding   A very gentle method of blending lighter, air-filled ingredients into heavier batters, usually with a rubber spatula, without losing air volume.

fondant   See *poured fondant* or *rolled fondant.*

food processor   A bowl with a blade inside it attached to an electric motor that is used for chopping and blending.

formula   The term used by professional bakers (especially bread bakers) for a recipe.

four-fold or **bookfold turn**   When a laminated dough is folded like a book to produce many layers of fat and dough.

French bread pan   See *baguette pan.*

French meringue   A mixture of egg whites and sugar beaten to stiff peaks, also known as a *common meringue*, it is the simplest type of meringue.

fresh currants   The small, shiny berries that grow on a prickly shrub in clusters like miniature grapes. They are tart in flavor and are red, black, or golden (also known as white). They are used to make jellies and fillings for cakes and pastries, and are used for garnishes. Not to be confused with dried currants.

fresh or **compressed yeast**   Fresh yeast mixed with a starch that is portioned or compacted into a small cube. Fresh yeast has a short shelf life and must be kept refrigerated. See *yeast.*

friction   The heat energy transferred to a yeast dough through the act of mixing.

friction factor   The difference between the temperature of the dough before and after mixing. The friction factor can be calculated for a particular mixer by taking the sum of the temperature of the room, the flour, and the water and subtracting it from the actual dough temperature multiplied by three.

frosting   A sweet topping or covering used to fill or coat the top or sides of cakes, cookies, and other pastries.

fudge-style frostings   A frosting based on fudge candy beginning with a boiled sugar syrup to which butter and flavorings are added.

galette   A French word referring to a free-form tart usually filled with fruit.

ganache   (ga-nosh) A versatile mixture of cream, chocolate, and flavorings used to make sauces, glazes, frostings, and candies.

ganache-style frostings   A rich frosting consisting of a mixture of simmering cream to which chocolate has been added and allowed to cool and thicken.

gelatin   A stabilizer that helps foods form a gel-like consistency, giving structure to desserts. It is derived from animal connective tissue or bones, or from plants. Left to bloom in a cold liquid, gelatin is then dissolved in hot ingredients or over a hot water bath and then chilled. It is available in sheets or as a granular powder.

gelatinization   The process that starch granules go through to ultimately thicken a liquid. Gelatinization occurs when a starch and a liquid are heated, the starch absorbs the liquid, then the starch swells and ultimately thickens the liquid.

gelatinization of starches   During the baking process, starches within the flour absorb moisture from the dough, swell, and become firm.

gelation   When gelatin firms up or sets up to become a solid-like gel.

genoise (jehn-waahz)   A type of egg-foam cake known as a *whole egg foam* or sponge cake in which whole eggs are warmed and beaten with sugar until thick and then folded into dry ingredients, usually with the addition of melted butter.

germ   The smallest part of the wheat kernel or other grain. The germ regenerates the plant and is the only part containing fat.

ginger   A spice that originates from a flowering tropical plant native to China. Ginger is part of an underground root system called a *rhizome* that grows horizontally. It has short, finger-like projections covered in a light brown skin. Once peeled, ginger can be sold fresh, pickled, ground, or crystallized. It has a sweet, almost peppery taste and is used extensively in baked goods.

glaze   A category of frosting used as a thin coating for cake layers, tarts, cookies, yeast breads, and coffee cakes, consisting of sugar syrups or thinned and melted preserves or jams.

gliadin   See *gluten*.

glutathione   A protein fragment (amino acid) found in milk and active dry yeast that weakens gluten in yeast doughs.

gluten   The network of fibers that is created when two proteins in wheat—glutenin and gliadin—are mixed with water. This web of fibers keeps gases trapped, causing baked goods to rise and providing strength and structure.

glutenin   See *gluten*.

granita (grah-nee-TAH)   An Italian sweetened, flavored, slushy ice that is scraped after freezing and scooped into glasses. It is served as a light dessert or a palate cleanser for in between courses of a meal. See *granité*.

granité (grah-nee-TAY)   The French version of a *granita*.

grater   A rectangular strip or box of metal with sharp holes of varying sizes cut out. Foods are passed up and down to allow small slivers to fall through to the other side. Graters can be used to remove the outer peel from citrus fruits, or to shred cheese, vegetables, or chocolate. It is similar to a microplane zester.

gum arabic   A gelatin-like stabilizer derived from the sticky sap of a tree that grows in Africa. Gum arabic is used for stabilizing emulsions in frostings and fillings.

gum tragacanth   A gum derived from a Middle Eastern shrub. It is used as a stabilizer for gum paste decorations and flowers when it is mixed into fondant. The decorations have the feel of bone china when they dry.

gums   The collective term for gelatin-like thickeners and stabilizers that are derived from plants, also known as vegetable gums. Examples of gums include *agar, carrageenan, gum arabic,* and *gum tragacanth*.

hard wheat   Wheat that is grown in harsher climates and contains greater amounts of protein and lower amounts of starch.

heavy cream   Cream that is pasteurized but not homogenized with a milkfat content of 36 to 40 percent and used to prepare whipped cream because of its ability to hold air. It is also used in frozen desserts such as ice cream and for ganache and in caramel and other rich sauces.

high-gluten flour   Flour containing a high level of protein (approximately 14 percent) used in yeast breads where a chewier texture is desired. High-gluten flour can also be used in combination with other flours that may lack gluten-forming proteins to strengthen them.

high-ratio cake   A cake that contains more sugar than flour by weight.

homogeneous   When different ingredients are thoroughly mixed together to form a uniformly blended mixture.

homogenization   A process whereby fat blobs are broken down into tiny particles so that they stay evenly dispersed in milk.

hull   The hard outside covering of the wheat kernel or other grain. See *bran*.

hydrogenation   When hydrogen is added to liquid fats such as oils to chemically and physically alter them to a solid form.

hygroscopic   When a substance absorbs water from the air, keeping the substance moist. Sugar is an example of a hygroscopic substance.

icing   See *frosting*.

icing comb   A rectangular or triangular piece of hard plastic or metal in which grooves or ridges have been cut out at regular intervals. It is dragged along the sides or top of a newly frosted cake where it leaves designs imprinted onto the frosting.

immiscible liquids   Two unmixable liquids (fat based and water based) that do not naturally stay blended together. Examples of two immiscible liquids are oil and water.

instant active dry yeast   A type of yeast that absorbs water instantly. It can be mixed directly in with the dry ingredients of a recipe. It produces more carbon dioxide gas per yeast cell, so a smaller amount can be used in yeast breads as compared to active dry yeast. Also known as *fast-rising dry yeast*. See *yeast*.

interfering agents   Ingredients added to a sugar syrup to keep sugar molecules from recrystallizing by preventing the sugar crystals from joining together. An example of an interfering agent is an acid or a different type of sugar such as corn syrup.

inversion   The process of preventing crystallization by adding an acid while heating a sugar syrup to break down the existing sugar into its component parts, thereby creating an impure state and controlling crystallization.

invert sugar   An invert sugar is sucrose that is chemically broken down into its two component parts (glucose and fructose) through the process of inversion.

Italian meringue   A type of meringue in which a hot sugar syrup is beaten into egg whites. Italian meringues are the most stable type of meringue.

kneading   A stage in yeast dough production in which the dough is pushed against a work surface and folded over onto itself until a smooth, elastic dough has developed. Gluten is developed during this process.

Lactobacillus   A species of bacteria that exist in a sour dough starter. These bacteria give off lactic acid, providing a slightly tangy taste to the finished bread.

laminated doughs   Rich doughs with or without yeast in which fat has been incorporated through a series of folds or turns. When baked, laminated doughs form hundreds of layers of flaky pastry such as in croissants, Danish pastry, and puff pastry. Also known as *rolled-in doughs*.

lean doughs   Yeast doughs that use little or no fat or sugar. They include breads that are prepared using few ingredients—French bread, Italian breads, pizza dough—and tend to have hard crusts.

leaveners   Ingredients that are added to batter and dough to help them rise. They include baking powder, baking soda, ammonium carbonate, yeast, and eggs. Natural leavening agents such as air and steam are added through the act of mixing and the addition of water-based ingredients.

levain   A sourdough starter made from wild yeast that is used to leaven a sourdough bread known as *pain au levain*.

litchis   Also spelled lychees. Litchis are small, round fruit native to Asia with a rough, leathery, inedible red skin and a delicate white flesh that encases a brown seed. The fruit has a light, sweet flavor with overtones of flowers, much like a perfume. It is available fresh or canned.

loaf pan   A rectangular pan, in different sizes, with high sides, generally used to bake quick breads and pound cakes.

low-fat milk   Milk from a cow that has had some of the milkfat removed so that it contains anywhere from 0.5 to 3 percent milkfat.

lychees   See *litchis*.

Maillard reaction   A reaction between amino acids and sugars that occurs between 300° and 500°F (149° and 260°C) causing the outer crust of breads and other baked goods to brown. This reaction contributes to crust formation and flavor.

makeup and panning   See *shaping*.

marshmallow   A light, foamy confection made from an Italian meringue that is stabilized with gelatin.

marzipan   A thick, sticky, dough-like paste made from ground almonds and sugar and used to coat cakes or in confections.

mascarpone cheese   Mascarpone cheese is known as the Italian cream cheese. It is rich like butter but it has a creamy consistency much like cream cheese.

meniscus   The level of a free-flowing liquid in a measuring container that marks the amount of the liquid the container is holding.

meringue   When egg whites are beaten with sugar to form an egg foam. The ratio of sugar to egg whites depends on what type of meringue is desired. Meringues are used to leaven cakes and soufflés or as a topping on pies or Baked Alaska. They may be baked until crisp and used as a base for cakes and tortes.

metal cake ring   A strip of stainless steel shaped into a circle much like a cake pan with no top or bottom. Metal cake rings are available in various diameters and heights and are used to mold layers of cake with fillings that need to firm up and set before being able to stand on their own. Cake rings can also be used as cake pans by placing them on aluminum foil on a sheet pan and filling them with batter. Also referred to as a *torte ring*.

microplane zester   A long, narrow, rectangular strip of metal, similar to a grater, with raised, sharp cuts, sometimes attached to a handle, used for grating hard cheeses, chocolate, and citrus peels. It is so named after the tool used by carpenters. Also known as a *rasp*.

mise en place (meez ahn plahs)   A French term that means "getting everything ready and in its place" to help a chef get organized by putting all ingredients, tools, and equipment together to get ready to bake.

mixing   The act of combining ingredients.

modeling chocolate   Melted chocolate mixed with corn syrup and kneaded together until a dough-like consistency forms. After several hours, it can be rolled out, cut, and shaped into decorations to top cakes and other pastries.

modified straight dough method   A variation of the straight dough method for mixing yeast breads whereby the fat and the sugar are combined before being added to the other ingredients to ensure their even distribution.

molded cookie   A category of cookie preparation in which a stiff dough is rolled into small balls and baked or flattened with the bottom of a glass, or criss-crossed with a fork and then baked.

mouthfeel   A term used to describe how a food tastes and feels in the mouth.

muffin method   A mixing method in which to prepare quick breads and muffins. The method consists of combining wet ingredients in a bowl before mixing them into a bowl of dry ingredients.

muffin tin   A muffin tin consists of round metal impressions in which muffin batter can be baked to form small cakes or muffins. Muffin tins come in professional sizes (holding 2 to 4 dozen muffins in standard, full, or half sheet pan sizes), miniature (12 muffins), standard (6 muffins), and jumbo (6 muffins) sizes. Also known as a muffin pan.

natural starter   A mixture of flour, water, and natural or wild yeast that is allowed to ferment. See *sourdough culture*.

neutralization reaction   A chemical reaction between equal amounts of an acid and a base that results in the formation of a salt and water. This results in a neutral pH of 7.0.

nibs   The kernel of the cocoa bean used in the preparation of chocolate.

nutmeg   A spice that originates as a hard seed from the tropical evergreen nutmeg tree. The hard outer coating of the nutmeg seed is covered with a red lace-like material, which when ground, becomes another spice called *mace*. Nutmeg has a strong, sweet aroma used in various sweet and savory dishes.

offset spatula   (1) A spatula with a wide metal blade having a slight bend in the blade just before the handle that is used to remove cookies and small pastries from a sheet pan. (2) A long, round-tipped knife with a slight bend in the blade just before the handle that is used to frost cakes, cookies, or spread fillings. Also referred to as a cake spatula.

old dough   See *pâte fermentée.*

one-stage method   The simplest method of mixing cakes in which all ingredients are added in one bowl, usually dry ingredients first, followed by the gradual addition of liquid ingredients.

osmosis   The tendency of water to go from a higher concentration of water, through a semi-permeable membrane, to a lower concentration of water in an attempt to balance the concentrations and form an equilibrium.

osmotolerant instant active dry yeast   A yeast used in rich sweet doughs. These doughs typically contain greater amounts of sugar, fats, and eggs. Osmotolerant yeast can be rehydrated without large amounts of water present in the dough and without being damaged because of its tolerance of osmotic changes within a dough. See *yeast.*

oven spring   The rapid rising of a yeast dough in the oven as the trapped gases within the dough expand.

overrun   The increase in volume caused by the incorporation of air during the freezing process of ice cream and other churn-frozen desserts.

paddle   A mixing tool used on an electric mixer. Shaped like a boat's oar with open spaces in between, the paddle is used primarily for creaming and blending.

palette knife   A tool similar to an offset spatula but without the bend near the handle. It is used to frost cakes and cookies, and to spread fillings.

parchment paper   Specially treated paper that is used to line cake and sheet pans to prevent foods from sticking. Parchment, also known as baking paper, comes in rectangular sheets 16⅜ by 24⅜ inches (41.5 by 62 cm), which fit perfectly into a full sheet pan. Parchment paper will not burn in a hot oven and can be used to make parchment cones for piping chocolate and thin icings.

passion fruit   A small, round, aromatic tropical fruit with a purple, wrinkled, hard, inedible skin. It has golden flesh with small edible seeds that taste both sweet and tart. The juice can be purchased without seeds in the frozen state and can be used to prepare fillings, mousses, and sauces.

pasteurization   The procedure in which a food substance is heated to a specific temperature for a specific amount of time in order to kill dangerous microorganisms that can cause foodborne illness.

pastry bag   A cone-shaped hollow bag made of various materials and of various sizes with a narrow opening at one end and a large opening at the other end. Frostings, cookie batters, doughs, and chocolate are squeezed through the narrow end of the bag, which is fitted with a pastry tip, to form decorative shapes and designs.

pastry blender   A tool consisting of five to six bent metal wires attached to a handle that is used to cut fat into flour for making pies and quick breads such as biscuits, shortcakes, and scones.

pastry brush   A brush resembling a paintbrush that comes in various sizes and used to lightly cover foods with glazes, butter, water, or egg washes. Also used to brush excess flour off doughs.

pastry cream   A stirred custard consisting of eggs, milk, sugar, and a starch such as flour or cornstarch that is used as a filling for cream pies, fruit tarts, cakes, and cream puffs.

pastry flour   A flour made from soft wheats containing more starch than protein. It is used for more tender pastries and quick breads.

pastry tip   A hollow metal cone shape with varying cuts at the smaller end that is fitted onto a pastry bag such that when frostings or batters are piped out, various designs and shapes are formed. Used to decorate or fill cakes, cookies, and various pastries with frostings, whipped cream, mousses, and pastry cream.

pasteurized fresh whole milk   Whole cow's milk is fortified with vitamin D and nothing artificial. It contains approximately 3.5 percent milkfat and has been heat treated (pasteurized) to 161°F (72°C) for a minimum of 15 seconds to kill bacteria that can cause foodborne illness. Whole milk can be dehydrated and dried to a powder.

pâte à choux (paht uh SHOO)   French for "cabbage paste." A steam-leavened specialty dough used to prepare cream puffs and éclairs that when shaped into small rounds resembles small cabbages. Also known as *éclair paste* or *choux paste.*

pâte brisée (paht bree-ZAY)   French for "broken pastry." Refers to a rich, flaky pastry dough containing flour, fat, and ice water.

pâte fermentée   French for "fermented dough." It is a piece of dough from the previous day's batch of bread dough that is incorporated into the next day's bread dough. Also referred to as *old dough.*

pâte sablée (paht SUB-lay)   French for "sandy dough." The richest and the most tender of the three types of pastry dough. It contains flour, fat, eggs, and more sugar than the other two types of pastry dough (i.e., *pâte brisée* and *pâte sucrée*).

pâte sucrée (paht soo-CRAY)   French for "sugar dough." Contains flour, butter, sugar, and egg yolks. Used as a rich, sweet, pastry dough for fruit tarts, pies, and other pastries.

patent flour   During the milling process, the finest particles of flour taken from the inner part of the endosperm.

pathogen   Microorganisms such as bacteria that cannot be seen by the human eye that can cause foodborne illness.

peanut butter   A ground paste made from peanuts, oil, and salt. By law, in order to be called "peanut butter," the paste must contain at least 90 percent peanuts. Peanut butter is commonly used in many desserts such as confections, cookies, cakes, frostings, and frozen desserts.

pectin   A type of thickener made from the natural sugars within the cell walls of plants, particularly unripened fruits. Pectin is used to thicken jam, jellies, preserves, and fruit glazes.

peel   See *baker's peel*.

persimmon   Small, glossy-skinned fruit resembling a tomato that varies from yellow to red in skin color. It has orange-red flesh with a jelly-like consistency. The two varieties found in the United States are hachiya and fuyu. Persimmons taste sweetest when fully ripened.

piped cookie   A category of cookie preparation in which a dough is pushed through a pastry bag fitted with a plain or decorative pastry tip into various shapes onto a sheet pan and baked. Also called a *pressed cookie*.

pirouette   A type of wafer cookie named after a ballet move. The cookie is rolled up tightly into a cylindrical tube while still warm and pliable.

plasticity   The ability of a fat to hold its solid shape at room temperature while still having the ability to be molded; refers to a fat that has a plastic consistency. An example of a fat with a high degree of plasticity at room temperature is solid vegetable shortening.

pomegranate   A medium, round fruit with a bright pink to red skin encasing an inedible yellowish flesh that is packed with edible seeds that hold a sweet-tart liquid covered with a membrane. Pomegranates are used as garnishes on fruit salads and tarts, and the juice can be purchased separately to prepare various fillings, sauces, and frozen desserts.

poolish   A French word for a thin sponge starter typically prepared with equal parts of flour and water by weight.

popovers   A steam-leavened quick bread made in deep muffin pans forming a puffy brown exterior with a hollow, eggy center.

popover pan   A baking pan similar to a muffin pan but with deeper, narrower impressions with which to bake popovers (puffy, eggy muffin-shaped puffs). The impressions are spaced farther apart to accommodate the rising of the popover.

porous   When a membrane or surface allows air or moisture through it. An eggshell is porous, allowing odors in and moisture to evaporate out.

poured fondant   A cooked sugar syrup that is allowed to crystallize enough to form a sugar paste. It is then melted down and used to glaze baked goods such as napoleons, petit fours, cakes, cookies, and other small pastries.

praline   A caramelized sugar syrup that is poured over nuts (usually almonds or hazelnuts) and allowed to harden. It is then chopped or ground and used to flavor cakes, icings, and candies or to coat cakes and other pastries.

preferment   Means "to ferment before." A mixture of flour, water, and yeast that is allowed to ferment before the actual dough is made. It is then added to other ingredients to form a dough. Preferments provide leavening and flavor to yeast breads.

pressed cookie   See *piped cookie*.

prickly pear   A barrel-shaped fruit from a species of cactus with sharp, prickly thorns. The flesh is a deep purple color. It has small black seeds, similar to a watermelon. Prickly pears have a spongy texture and a mildly sweet flavor. Also known as cactus pears.

profiterole   A small cream puff made from pâte à choux dough.

proof box   A room or cabinet-like box in which humidity and temperature can be controlled to ferment and proof yeast doughs.

proofing   (1) A stage in yeast dough production in which the dough is shaped into rolls, braids, or loaves and allowed to ferment one last time before being baked. (2) A procedure to determine if yeast is alive. The yeast is dissolved in warm water with or without a small amount of sugar. If the mixture becomes foamy after 5 to 10 minutes the yeast is alive and can be used to leaven yeast dough.

protease   An enzyme occurring naturally in dairy products and certain fresh fruits that breaks down proteins. Protease can have a negative effect on gluten formation in yeast doughs and can prevent gelatin from gelling.

protein   Chains or strands of amino acids chemically linked together.

puff pastry   A laminated dough made by enclosing fat into a non-yeasted dough that bakes into a light, flaky pastry with multiple layers leavened by steam.

punching   Also referred to as *degassing*. A stage in yeast dough production in which, after the dough has fermented, the edges of the dough are pulled over into the center to release carbon dioxide and redistribute the yeast.

purifying   Part of the flour milling process whereby air currents are used to blow away any remaining bran pieces on the endosperm after breaking.

quick breads   Refers to a category of breads, scones, biscuits, muffins, and popovers that are quick to make and use chemical leaveners instead of yeast.

quince   A pear-shaped fruit with yellow skin that is always served cooked. It tastes similar to a tart apple or pear.

ramekin   A small baking dish usually made of ceramic or heat-resistant glass used to bake individual soufflés, custards, and cakes. Available in various sizes.

reducing   Part of the milling process whereby endosperm are ground into flour.

refrigerator cookie   A category of cookie preparation in which a stiff dough is shaped into logs, wrapped well, and refrigerated or frozen until firm. The logs are then cut into slices and baked.

resting   A stage in yeast dough production in which, after the dough is scaled and rounded, it is covered and allowed to relax for a short period of time. This allows gluten to relax before shaping. See *benching*.

retarding   A slowing down of the fermentation process by placing dough in a special temperature-controlled box called a retarder or in the refrigerator for several hours.

rhubarb   A long, purplish-pink plant with celery-like stalks used in fillings, cakes, dessert sauces, and pies. The leaves are poisonous and should be carefully trimmed. Rhubarb is usually cooked with sugar to decrease its tartness.

rich dough   A yeast dough that contains greater amounts of fat and sugar. It may also include eggs. The crust of breads made with rich doughs tends to be softer than those made from lean doughs. Some rich yeast breads include brioche, coffee cakes, and cinnamon rolls.

ricotta cheese   Cheese that is made from the reheated liquid whey that is left over after whole cow's milk cheese is made. Curds are formed and then drained. (The Italian variety is made from sheep's milk.) Ricotta is used for such Italian pastries as cheesecake and cannoli.

rolled cookie   A category of cookie preparation whereby a dough is refrigerated until firm and then rolled thin and cut into shapes before baking.

rolled fondant   A cooked sugar paste that is cooled, beaten, and kneaded like a dough. It is rolled out and used to cover cakes and other pastries.

rolled-in doughs   See *laminated doughs*.

rolled-out frostings   Dough-like frostings made up of various ingredients such as fondant, modeling chocolate, or marzipan. Rolled-out frostings can be molded and shaped into decorations or rolled thin to cover cakes and other pastries.

rolling cutter   A tool that resembles a row of small pizza cutters that are joined together. It can expand and contract like an accordion to make different size cuts. It can be used to cut croissant dough, cakes, and cookies.

rounding   A stage in yeast dough production where pieces of yeast dough are rounded into smooth balls, after scaling, forming a smooth, elastic skin of gluten around the outside. Rounding makes the final shaping of the dough easier.

royal icing   A fluffy, uncooked, decorative icing that consists of confectioners' sugar, an acid such as lemon juice or cream of tartar, and egg whites. Resembling a meringue, royal icing dries very hard and is used to create piped decorations or flowers on cakes, cookies, and other small pastries.

rubber spatula   A tool consisting of a soft, rubber, scoop-like spoon attached to a wooden or plastic handle that is used to gently blend together or fold ingredients. Also used to scrape down ingredients from the sides of a mixing bowl and to remove batters and doughs from a spoon.

sabayon   A rich, foamy French custard sauce consisting of egg yolks whisked with sugar or corn syrup over a hot water bath until thickened and a pale yellow. Sometimes white wine is added. The Italian version is known as *zabaglione*.

saffron   A spice from the dried yellow-orange inner threads (the stigma) of the purple crocus plant. It gives foods a beautiful color and it has a pungent aroma. One of the most expensive spices in the world, saffron is handpicked and takes more than 75,000 flowers to produce 1 pound (455 g) of saffron. Besides being used in Middle Eastern and Spanish cuisine, saffron is used to color and scent certain yeast breads and rolls.

Salmonella   A bacteria associated with eggs and poultry that can cause foodborne illness.

saturated fat   Fats directly derived from animals (with the exception of tropical plant oils and cocoa butter) and having a chemical structure wherein there are as many hydrogen atoms as possible bonded to the carbon atoms, and all of the bonds are single. Saturated fats are generally solid at room temperature.

scalding   To heat a liquid to just below the boiling point.

scaling   (1) The act of weighing ingredients. (2) A stage in yeast dough production when the dough is divided and weighed into portions after fermentation and punching.

scoring   See *slashing*.

seizing   When melted chocolate thickens to a dried out, clumpy mass after being exposed to a small quantity of water or moisture.

self-rising flour   All-purpose flour that contains baking powder and salt.

separated egg-foam cake   A type of egg-foam cake in which the eggs are separated and the yolks are beaten with a portion of the sugar until thick and the beaten whites, along with sifted dry ingredients, are folded in alternately.

shaping   A stage in yeast dough production in which scaled pieces of dough are formed into the desired shapes that will be placed into the oven. Also called *makeup and panning*.

sheet cookie   The category of cookie preparation in which a batter is spread into a sheet pan with sides and baked. The baked sheet is then cut into squares or other shapes.

sheet pans   Sheet pans are rectangular metal baking pans. They come in two sizes: full and half. The full sheet pan measures 18 by 26 inches (45 by 65 cm) and has sides that are 1 inch (2.5 cm) high. The half sheet pan measures 13 by 18 inches (32.5 by 45 cm). They are so named because two half sheet pans put together would equal one full sheet pan. Full sheet pans are used mostly in commercial kitchens because they are too large to fit into a standard sized non-commercial oven. Most half sheet pans fit into noncommercial ovens.

sieve   A small tool used to separate finer particles from coarser ones. It consists of a handle attached to a metal bowl with screen-like openings. The sieve is used to sift out lumps from dry ingredients like flour or confectioners' sugar, to strain lumps from desserts like pastry cream or custard sauces, or to separate out solids from liquid ingredients. Also known as a *strainer*.

sifter   A metal cup with a screened bottom and rotating metal wires inside to help separate out lumps in dry ingredients like flour, confectioners' sugar, and cocoa powder. Used for dry ingredients only, the sifter resembles a sieve.

sifting   (1) Part of the flour milling process whereby particles of flour are sorted by size. (2) The act of putting dry ingredients through a sifter.

silicone baking mats   Reusable, flexible, plastic rectangular sheets coated with silicone and able to withstand extreme temperatures, both cold and hot. Resembling rubber placemats, they are placed into full sheet or half sheet pans instead of parchment paper with no need to grease or flour them. Wiped clean with a damp sponge or cloth, silicone baking mats are reusable.

simple icing   An easy icing consisting of confectioners' sugar and water, cream, milk, citrus juice, or corn syrup added with flavorings to make a thin, pourable icing that can be drizzled over coffee cakes, scones, cookies, or sweet yeast breads.

simple syrup   A sugar syrup consisting of equal parts of water and granulated sugar by weight that is brought to a boil until the sugar is dissolved. Simple syrups can be flavored and brushed onto cake layers to keep them moist, used in a base for frozen desserts, or brushed onto the tops of pastries as a glaze.

single-acting baking powder   A chemical leavener that requires moisture to produce carbon dioxide gas to leaven baked goods. It is rarely used because all leavening power would be gone as soon as the ingredients were moistened.

skim milk   Whole cow's milk that has had most of the milkfat removed so that it contains up to 0.5 percent milkfat. Also referred to as *fat-free* or *nonfat milk*.

slashing   Shallow cuts made with a razor on the surface of an unbaked loaf of yeast bread to allow the bread to expand during baking. The cuts may also be decorative. Also referred to as *scoring*.

smoked gouda   A mild cheese made from cow's milk originating from Holland that has been exposed to smoke, giving it a brownish color and smoky flavor.

soft peaks   The stage of beating a meringue until the beater or whip is held up and the meringue curls over on itself.

soft wheat   Wheat that is grown in milder climates and contains less protein and a higher starch content.

solute   The substance that is dissolved in a solution.

solution   A *solute* and *solvent* that are evenly distributed when mixed together.

solvent   The liquid that a substance is placed in to help it to dissolve.

soufflé   A light, airy French dish consisting of a base for structure and added flavorings, to which beaten egg whites are folded in to provide leavening. Soufflés can be baked or frozen in small ceramic dishes called *ramekins* or soufflé cups.

sourdough culture   A live, bubbly mixture used to leaven bread doughs that uses flour, water, and wild or natural yeast and bacteria that produce flavorful by-products through the process of fermentation. Breads baked with sourdough cultures have a tangy, acidic flavor. See *natural starter* and *starter*.

sourdoughs   A type of preferment that uses a natural or wild yeast starter. Sourdough starters are used to leaven bread doughs. Unused starter can be replenished with flour and water and maintained for long periods of time.

sour cream   Cream that has been soured or fermented with harmless bacteria to give it a thicker consistency and a tangy flavor. It may contain other ingredients such as gelatin or enzymes to help it thicken. It contains between 18 and 20 percent milkfat.

spices   Seasonings derived from aromatic dried plants containing essential oils that impart deep flavor to foods. The various parts of a plant that can be used for spices include the bark, flower buds, berries, seeds, and roots. They are available whole or ground. The quality of a spice depends on the method of harvesting, processing, and climatic conditions.

spiking   When a small amount of commercial yeast is added to a natural starter to create more leavening power.

sponge   A type of preferment consisting of a mixture of flour, water, and yeast that is allowed to ferment for 30 minutes to several hours before other ingredients are added to make a yeast dough. Sponges add flavor to yeast doughs and give the leavening process a head start.

sponge method   A method of yeast dough preparation in which the dough is prepared in two stages. A portion of the flour, water, and yeast are mixed together and allowed to ferment before other ingredients are added to make a dough. The sponge or "pre-dough" is referred to as a preferment and imparts great flavor and leavening power to yeast breads.

springform pan   A springform pan is a round pan of variable size with a removable bottom. It resembles a cake pan but has higher sides for baking cheesecakes and other cakes and desserts. The sides of the pan can be detached from the bottom for ease of removal.

star fruit   A fruit native to Asia and resembling a five-pointed star when cut crosswise. The taste ranges from tart to sweet, similar to a mild Granny Smith apple. Also known as *carambola*.

starch retrogradation   The process of staling whereby chemical changes occur in the molecular structure of a baked good causing the starches to bond more closely over time, forming a drier texture. Also known as chemical staling.

starches   Long chains of sugars chemically bonded to one another in the form of semicrystalline shapes. Referred to as starch granules, the inside structure consists of amylose and amylopectin. Used to thicken desserts such as pastry creams and pie fillings.

starter   A mixture of flour, water, and yeast (either commercial or wild) that is allowed to ferment before a portion of it is added to other ingredients to make a yeast dough. The remaining starter is saved and fed additional flour and water to maintain it over a period of time until it is used again. Starters help leaven yeast doughs and impart complex flavors to yeast breads.

stiff peaks   The stage of beating a meringue until the beater or whip is held up and the meringue stands straight up in a vertical peak.

still-frozen desserts   Frozen desserts that lay still in the freezer with no agitation or churning as they are being frozen. Still-frozen desserts include mousses, semifreddos, bombes, and parfaits.

stippling   See *docking.*

stirred custard   A mixture of eggs and milk or cream stirred on top of the stove with or without the addition of a starch and cooked until thickened.

straight dough   A yeast dough in which most of the ingredients are mixed together in one bowl.

straight dough method   The simplest mixing method for yeast breads in which all the ingredients are added in one bowl and mixed.

straight flour   The particles of flour from the entire endosperm.

straight spatula   A long, rounded metal knife used to spread frostings and fillings. Also known as a *palette knife.*

streams   Particles of milled flour that are sorted by size in order to be classified.

sucrose   The chemical name for table or granulated sugar; derived from the sugar cane or sugar beet plant. Sucrose is composed of two simple sugars—glucose and fructose—bound as one molecule.

sugar bloom   A recrystallization of sugar that forms on the surface of chocolate that has been exposed to moisture giving it a whitish coating.

sugar syrup   When one or more sugars are dissolved in water. Sugar syrups cooked to specific temperature are the foundation of many desserts including caramel, Italian meringues, marshmallows, fudge, and pulled sugar decorations.

surface tension   The natural tendency for two immiscible (not mixable) liquids to separate.

suspension   Tiny liquid drops or solid particles floating freely in another liquid.

sweetened condensed milk   Whole milk that has 60 percent of the water evaporated from it and with extra sugar added, leaving a sweet, thickened liquid that is sold in cans. Sweetened condensed milk is used in desserts such as confections, cheesecakes, and custards.

Swiss meringue   A type of meringue in which egg whites are warmed with sugar over simmering water before being beaten.

temperature danger zone   The temperature range between 41° and 135°F (5° and 57°C) in which bacteria grows very quickly in foods. This temperature range may vary from state to state.

tempering (1) The act of bringing eggs up to the proper temperature by slowly whisking in a hot liquid (e.g., hot milk) to prevent curdling. (2) The stabilization of fat crystals within chocolate through a heating and cooling process.

thickener A food that helps ingredients to become less fluid and more dense.

three-fold or **letterfold turn** When a laminated dough is folded into thirds like a letter to incorporate the fat and to produce multiple layers of fat and dough.

torte ring See *metal cake ring.*

trans-fats Liquid fats such as oils that are partially hydrogenated to a solid or partially solid state at room temperature. The chemical structure of the fat is such that the hydrogen atoms sit diagonally across the double bond. Trans-fats are associated with an increased risk of heart disease and certain cancers.

truffles A rich chocolate confection prepared from a mixture of heavy cream and chocolate (ganache) flavored with extracts, alcohol, fruits, nuts, or purées. The mixture is chilled and rolled into balls where they can be served as is or dipped into melted chocolate.

tube pan A deep, round cake pan with a hollow tube in its center used to bake angel food, sponge, and pound cakes. Some tube pans come with a false bottom or have metal tabs that stick out from the top edge of the pan to allow cakes to balance upside down while cooling.

tuile (tweel) French for "roof tile." A type of crisp, thin *wafer cookie* that can be molded into various shapes while warm before cooling and hardening.

tunneling The result of too much gluten produced by overmixing that results in large holes or cavities "tunneling" through the inside of a quick bread.

turn The sequence of rolling out and folding a laminated dough to create multiple layers.

turntable A rotating, elevated plate used to facilitate frosting a cake.

two-stage method A cake mixing method in which the liquids are added in two stages. High-fat cakes that contain more sugar by weight than flour (high-ratio cakes) are mixed using this method.

ultrapasteurization When milk or cream is held at an even higher temperature (280° to 300°F; 138° to 150°C) than regular pasteurization requires. The product has a longer shelf life but the cream will not attain the same volume when whipped. Also known as ultrahigh-temperature pasteurization (UHT).

unsaturated fat A type of fat derived from plants that tends to be liquid at room temperature (with the exception of partially hydrogenated vegetable shortenings). In the chemical structure, fewer than the maximum number of hydrogen atoms are bonded to each carbon atom, creating some double bonds between carbon atoms.

unsweetened chocolate Chocolate liquor that has been cooled and molded into bricks or disks. Also referred to as *baking chocolate* or bitter chocolate.

wafer cookie A category of cookie preparation in which a batter is spread onto a sheet pan or over a stencil before being baked. Wafer cookies are soft when hot and harden as they cool. See *tuile,* a type of wafer cookie.

wash A liquid used to brush over food such as a yeast dough before baking to add color or shine, or to help toppings adhere.

wet method A method of preparing caramelized sugar by heating sugar and water and boiling it to the caramel stage.

wheat berry The whole wheat kernel before processing or milling consisting of the *hull* or *bran, endosperm,* and *germ.*

wheat kernel See *wheat berry.*

whip A tool used on an electric mixer that is shaped like a wire whisk and used to beat air into eggs or cream.

whisk A tool having thin, curved metal loops attached to a main handle. Whisks are used to combine dry or wet ingredients and to whip air into ingredients such as egg whites or heavy cream.

white chocolate Consists of cocoa butter, milk solids, sugar, and flavorings, but because it contains no chocolate liquor, it is not really chocolate. There are two qualities of white chocolate: those that contain cocoa butter and are of a higher quality and those that contain little or no cocoa butter.

whole egg-foam cake A type of egg-foam cake in which whole eggs are beaten with sugar and then sifted dry ingredients are folded in.

whole milk Milk from a cow containing approximately 3.5 percent milkfat.

whole wheat flour Flour that is milled from all three components of the *wheat kernel*: the *bran* or *hull*; the *germ*; and the *endosperm*.

yeast A one-celled microscopic living fungus that undergoes fermentation and that is used as a leavening agent for yeast breads and in the production of cheese, beer, and wine. Available as *fresh* or *compressed,* active dry, *instant active dry,* and *osmotolerant instant active dry.*

yogurt Milk that is heated with special bacteria until it ferments and becomes thick and tangy. The fat content varies depending on how much butterfat the milk contains. Yogurt can be purchased as plain and unsweetened, which is best used for baking, or sweetened and flavored, which can be eaten on its own. Mildly flavored yogurts (e.g., vanilla and lemon) can be easily substituted for sour cream in baked goods.

zabaglione (zah-bahl-YOH-nay)　The Italian version of a sabayon, a sweet custard sauce that uses sweet Marsala wine. See *sabayon.*

zester　A tool consisting of five, small, angled holes with sharp edges attached to a handle used to remove long, narrow strips from the outermost peel or rind of citrus fruits.

Amendola, J., & Rees, N. (2003). *Understanding baking: The art and science of baking* (3rd ed.). New York: John Wiley & Sons.

Aslam, M., & Hurley, W. L. (1996). www.aces.uiuc.edu/~ansystem/dairyrep96/Aslam.html. Accessed 8/2/04.

Beranbaum, R. L. (2003). *The bread bible.* New York: W. W. Norton & Co.

Beranbaum, R. L. (1988). *The cake bible.* New York: William Morrow & Co.

Beranbaum, R. L. (1998). *The pie and pastry bible.* New York: Scribner.

Brown, A. (2004). *Understanding food, principles and preparation.* Belmont, CA: Wadsworth, A Division of Thomson Learning.

Card, M. (2003). "Eggs 101." *Cook's Illustrated,* March and April, p. 16.

Child, J. (1989). *The way to cook.* New York: Alfred A. Knopf.

Collin, J. (2003). "A Rustic Italian Loaf." *Cook's Illustrated,* January and February, p. 10.

Corriher, S. O., (1997). *Cookwise: The hows and whys of successful cooking with over 230 great-tasting recipes.* New York: William Morrow & Co.

Figoni, P. (2004). *How baking works.* Hoboken, NJ: John Wiley & Sons.

Friberg, B. (2002). *The professional pastry chef: fundamentals of baking and pastry* (4th ed.). New York: John Wiley & Sons.

Gisslen, W. (2001). *Professional baking* (3rd ed.). New York: John Wiley & Sons.

Gisslen, W. (1999). *Professional cooking* (4th ed.). New York: John Wiley & Sons.

Glezer, M. (2000). *Artisan baking across America.* New York: Artisan.

Healy, B., & Bugat, P. (1999). *The art of the cake.* New York: William Morrow & Co.

Heatter, M., (1999). *Maida Heatter's book of great desserts,* Kansas City: Andrews McMeel Publishing.

Labensky, S., Ingram, G., & Labensky, S. (2001). *Webster's new world dictionary of culinary arts* (2nd ed.). Upper Saddle River, NJ: Prentice-Hall.

Leader, D. & Blahnik, J. (1993). *Bread alone: bold fresh loaves from your own hands.* New York: William Morrow & Co.

Lynch, F.T. (2000). *The book of yields accuracy in food costing and purchasing* (5th ed.). Hoboken, NJ: John Wiley & Sons.

McGee, H. (1984). *On food and cooking: the science and lore of the kitchen.* New York: Scribner.

Medrich, A. (1990). *Cocolat.* New York:Warner Books Inc.

National Restaurant Association Educational Foundation (2004). *ServSafe coursebook* (3rd ed.). Hoboken, NJ: John Wiley & Sons.

Reinhart, P. (2001). *The bread baker's apprentice.* Berkeley: Ten Speed Press.

Severson, K. (2003). *The trans fat solution.* Berkeley: Ten Speed Press.

Sherber, A. & Dupreé, T.K. (1996). *Amy's bread.* New York: William Morrow & Co.

Teubner, C. (1997). *The chocolate bible.* New York: The Penguin Group, Penguin Putnam.

Wolke, R. (2002). *Kitchen science explained, what Einstein told his cook.* New York: W.W. Norton & Co.

The following Web sites were used as resources:

http://www.flour.com, accessed 9/29/2003 "The World of Flours & Flour Milling,"

http://www.ynhh.org/online/nutrition/advisor/sugar_alcohol.html, accessed 12/7/2004 "Eat any sugar alcohol lately?"

Note: As with any dynamic informational tool, Web sites will change and even disappear from time to time. Any Internet user must be aware of this fact and be prepared to investigate and discover other comparable Web sites.

RECIPE INDEX